A Guided Tour of the Living Cell

A GUIDED TOUR OF THE LIVING CELL

VOLUME ONE

Christian de Duve

Illustrated by Neil O. Hardy

This book is published in collaboration with
The Rockefeller University Press

**SCIENTIFIC
AMERICAN
LIBRARY**

An imprint of Scientific American Books, Inc.
New York

Library of Congress Cataloging in Publication Data

De Duve, Christian.
A guided tour of the living cell.

(Scientific American library)
"This book is published in collaboration with the Rockefeller University Press."
Includes index.
1. Cells. I. Hardy, Neil O. II. Title.
QH581.2.D43 1984 574.87 84-5534
ISBN 0-7167-5002-3 (v. 1)
ISBN 0-7167-5006-6 (v. 2)

Printed in the United States of America

Book design by Malcolm Grear Designers

Scientific American Library is published by Scientific American Books, Inc., a subsidiary of Scientific American, Inc.

Distributed by W. H. Freeman and Company, 41 Madison Avenue, New York, New York 10010.

1 2 3 4 5 6 7 8 9 KP 2 1 0 8 9 8 7 6 5 4

For Thierry, Anne, Françoise, and Alain

To their mother

What wonders would he discover, who could so fit his eyes to all sorts of objects, as to see, when he pleased, the figure and motion of the minute particles in the blood, and other juices of animals, as distinctly as he does, at other times, the shape and motion of the animals themselves!

JOHN LOCKE,

AN ESSAY CONCERNING HUMAN UNDERSTANDING

Contents

Volume One

Preface

Every year at Christmas time, The Rockefeller University in New York invites some 550 selected high-school students to a series of four lectures by one of its professors. In 1976, my turn came to deliver the Alfred E. Mirsky Christmas Lectures on Science, as they are now called in memory of the distinguished biologist—an expert on the cell nucleus—who founded them in 1959.

To give such lectures is both a rewarding experience and a demanding challenge. One is faced with an audience that combines in a unique way the eager receptiveness of youth with a daunting degree of sophisticated knowledge in certain areas. Reflecting on how to give these youngsters, who probably knew all about DNA but little about other cell components, some sort of balanced view of cellular organization, I came upon the idea of taking them on a tour. We would shrink to the size of bacteria or, alternatively, blow up the cell a millionfold, which amounts to the same thing. We could then conveniently walk—or, rather, swim—around, look at the different cell parts to observe their structure, and watch them in action to understand their function. Once I got into the spirit of the game, I found it most enjoyable. And the audience was appreciative.

What started more than seven years ago as a four-hour fantasy has become an unwieldy nineteen-chapter opus. Writing it, like preparing the Christmas lectures, has been great fun. So much so, in fact, that the reader came to be somewhat forgotten in the process. The Tour is not a textbook organized so as to offer students an equitable coverage of a topic pitched at their required level of comprehension. It is even less the kind of critically documented scholarly volume that is written for the benefit of experts. But neither is it a popularization work of the sort that covers difficulty with a deceptive cloak of simplicity. It is just what the title says—a guided tour—with all that such an appellation implies in arbitrariness, including the exasperating privilege that tourist guides arrogate to themselves to linger in certain places in order to tell a story or propound some private opinion, only to rush through the next part of the tour to make up for lost time. It is a sharing of a very personal view, gained in decades of roaming through what Albert Claude has called the mansion of our birth and musing over its wonders, a view unavoidably colored, therefore, by preference, prejudice, and familiarity (or the lack of it). As it is written, it is not specifically directed at the student, the expert, or the layman. My hope is that each may find in it some parts or aspects that will make up for its shortcomings as a whole.

From the original "guided tour," I have retained the idea of reducing the observer or expanding the cell a millionfold, thereby making the cell components visible to the naked eye, aided solely, in certain cases, by "molecular eyeglasses." To readers who object to such a device in a serious scientific context and who resent being turned into minuscule "cytonauts," I offer my sympathy but no excuses. There is nothing in the nature of science that demands it to be solemn. There is, however, a danger in metaphor as a vehicle for concepts not easily defined in ordinary language, and even more so in an anthropomorphic description of processes such as those that occur in living cells, which we too readily are tempted to endow with our own brand of purposiveness. I hope the playful imagery adopted in parts of this tour will not prove misleading.

Cells contain a wealth of intricate structures. To try to make these structures visible by magnification is perfectly legitimate. We do not do otherwise with our optical instruments and electron microscopes. Even molecules have an anatomy, and a good part of modern biochemistry is concerned with their faithful three-dimensional representation. In the tour, a special effort has been made, thanks to the untiring and enthusiastic collaboration of Neil Hardy, to depict these structures as accurately as possible and at scale, within the limits imposed by the available information and by the demands for readability. In a number of cases, unfortunately, imagination had to be enlisted, and controversy had to be muted. In spite of the "biological revolution," knowledge of the cell remains fragmentary. It is also advancing rapidly, and the best one can do is to capture as late a stage as possible, but hardly the final one, in this progression. For factual illustration, I have tried to assemble a collection of pictures that have some historical interest, to the point of sometimes choosing an old master over a technically superior, more recent document. But I have also enlisted the collaboration of a few "young masters."

Anatomy, even animated by descriptions of moving parts, rarely reveals function and hardly ever elucidates mechanisms. In the living cell, this can be done only with the help of biochemistry. Not wishing to burden the reader with arrays of complex molecules and strings of intricate reactions, yet loath to resort to the kind of superficial and evasive language often used as a substitute, I have tried to focus on what I see as the main dynamic lineaments of life, stripped of the many individual shapes and forms that clothe and, to some extent, hide them. My guide in this attempt has been energetics, rather than organic chemistry. This may not be to the taste of those more interested in real mechanisms than in abstract concepts, especially since I have gone so far as to invent some rather outlandish terms, such as "oxphos unit" and "Janus intermediate," to avoid entering into specifics. I can only ask those who might be antagonized by this treatment to overcome, or at least to delay, their reaction and bear with me for a while. They may discover that this sort of analysis offers a certain global vision of the metabolic forest that does not demand a knowledge of any individual trees. However, I realize that I may have succumbed to a Gallic tendency that does not export well to other shores. To help the reader wade through the more biochemical parts of the book, I have collected the most important notions of descriptive biochemistry and bioenergetics in two appendices.

A work such as this does not see the light of day without much outside help and support. My first thanks go to the 550-odd high-school students whom I took on the original "guided tour of the living cell" on December 27–28, 1976. Their enthusiastic response and eager questions have been a major incentive to me. Yet I would probably not have started what turned out to be a major undertaking without the kind insistence of Bill Bayless and his colleagues of The Rockefeller University Press and without the encouraging interest of Jaime Etcheverry, of Buenos Aires, in the recordings of the lectures.

Two persons have proved invaluable collaborators, not only by the quality and importance of their contributions, but also because, with them, what is often drudgery, sometimes even agony, turned into genuine fun. I have already mentioned Neil Hardy, a gifted artist, whose professional conscience led him to learn more biochemistry than I could ever teach him and whose good nature I ruthlessly exploited in my groping search for clarity and intelligibility. The other is Helene Jordan Waddell, former director of The Rockefeller University Press, who tirelessly went through innumerable versions of the manuscript, weeding out faulty constructions, stylistic abominations, and other monstrosities with uncompromising firmness, yet always maintaining an unobtrusive and probably often mistaken respect of the author's idiosyncrasies. Both have become great friends. In later stages, I have received much helpful and sympathetic support from the publishers at Scientific American Books: Neil Patterson, Linda Chaput, and in particular Patty Mittelstadt, who has carried out with unfailing devotion, strict professional competence, and admirable forbearance the arduous task of giving the book its final polishing and putting it together for production under what at times proved rather stressful conditions.

Many friends and colleagues have responded generously to my sometimes intemperate requests for illustration material. They include Pierre Baudhuin, Wolfgang Beermann, Marcel Bessis, Robert Bloodgood, Daniel Branton, Ralph Brinster, John Cairns, Pierre Chambon, David Chase, Isabelle Coppens, Richard Dickerson, David Dressler, Marilyn Farquhar, Walter Fiers, Brian Ford, Werner Franke, Yukio Fujiki, Joseph Gall, Ian Gibbons, Jerome Gross, Pierre Guiot, Françoise Haguenau, Etienne de Harven, John Heuser, James Hirsch, Hans-Peter Hoffman, David Hogness, Hugh Huxley, James Jamieson, Morris Karnovsky, John Kendrew, Richard Kessel, Ulrich Laemmli, Emmanuel Margoliash, Arvid Maunsbach, Oscar Miller, Eldon Newcomb, Alex and Phyllis Novikoff, George Palade, Donald Parsons, Keith Porter, Evans Roth, Helen Shio, Samuel Silverstein, Sidney Tamm, Herman Van den Berghe, Marten Veenhuis, Leo Vernon, Eugene Vigil, Luc Waterkeyn, James Watson, Maurice Wilkins and Heinz Günther Wittmann. I thank them all most heartily, as I do also the Arp Foundation in Paris, the late Buckminster Fuller, Dan Dixon, Karl Koopman, Henry Moore, Enrico di Rovasenda, the Metropolitan Museum of Art and the Museum of Modern Art of New York, and the Vatican Museum, for their artistic contributions.

Finally, I owe a special debt of gratitude to what, in retrospect, looks like an army of secretaries endlessly committing new drafts to the memories of their word processors. I cannot name them all, but I must mention Norma Musiek and Patricia Lahy in Brussels and especially Karrie Polowetzky, who, in New York, may be said to have seen it all and, thanks to a happy blend of shockproof resilience, unflappable equanimity, and ever-obliging dedication, survived.

My thanks also go to Miklós Müller, who read the whole manuscript and made many helpful suggestions, to Jacques Berthet, who looked over the appendix on bioenergetics, and to a number of anonymous reviewers who, often in no uncertain terms, pointed out mistakes and inaccuracies, as well as what they considered objectionable terms or modes of presentation. I have tried to correct all factual errors as much as possible. But I have not always followed my critics' advice on style and terminology. For all the many blemishes, minor and major, that remain, I must claim sole responsibility.

To this long list of creditors I must finally add my wife, my children, my friends, and my colleagues on both sides of the Atlantic, who have had to suffer with increasing patience the growing and seemingly never-ending demands that "the book" came to make on my time, my attention, and my temper. There is clearly no way for me to make up for all these cumulative derelictions. If, at least, I have succeeded in conveying some of the feeling of wonder and reverence that the living cell inspires in those who are fortunate enough to busy themselves with its exploration, and also some of the joy and excitement that the explorers derive from their expeditions, I will consider my own time and efforts not entirely wasted.

CHRISTIAN de DUVE
New York and Nethen
April, 1984

A Guided Tour of the Living Cell

1 Preparing for the Tour

We are about to enter a strange world, fascinating, mysterious, but very far removed from our everyday experience. It is a world that exists in each of us, multiplied more than 10,000 billion times, and in every other living being. All are made of one or more units of microscopic size called cells, each of which is capable of leading an independent life if given a suitable environment. In touring the cell, we will in fact be looking at life itself, in its most elementary and basic form. Before we start, some briefing is called for.

The World of Cells

To begin with, let it be clear that there is no such thing as *the* living cell. There are only living cells, innumerable varieties of them. Should we look simply at their more obvious features—such as size, shape, pattern of movement, and other external manifestations—we would find them so diverse that the fundamental kinship between all cells might well escape us, as it escaped the early microscopists for more than 150 years. But when we probe deeper, as we will in this tour, unity reveals itself. By the time we reach the world of submicroscopic structures and of molecular functions, the differences between cells are largely erased. When we speak of the living cell as the object of our tour, we mean some sort of common denominator of all living cells that shares with them the main attributes of life.

Much of our tour will be conducted through this composite entity, but we will make reference also to real cells.

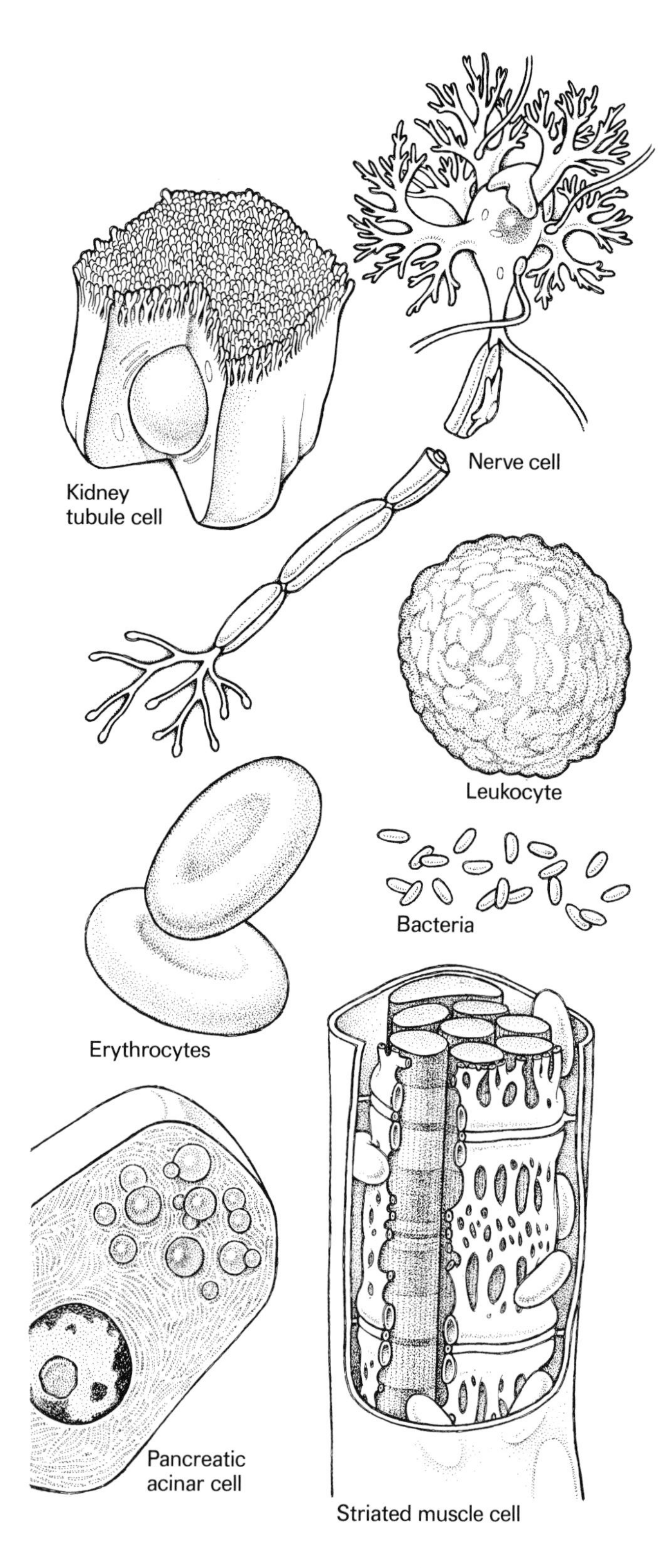

The diversity of living cells.

It is therefore necessary, before we start, to get some idea of the different kinds of cells that can be encountered.

Our own bodies are composed of some few hundred cell types, each represented by up to thousands of billions of individuals. They are the nerve cells, the muscle cells, the gland cells, the blood cells, and so on, of which in each case a number of different subtypes exist. As these names indicate, cells of a given type tend to be grouped into organs, or systems, to perform specific functions. The manner in which they are grouped often plays a decisive role in the expression of these functions. For instance, all striated muscle cells look alike and share the ability to contract. How they are associated makes the difference between the hundreds of different muscles found in the human body. Association patterns between cells reach their highest degree of complexity in the central nervous system, which is made up of tens of billions of cells, each of which may be connected with as many as ten thousand of its congeners. Such connections are established by cells that somehow seek and recognize each other, and then join together. Their associations are stabilized and supported by a variety of extracellular structural elements, which are mainly responsible for maintaining the characteristic architecture of each tissue.

Our close cousins, the other mammals, are constructed very much like us, with essentially the same types of cells. In fact, similar cells are encountered far down the animal scale. Typical muscle or nerve cells are found in fishes, in insects, in mollusks, in worms, but their arrangements become progressively simpler. Down at the level of the lower invertebrates, such as the sponges, the pattern itself begins to change—from that of a multicellular organism to that of a colony of semi-independent cells. At the bottom of the ladder are the fully independent protozoa, such as amoebae, which consist of single cells.

Plants also are made up of different cell types held together by structural elements. But the organization of plants is different from that of animals. They are built entirely around a solar economy conditioned by the occurrence in their cells of a special kind of light-powered factory, the green chloroplasts. Take these away, and you are left with something much like an animal cell. As with animals, there are different degrees of complexity in the organization of plants, from the highly complex flowering plants and trees, down to the lowly unicellular algae. Their nonphotosynthetic relatives, the fungi, display a similar range of complexity, descending from mushrooms to molds and yeasts.

All these cells, which make up the animal and the plant kingdoms, are constructed according to the same general blueprint. In particular, their bodies house a voluminous central structure of characteristic shape, called a nucleus, and are subdivided into many distinct compartments by membranous partitions. Such cells are called eukaryotes, which in Greek means that they have a good (*eu*) nucleus (*karyon*). They are the cells we plan to visit, remaining mostly in our own animal kingdom, with an occasional excursion into the plant world.

Next to the eukaryotes, there is a simpler form of living cells, called prokaryotes because they have only a primitive sort of nucleus. The prokaryotes are the bacteria. They are much smaller than eukaryotes, live singly or in very crude colonies, and show little internal organization. Yet, they occur in an enormous variety of species and have succeeded in invading the most inhospitable of environments, including the steaming ponds of hot springs and the almost solid brine of drying seas. There are bacteria everywhere and they carry out many of the essential mechanisms whereby the constituents of dead organisms are recycled back into forms that can again sustain life. Without bacteria, eukaryotic life would soon become extinct. Some bacteria are harmful, however, through their ability to invade higher organisms and cause diseases.

Finally, if we go one order of magnitude lower again, we encounter the viruses. These are no longer considered cells, as they do not have the capacity for independent life. Nevertheless, they possess one key property of life, which is the ability to supply the instructions for their own reproduction when provided with the required machinery. This they find by the simple device of entering an authentic cell, either eukaryote or prokaryote. Once inside, they appropriate the cell's copying devices and are thereby multiplied manifold, most often causing the invaded cell to degenerate and die. We will have more to say about the viruses when visiting the main information and duplication centers of the cell.

The World of Molecules

If we wish to understand how cells are constructed and how they function, we must use the language of chemistry. We must even use a highly sophisticated form of this language, in view of the exceptional degree of chemical wizardry displayed by living cells. Reflecting this complexity and the progress in our understanding of it, the science of biochemistry has developed enormously in recent years.

Not all tourists can be scholars, however. And it would be a great pity if the beauty and fascination of the world of cells should be reserved to that small minority of cognoscenti who are familiar with the world of biomolecules. Certainly, we would like to take along as many people as we possibly can, and every effort will be made to facilitate their participation. But some chemistry is going to be needed. Without it, much of the tour will be wasted.

It will be assumed, therefore, that everybody has some familiarity with the concepts and laws of chemistry. When possible, images and models will be used to convey key notions. Scientific rigor and accuracy will not, however, be sacrificed to the requirement for simplicity. A more systematic survey of the main constituents of living cells and of their chemical structures is given in Appendix 1. Essential physicochemical concepts are recalled in Appendix 2.

It is hoped that the tour will be accessible to many in this way. Even so, all of you are urged to delve deeper into the world of molecules, since it will heighten your enjoyment of the world of cells. You need not be an Egyptologist to enjoy the pyramids and the treasures of King Tut. But the more you know of the history of these famous objects, the more benefit and pleasure you derive from seeing them.

The Problem of Size

Cells are measured in micrometers (1 μm = one-millionth of a meter), molecules in nanometers (1 nm = one-billionth of a meter). Such small dimensions are very difficult to visualize. Take an average eukaryotic cell, for example. Roughly spherical in shape, it has a diameter of about 25 μm, or one-thousandth of an inch, which means that one billion cells fit snugly within one cubic inch. Bacteria are about 1 μm in diameter, which means that more than 10,000 can fit into a single eukaryotic cell. Many viruses are so small that thousands can be accommodated in a single bacterial cell, or tens of millions of billions within one cubic inch. Imagination refuses to follow.

In our tour, we will solve this difficulty by shrinking to the size of bacteria, say by one million times in each of the three dimensions. This is equivalent to saying that we remain as we are and blow up everything around us one million times. Magnified in this way, the earth would reach far beyond the present location of the sun, to the point where it would take more than 18 hours for a ray of light to travel from one pole to another, and the cell would grow to the convenient size of a large auditorium. We can now stop by any part that catches our fancy and discern every individual detail, down to single molecules.

The Fourth Dimension

Biology, like geology and cosmology, also concerns itself with historical events. Its objects cover a span of several billion years. This fourth dimension became evident some 200 years ago, when fossils were discovered and recognized for what they are: not, as some would have had it, the remains of victims of the Flood or the creation of a playful deity that planted a few dead species among the living ones, but the bones and shells of animals long extinct, the petrified imprints of plants that flourished eons ago. As geological dating progressed, a historical pattern began to reveal itself: the more ancient

The history of living forms (the fourth dimension).

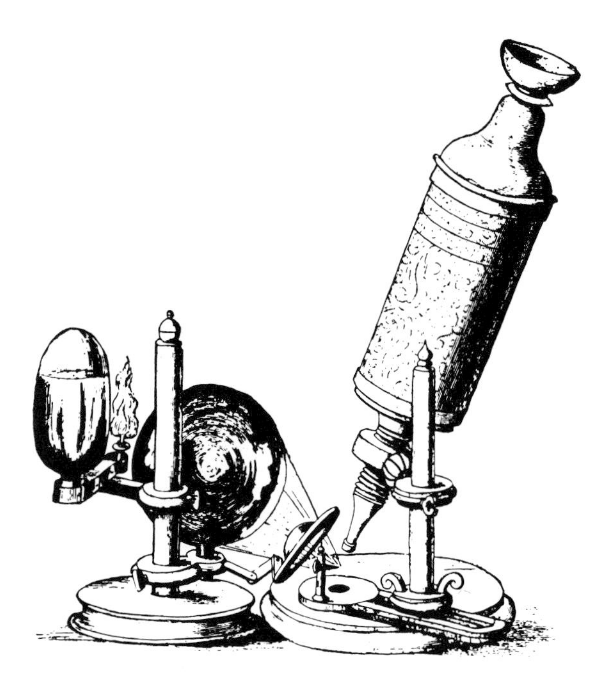

Robert Hooke's microscope, together with the oil lamp that he used for illumination, as drawn by the inventor in his *Micrographia.*

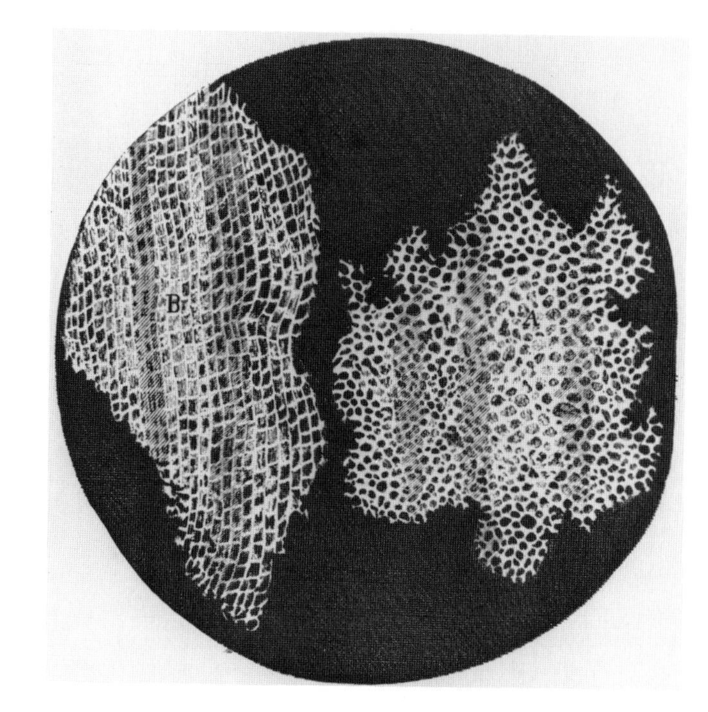

Historical drawing by Robert Hooke of the microscopic structure of a thin slice of cork, illustrating the small cavities that Hooke referred to as cells.

the fossil remnant, the more rudimentary its degree of organization. Mollusks went back farther than fishes, which themselves antedated reptiles. Birds and mammals came later, to be succeeded finally by the early humanoids. Out of this pattern, the concept of evolution emerged during the first half of the nineteenth century, culminating in the publication in 1858 of Darwin's seminal work, *On the Origin of Species by Means of Natural Selection*.

Although the fossil record holds few clues to the evolution of cells, recent advances in biochemistry and molecular biology have provided powerful new means of reconstructing the past by probing the present. The discoveries made in this way have generated considerable excitement, and the fourth dimension has invaded cell biology, permeating our concepts of the living cell and of its contents. We cannot ignore it on a tour like this. Occasionally, as we stop for a break, we will reflect on the origin and evolutionary history of what we see.

Tools and Their Development

Hardly more than 300 years have elapsed between the day when a living cell was first glimpsed and the present era of massive tourist invasion and media popularization. Every milestone on the way that led the early explorers deeper into the cell bears the name of a new tool or instrument. Their main steps are worth recalling.

The Morphological Approach

The world of cells is invisible to the unaided eye. It remained entirely unknown and unexplored until the middle of the seventeenth century, when men of an inquisitive mind served by skilled hands started grinding lenses and using them to extend their power of vision. One of the first makers of microscopes was the Englishman Robert Hooke—physicist, meteorologist, biologist, engineer, architect—a most remarkable product of his

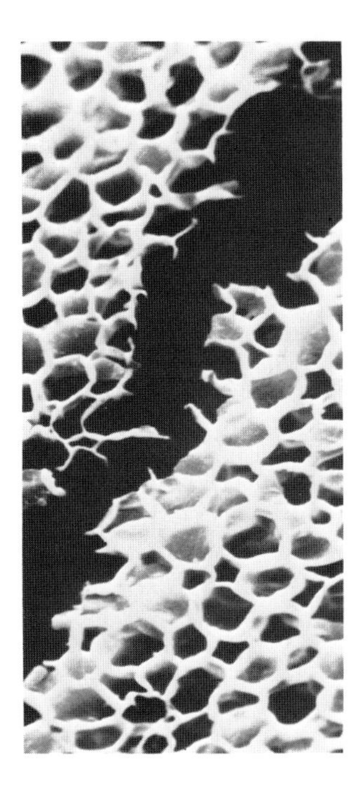

This picture was taken by the British biologist Brian J. Ford, at Cardiff University, by means of a modern instrument known as the scanning electron microscope, which emphasizes the relief of objects. It shows a section of cork similar to, and contemporary with, that drawn by Robert Hooke. The section was cut by Antonie van Leeuwenhoek in 1674 and discovered by Ford in 1981 in a file of Leeuwenhoek's letters to the Royal Society of London (of which Hooke was the secretary).

A replica of Leeuwenhoek's microscope. Objects were impaled on a needle and examined through a glass bead inserted in a copper plate.

time. In 1665, he published a beautiful collection of drawings called *Micrographia*, describing his microscopic observations; among them was that of a thin slice of cork showing a honeycomb structure, a regular array of "microscopic pores" or "cells." In this description Hooke used the word cell in its original meaning of small chamber, as in the cell of a prisoner or a monk. The word has remained, not to describe the little holes that Hooke saw in dead bark, but rather to designate the little blobs of matter that are the inmates of the holes in the living tree.

One of Hooke's most gifted contemporaries was the Dutchman Antonie van Leeuwenhoek, who made more than two hundred microscopes of a very special design. They consisted simply of a small bead of glass inserted in a copper plate. By holding this contraption close to his eye and peering through the glass bead at objects held on a needle that he could manipulate with a screw, Leeuwenhoek succeeded in obtaining up to 270-fold magnifications and made remarkable findings. He was able to see for the first time what he called "animalcules" in blood, sperm, and the water of marshes and ponds. Amazingly, he even saw bacteria, which he drew so accurately that specialists can identify them today.

Not all early users of microscopes were as perceptive. Especially when it came to objects as small as living cells, the images they were able to observe with their simple instruments were so blurred that most details had to be filled in by the imagination. Many showed admirable restraint in the use of this faculty. Others took full advantage of it, to the point of distinguishing, as did the Frenchman Gautier d'Agosty, an enthusiastic adept of the "preformist" theory, a fully formed baby within the head of a sperm cell.

And so, for a long time, microscopy did little more than hover around the world of cells until, in 1827, the Italian physicist Giovanni Battista Amici succeeded in correcting the major optical aberrations of lenses. The increase in the sharpness of the images was dramatic; so

This is one of the plates illustrating Theodor Schwann's *Mikroskopische Untersuchungen*, published in 1839. It is shown in support of the theory that plants and animals are made of similar cells. Figures 1, 2, 3, and 14 are from plants, the others from animals. Schwann mentions that Figures 2 and 3 were given to him by Mathias Schleiden, the botanist who jointly with Schwann is generally given the credit for first enunciating the generalized cell theory.

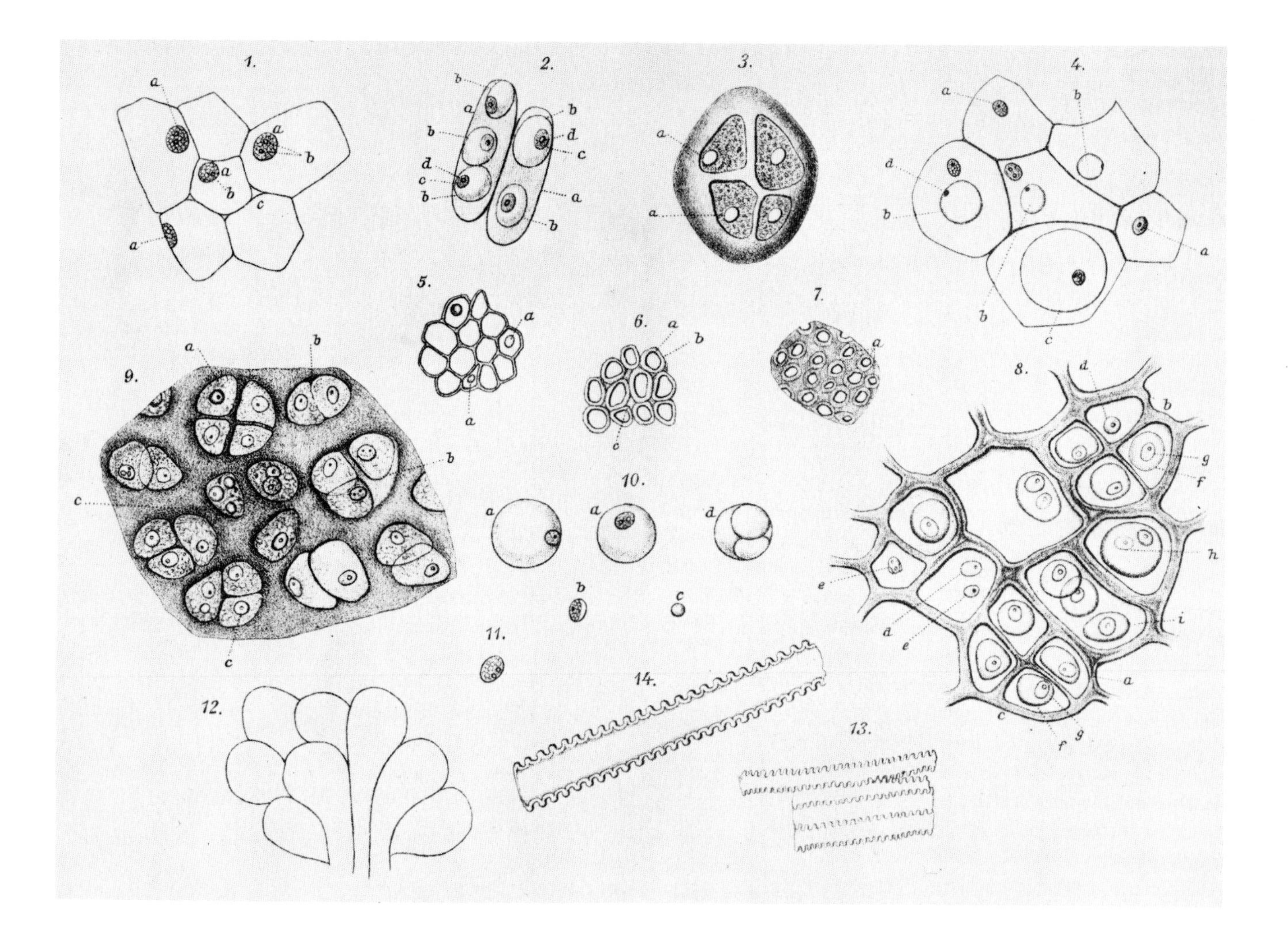

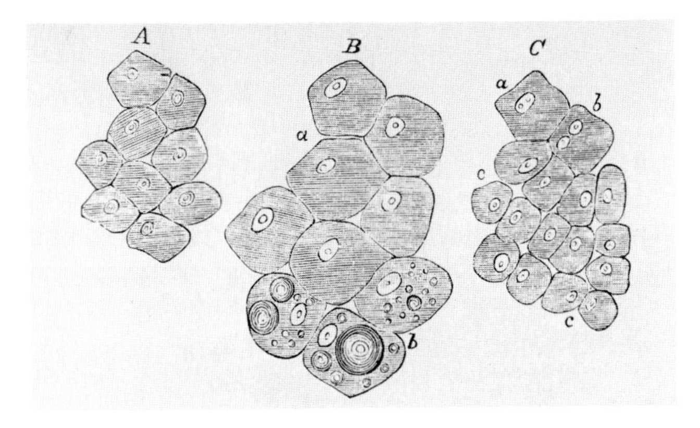

One of the 144 engravings illustrating *Cellularpathologie*, which was published in 1858 by Rudolf Virchow, who first recognized clearly that cells originate from cells, not by crystallization from some amorphous plasma, as was believed by Schwann.

much so that only a few years later the generalized theory could be formulated that all plants and animals are made of one or more similar units, the cells. This theory was proposed for plants in 1837 by the German botanist Mathias Schleiden, and was extended to animals by his friend, the physiologist Theodor Schwann. It was subsequently completed by the pathologist Rudolf Virchow, when he proclaimed in 1855: "*omnis cellula e cellula,*" every cell arises from a cell—actually a paraphrase of "*omne vivum ex ovo,*" every living being arises from an egg, an affirmation made by William Harvey, the English physician who discovered blood circulation and who died just a few years before Robert Hooke's discovery. Virchow also championed the extension of the cell theory to pathology, as witnessed by the title of his *Cellularpathologie*, published in 1858. By the middle of the nineteenth century, the cell theory was firmly established, and the science of cells, or cytology (Greek *kytos*, cavity), started to flourish. The first journal devoted exclusively to cell biology was started in 1884 by Jean-Baptiste Carnoy, at the Catholic University of Louvain, Belgium, under the name *La Cellule*. By the turn of the century, a number of important cell parts had been described and named.

Progressively, however, investigators ran into a new obstacle, apparently insurmountable, as it was set by the very laws of physics. Even with a perfect instrument, no detail smaller than about half the wavelength of the light used can be perceived, which puts the absolute limit of resolution of a microscope utilizing visible light at about 0.25 μm. In the world of cells, such a dimension is quite large, relatively speaking. Just think of what we would miss of our own world if no detail smaller than ten inches could be distinguished. That is all the classical microscop-

LA CELLULE

RECUEIL

DE

CYTOLOGIE ET D'HISTOLOGIE GÉNÉRALE

PUBLIÉ PAR

J. B. CARNOY, PROFESSEUR DE BIOLOGIE CELLULAIRE, G. GILSON, PROFESSEUR D'EMBRYOLOGIE,
ET J. DENYS, PROFESSEUR D'ANATOMIE PATHOLOGIQUE,
A L'UNIVERSITÉ CATHOLIQUE DE LOUVAIN,

AVEC LA COLLABORATION DE LEURS ÉLÈVES ET DES SAVANTS ÉTRANGERS.

TOME I

ÉTUDES SUR LES ARTHROPODES

I. Étude comparée de la spermatogénèse chez les arthropodes,
par G. GILSON.

II. La cytodiérèse chez les arthropodes,
par J. B. CARNOY.

LIERRE
JOSEPH VAN IN & Cie,
IMPRIMEURS-ÉDITEURS.

GAND
J. ENGELKE, LIBRAIRE,
rue de l'Université, 24.

Title page of the first issue of the first journal of cytology, published in 1884.

Portrait of Lavoisier and his wife, painted by Louis David in 1788, six years before the French physicist was condemned to death by guillotine by a judge who commented that "La République n'a pas besoin de savants" [the Republic has no need for scientists]. This historic painting hung for many years in the library of The Rockefeller Institute for Medical Research, now The Rockefeller University, where many of the discoveries in modern cell biology were made. It is now at The Metropolitan Museum of Art in New York.

ists would have seen if they had toured the living cell magnified a millionfold, as we are about to do.

The Chemical Approach

In the meantime, however, while microscopists struggled to improve their instruments, a second type of exploration of the cell was started, thanks to the discoveries by such scientists as the Frenchman Antoine de Lavoisier, the Englishman Joseph Priestley, and others, who, in the later part of the eighteenth century, founded the new science of chemistry. In contrast with morphology, which progresses from larger to smaller entities, chemistry moves from the smaller to the larger. It started by identifying the elements, the atoms, and then went on to characterize some of their simpler molecular combinations. A historical landmark in the penetration of the living world by chemistry is the first synthesis of a biological molecule, urea, by the German Friedrich Wöhler in 1828. The boundary between mineral and organic chemistry, believed by many to be passable only with the help of a special vital force, was crossed.

During the next hundred years, considerable advances were made in our understanding of the chemical composition of living cells. The amino acids, the sugars, the fats, the purines, the pyrimidines, and other relatively small natural molecules were recognized, purified, analyzed structurally, and reproduced synthetically. Some insight was gained into the metabolic transformations they undergo in the organism, as well as into their mode of association into the major biological macromolecules—the proteins, the polysaccharides, and the nucleic acids. But here again the obstacles to progress became increasingly formidable. When structural complexity reaches the degree found in these large molecules, the tools of classical chemistry become almost powerless.

The Experimental Approach

For a long time cells were studied mostly by observation. But, as the experimental method became progressively developed in the physical sciences, it began to be applied to living organisms. This move was aided powerfully by the great upsurge of biomedical exploration in the second half of the nineteenth century. Physiology, pharmacology, genetics, bacteriology, immunology, experimental embryology, comparative and evolutionary biology—all made important inroads into the world of cells. An especially significant development came in the early part of the present century, when the American Ross Harrison and the French-American Alexis Carrel showed that animal cells could be cultured in the test tube, like unicellular microorganisms. They thereby demonstrated the cells' capacity for independent life and set up a technique that is still rendering major services today.

Powerful as they were, all these advances utilized means that, perforce, had to remain indirect and circui-

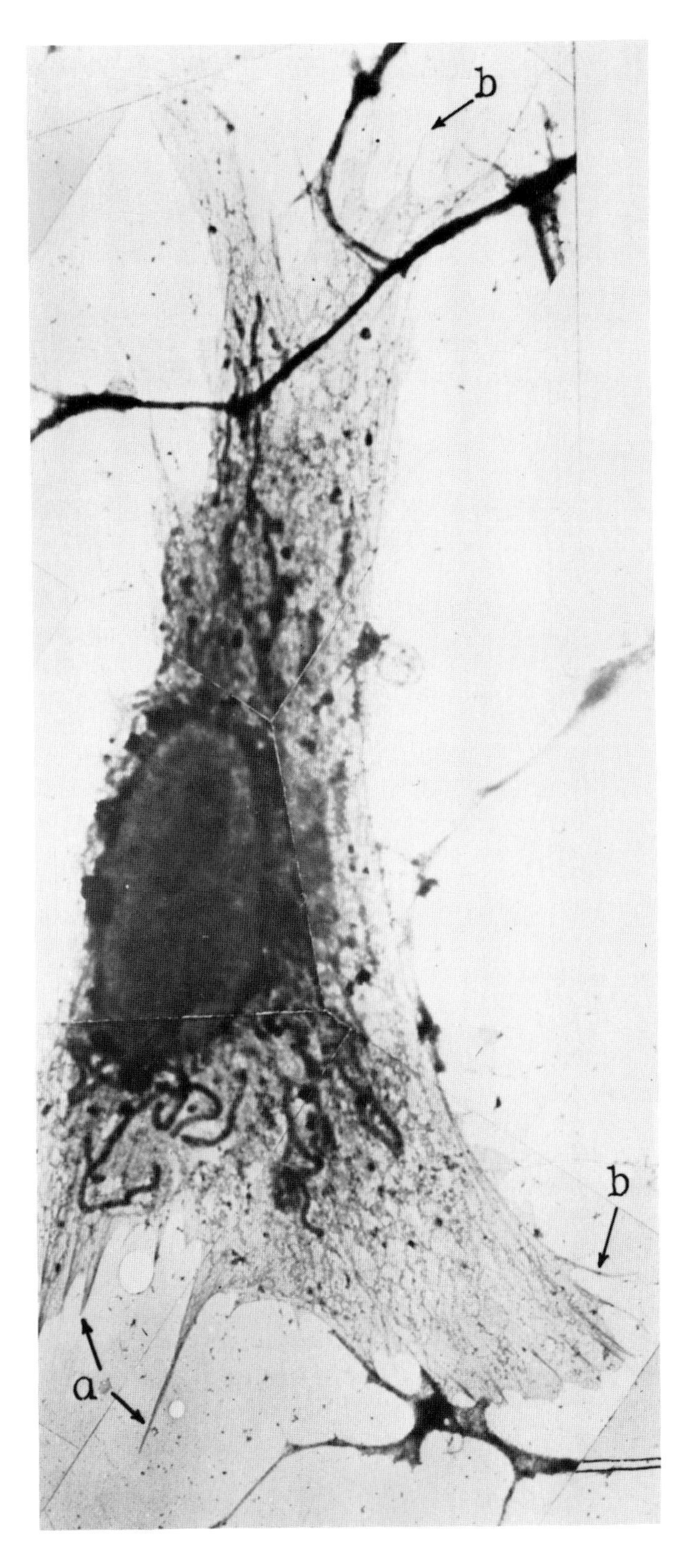

The first electron-microscope picture of a cell was taken by Keith R. Porter, Albert Claude, and Ernest F. Fullam and published in 1945 in the *Journal of Experimental Medicine.*

tous, leaving the cells themselves as virtually unopened "black boxes." And so, in spite of much progress on all fronts, there remained, between the smallest entity discernible in the light microscope and the largest molecular size accessible to chemistry, an unexplored no-man's-land extending over two orders of magnitude, a vast region that had to be labeled terra incognita on the map of the living cell. Scientists knew that this mysterious territory held major notions and concepts without which the life of cells would forever remain ununderstandable. But they could only stare in frustration at its seemingly impenetrable boundaries. Some did not give up, however. Like their predecessors, they put ingenuity to the service of inquisitiveness, and searched for the only solution: better tools.

The 1945 Breakthrough

This long, continuous effort came to sudden fruition at the end of World War II, when, through a remarkable combination of circumstances, a battery of powerful new instruments and techniques became available at more or less the same time. The morphologist's share of this boon was the electron microscope. This instrument, which was invented in the 1930s, has a sufficient resolution to traverse the whole unknown cellular territory, down to the nanometer level. But the weak penetrating power of the electron beam requires specimens to be extremely thin—a few millionths of an inch at most—and to be examined under high vacuum. These requirements posed great technical difficulties, which discouraged many. But a few held on, fired by the perspectives of progress opened by the new instrument. In a surprisingly short time, they succeeded in developing techniques for the preparation of specimens and constructed instruments that would slice the specimens into ultrathin films. Images of increasing quality were obtained, so that by the early 1960s much of the unknown territory had already been mapped out morphologically.

At the same time, biochemistry also acquired a number of incisive new tools. The most important ones among these were chromatography and isotopes. Of the two,

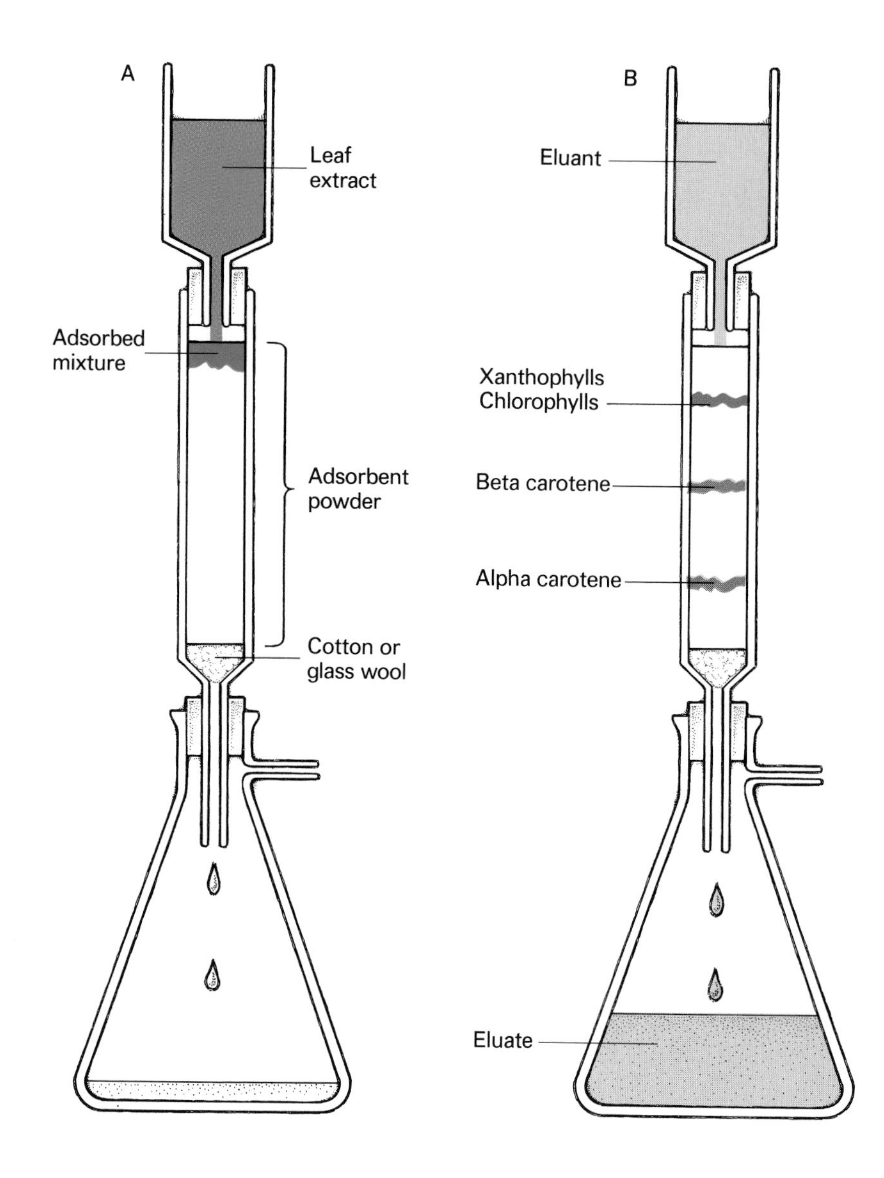

Chromatography. A simple device similar to that diagrammed here was used by Tswett to separate the main pigments in a leaf extract. It consists of a column of adsorbent powder, mounted so as to allow fluid to seep through.

A. A leaf extract is allowed to run through the column. All pigments remain bound on top of the adsorbent column.

B. After the column has been charged, a suitable fluid (eluant) is let through, displacing the different pigments at different rates, depending on the strength of their binding to the powder.

chromatography is especially remarkable, as it utilizes a very simple phenomenon, familiar to everyone who has ever seen a drop of ink spread on a piece of blotting paper or tried to remove a stain with cleaning fluid: the fringe, or halo, surrounding the spotted area. It is explained by the fact that different dyes do not move at the same rate with the spreading fluid. Some may move together with the solvent front, but many are retarded to a greater or lesser degree by binding to the paper or cloth fibers. They thus form concentric rings. In the beginning of this century, a half-Italian, half-Russian botanist, Mikhail Semenovich Tswett, became the first to make use of this phenomenon. By passing a leaf extract through a vertical tube packed with some adsorbent powder, he was able to separate the leaves' main green and orange pigments, which appeared as distinct colored bands or rings on the column. Hence the name "chromatography" that he gave to his technique (Greek *khroma*, color; *graphein*, to write).

Tswett died relatively young, and the potentialities of his remarkable technique remained largely unexploited until it was revived in the early 1940s. There are now innumerable variants of chromatography, not restricted to pigmented molecules, of course, but applicable to all

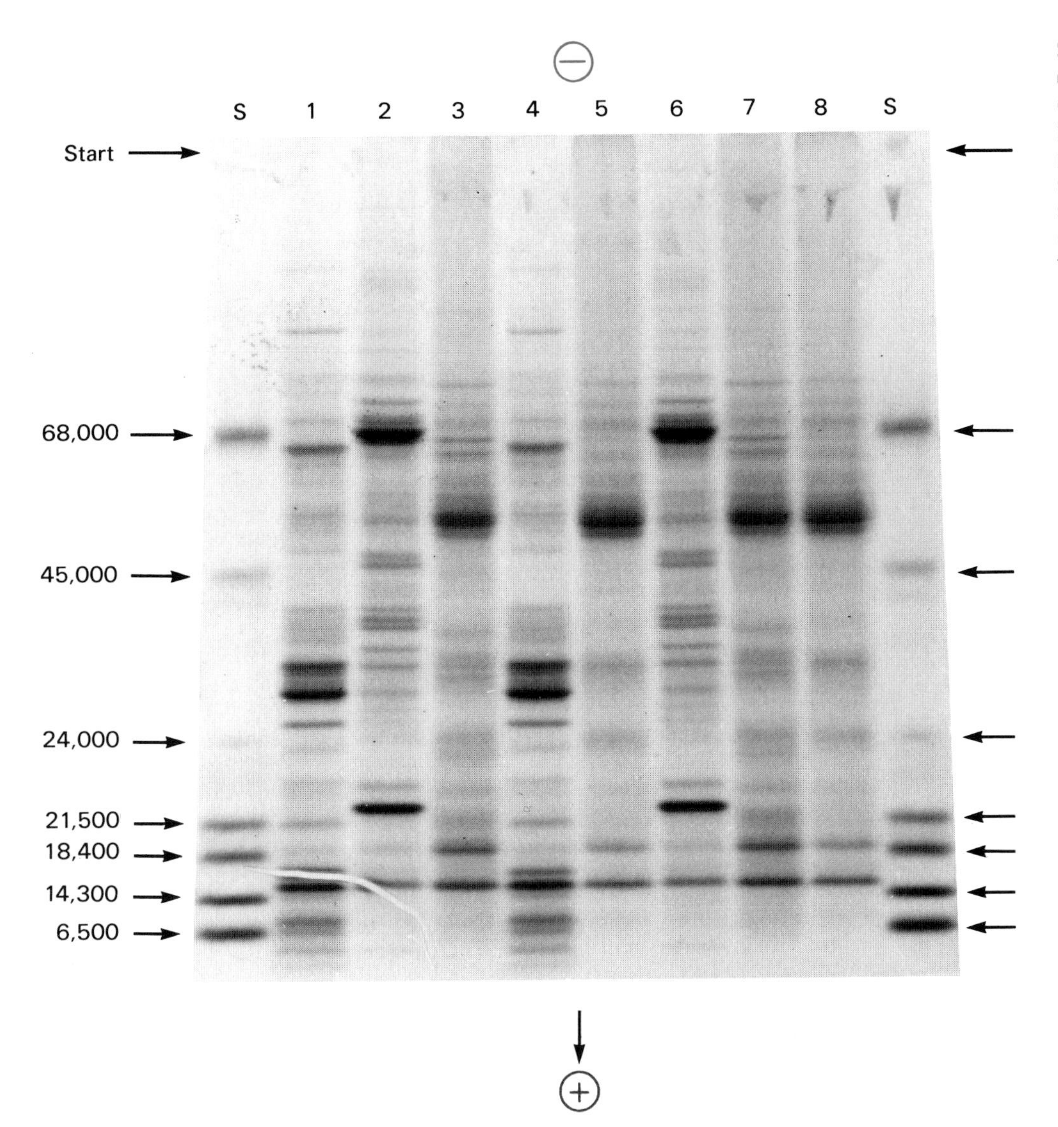

Separation of proteins by sodium dodecyl sulfate polyacrylamide gel electrophoresis (SDS-PAGE). Like most modern versions of chromatography, this technique uses a flat substrate—looking somewhat like a slab of jelly—instead of a cylindrical column. However, it relies on an electric field, not on a flow of fluid, to cause the different kinds of electrically charged protein molecules to move apart from each other. Mixtures are deposited as narrow bands at the start position. Electromotive force is applied and causes different protein species (all negatively charged by binding of SDS) to move toward the positive electrode, at rates inversely related to their molecular masses. At the end of the experiment, protein bands are revealed by staining with a dye. Lanes marked S contain standards of indicated molecular masses. Other lanes contain membrane proteins extracted from different organelles: mitochondria (1 and 4), peroxisomes (2 and 6), rough-surfaced endoplasmic reticulum (3 and 7), and smooth-surfaced endoplasmic reticulum (5 and 8). Note the differences between samples of different origin, enhanced by the reproducibility of duplicates.

substances that can be recognized by some analytical procedure. Related to chromatography is the technique of gel electrophoresis, in which an electromotive force, rather than a flow of solvent, moves the (electrically charged) components to be separated. These developments have, in one sweep, revolutionized the whole field of chemical separation and analysis. What in earlier days either could not be accomplished at all or required laboriously repeated extractions, precipitations, or crystallizations, to be performed on large amounts of starting material, can now be done without effort on trace quantities of almost any kind of mixture.

The second tool that has radically changed the chemical exploration of living cells is the isotopic-tracer method. Isotopes are species of the same element that have different atomic weights. Some exist naturally, and many more can be made artificially by nuclear reactions. For instance, in addition to the hydrogen atom ^{1}H, of atomic weight 1, by far the most abundant in nature, there is a heavy natural isotope, ^{2}H (deuterium), and an even heavier artificial one, ^{3}H (tritium). Chemically, these three forms of hydrogen have closely similar properties. They all combine with oxygen to form water, with carbon to make hydrocarbons, and so on. But they can be distinguished from

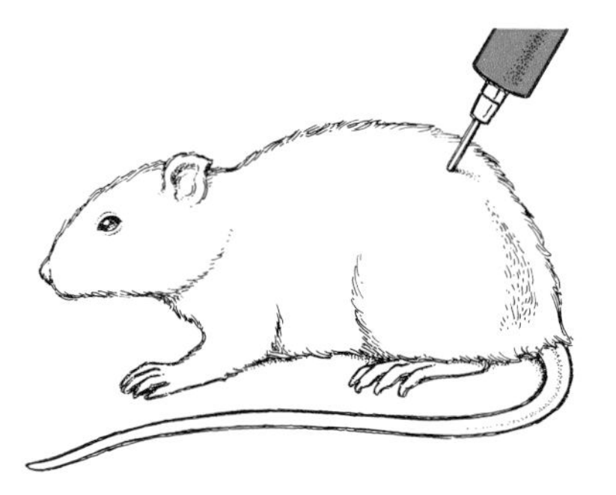
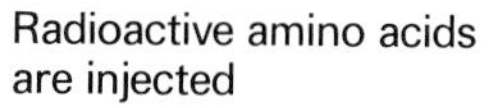
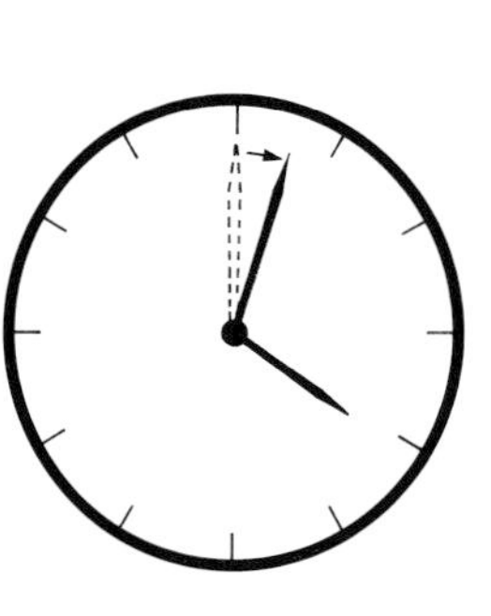
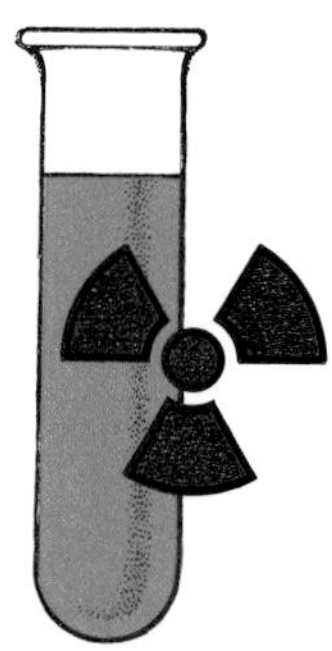

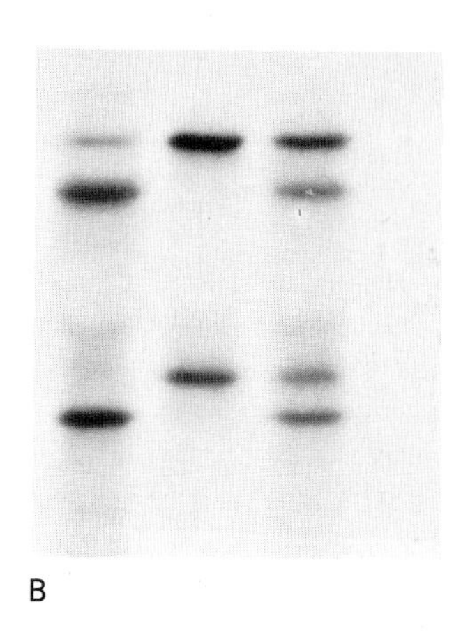

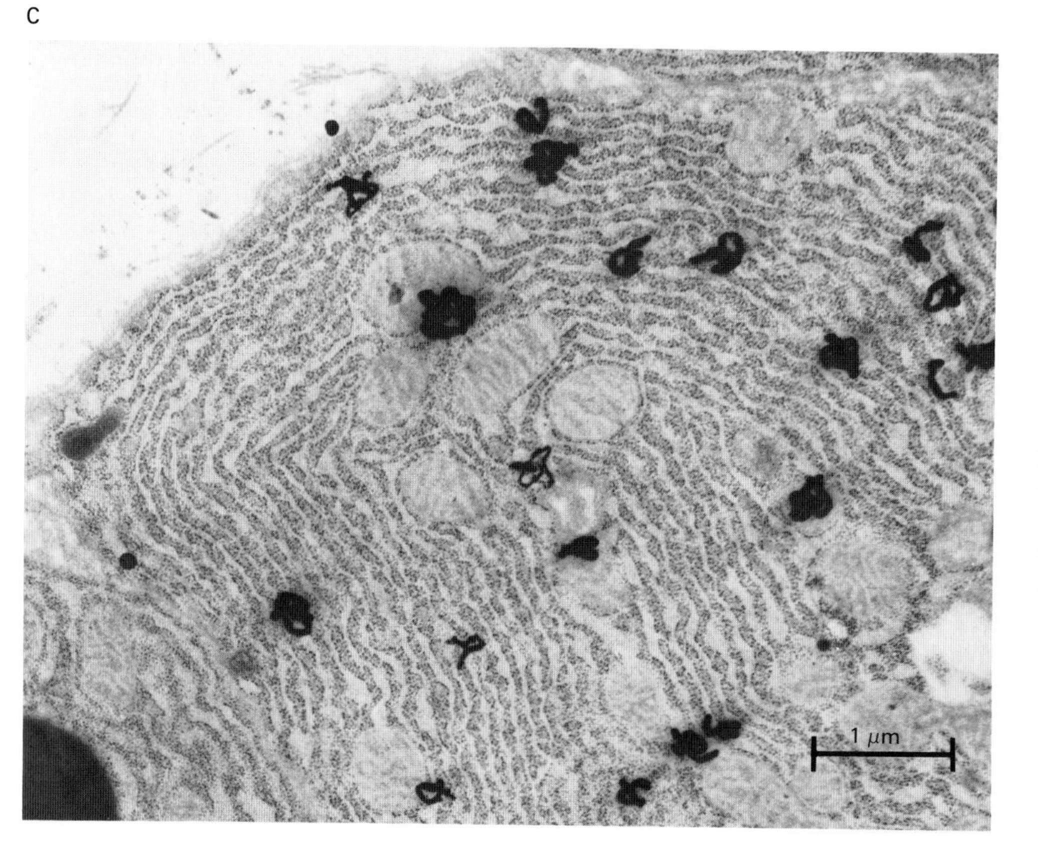

Three examples illustrating the use of radioactive isotopes.

A. Use of labeled amino acids as substrates allows the accurate measurement of the amount of protein synthesized from them, even though this amount is exceedingly small with respect to the total amount of protein present.

B. Impression of a radiographic film by emitted radiation reveals position of proteins synthesized from radioactive amino acids (as in part A), after SDS-PAGE separation (see the figure on p. 13).

C. High-resolution autoradiography combined with electron microscopy serves to localize newly made proteins in the part of the cell occupied by rough-surfaced endoplasmic reticulum. In this section through a cell from the pancreas of a guinea pig that received an injection of a tritium-labeled amino acid (^{3}H-leucine) 3 minutes before killing, each dense curlicue represents a track left in the photographic emulsion by a single high-energy electron emitted locally by a radioactive protein.

each other by a mass spectrograph, which, as the name indicates, separates atoms according to mass. Tritium is particularly easy to detect, because, in addition, it is radioactive, as are most of the isotopes used as tracers.

The advantage of isotopes is that they can be used to label specifically certain molecules, or parts of molecules, so that these can be distinguished and recognized from their congeners with almost no disturbance of the system. One of the many fruitful applications of this technique has been in the analysis of biosynthetic processes, most of which could not have been unraveled otherwise. For instance, as soon as labeled amino acids became available it was possible to study their assembly into proteins in the living animal or in the test tube, even though the amount of new protein manufactured was infinitesimally small relative to the amount of protein present. Although undetectable chemically, the newly made proteins could be recognized and even measured accurately, thanks to their radioactivity. In actual fact, the use of isotopes for studies of this sort had started before World War II, with the few natural (^{2}H, ^{15}N) or artificial (^{32}P) isotopes then available. But only after nuclear reactors were developed and a wide variety of radioisotopes could be manufactured on a large scale at a reasonable price did the technique truly come into its own. Without it, the revolutionary advances of the past decades in cellular and molecular biology would have been impossible. It represents one major peaceful benefit of nuclear power.

With morphology and biochemistry thus both becoming immeasurably strengthened, there still remained the need for a bridge between the two. This was provided by the development of procedures for separating the cell parts in such a way that they could be analyzed and their properties and functions recognized. For this to succeed, biochemists first had to learn to break cells open as gently as possible, so that the fragile cell constituents would be simply let loose but not significantly harmed. Once this is accomplished, one can take advantage of the differences in physical properties, mostly size and density, among the various cell constituents to separate them from each other. The procedures used for such fractionations depend mostly on centrifugation. In this respect, the development of centrifugation as an analytical tool and the construction of high-speed ultracentrifuges, which took place in the 1920s and 1930s, played an important role.

By some remarkable historical coincidence, all these new tools became available more or less simultaneously in the mid-1940s. Among the many names associated with these events, that of Albert Claude, who died in May 1983, deserves special mention. Born in Belgium at the turn of the century, Claude spent the years from 1929 to 1949 at The Rockefeller Institute for Medical Research in New York and there developed almost single-handedly the first applications of electron microscopy to the study of cells, as well as the main techniques for fractionating cells by centrifugation. He thus led both the morphological and the chemical armies into the cell's virgin territories and saw them effect their fateful junction. Today the ground that he opened is so familiar as to allow even the kind of excursion we are about to undertake. One part of the cell, however—actually its most centrally important part, the nucleus—would have remained largely impenetrable had not one additional development taken place.

The Molecular Biology Boom

This development started at the same time as the main invasion of the eukaryotic cell, but at the very fringe of the living world, with an attempt to analyze by means of genetic methods the properties of some simple viruses that infect bacteria and are called bacteriophages—literally bacterium eaters—for this reason. This humble beginning turned out to be the right approach to the problem of genetic organization, which even in these simplest of all organisms has proved to be of enormous complexity. For a long time, the new discipline, now known as molecular biology, remained confined to the study of viruses and of bacteria. But in the last few years, it has broken into the domain of eukaryotes and provided investigators with incisive new tools with which they are now exploring the most hidden recesses of the cell. It has also spawned the powerful new recombinant DNA technology, which may have profound influences on the future of humanity.

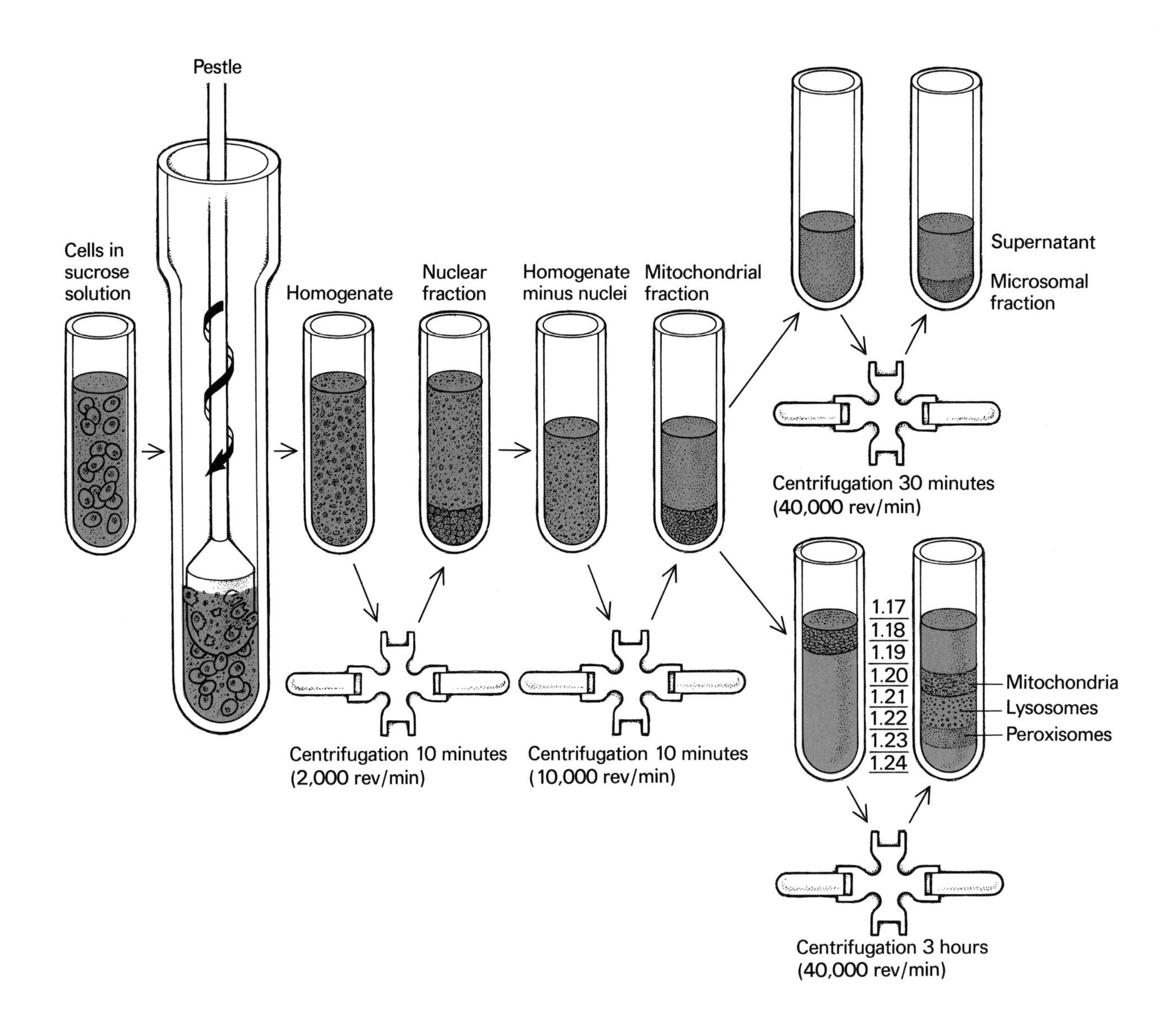

The fractionation of cells by differential centrifugation as first developed by Albert Claude and his pupils Walter Schneider, George Hogeboom, and George Palade. Cells are first ground in a solution of sucrose. (The grinder consists of a heavy-walled glass tube and of a tight-fitting pestle that is simultaneously rotated and moved up and down.) The resulting "homogenate" (an inappropriate term, as the main virtue of the preparation is its heterogeneity) is spun at low speed in a centrifuge, so that only the heavier components (mostly nuclei) are sedimented. The supernatant is then centrifuged at higher speed to separate the mitochondrial fraction. Very high centrifugal speeds (40,000 or more revolutions per minute) are used to bring down the small cell particles that make up the microsomal fraction. Soluble components and very small particles remain in the final supernatant. Altogether, four fractions are separated by this technique of differential centrifugation. Each fraction is grossly impure and heterogeneous and requires further subfractionation by more refined methods. One such method, called density-gradient centrifugation, is illustrated at the lower right. A crude mitochondrial fraction is layered above a density gradient. High centrifugal force is then applied to cause the particles to move down in the gradient until they reach a position at which their density equals that of the medium (equilibrium position). This technique therefore separates the cell constituents according to their density. In this experiment, particles present in the crude mitochondrial fraction have been separated into three distinct groups: mitochondria, lysosomes, and peroxisomes.

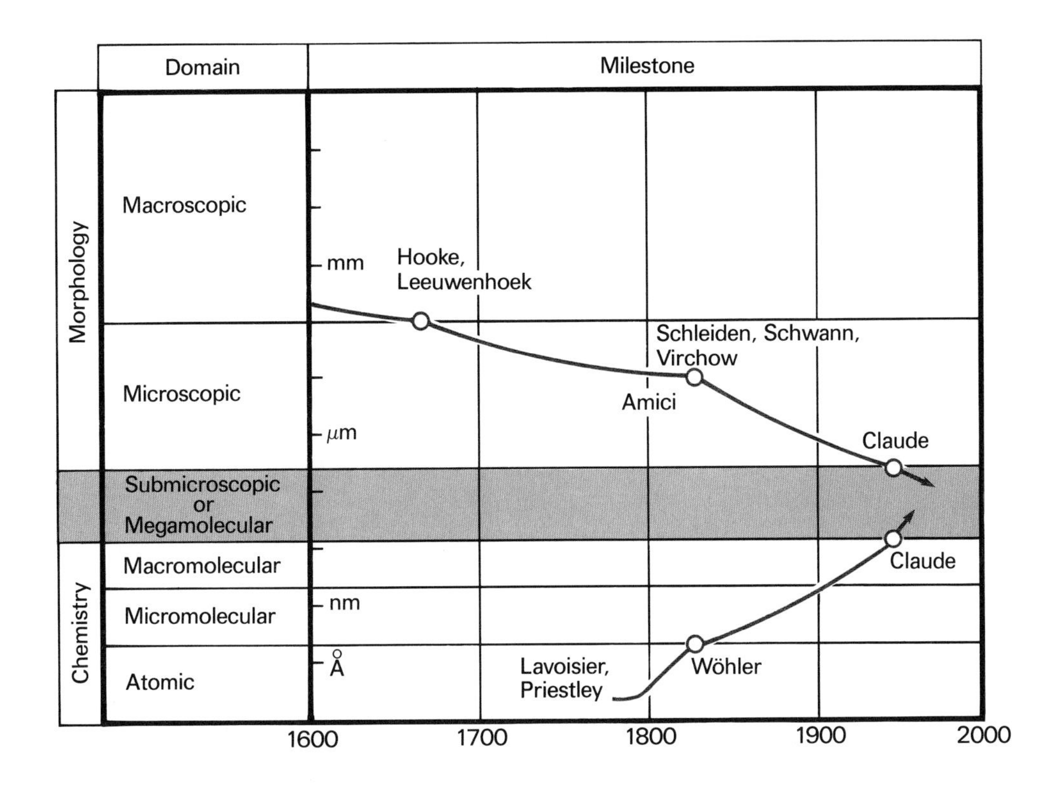

Some milestones in the invasion of the living cell. This diagram shows, as a function of time, the progress of morphology in the resolution of objects of decreasing size and that of chemistry in the characterization of molecules of increasing size. The no-man's-land between the microscopic and the macromolecular domains was invaded simultaneously from both sides immediately after the end of World War II.

An important lesson to remember from this brief historical survey is the crucial role of instruments and techniques in the progress of science. Not that the creative faculties of intellect, intuition, imagination, sometimes even genius, may not be of critical significance at some stage. But these faculties are powerless without the means of establishing contact with reality. As has been said by Claude, "In the history of cytology, it is repeatedly found that further advance had to await the accident of technical progress." Many such "accidents" have happened since Robert Hooke first turned his microscope on a slice of cork. Their consequences have been increasingly far reaching, culminating in the major discoveries of the last decades. Although it is always difficult to judge one's own time in historical perspective, one cannot help the feeling that the second half of this century will be remembered for one of the great breakthroughs of human knowledge—perhaps the greatest to date, as it concerns the basic mechanisms of life.

The Personal Factor

Scientists, it has been said, do not read the book of Nature. They write it. This does not mean, of course, that works of science are works of fiction. On the contrary. Scientists do the best they can to stick to the facts, and the collective process whereby the facts are recorded and put together aims at objectivity.

The point is that, however hard one may try, perfect objectivity is unattainable. There is no such thing as an isolated fact. There are only recordings and interpretations of facts by individuals. Even the simplest observations come to us through our senses, which act as highly selective, as well as subjective, filters that let through only certain limited aspects of reality. We have greatly extended the range of our senses by means of instruments, but only at the cost of interposing additional filters. A lens may reveal more than the eye does, but not without distortion. We can probe even deeper with an electron microscope, reach atoms by X-ray diffraction, detect subatomic particles with a high-energy accelerator. But the resulting picture increasingly depends on elaborate theories, complex machines, and delicate manipulations.

The choice of facts to be collected introduces an additional subjective element. This is true even in the observational sciences. An astronomer has no power over the stars he looks at, but he nevertheless decides in what direction to point his telescope and thereby influences the development of astronomy by his personal preferences. The individual's decision plays a much more important role in the experimental sciences, where facts are provoked. Every experiment is a question asked of Nature. As in all interrogations, the answer depends to some extent on how the question is asked. Scientists are well aware of this danger but, try as they may, cannot avoid it entirely.

Finally, and most importantly, we must remember that the collection of facts is only a small part of the scientific process. What counts is the way in which the facts are interpreted and organized into theories that advance our *understanding*. The construction of a theory is a highly creative process that bears the imprint of its author. That is why we speak of Einstein's relativity theory as we speak of Leonardo's *Mona Lisa*. We do so, not only to give Einstein his well-deserved credit, but also to indicate that the theory of relativity may, for all its universal significance, include lingering traces of the mind that conceived it. Science aims at perfect objectivity, but it progresses toward this goal by successive approximation. We may at any given stage have the impression that we have reached the goal, but subsequent events prove this to be an illusion.

This warning should be kept in mind throughout our tour. Every object, every site, every happening, every process, every mechanism that will be pointed out as though it were there to be seen is actually a product of individual human brains mulling and churning over collections of images and sets of figures, themselves the products of recordings made by intricate instruments on biological materials subjected to complex experimental manipulations. There is no other way of entering the cellular microcosm, and scientists are justifiably proud of having already reached as far as they have. But the risks of distortion are considerable. And so is the amount of guesswork needed to fill in all the gaps in our knowledge. As a result, even fairly well established facts may be presented very differently by different scientists, depending on their personal biases and particular areas of expertise.

Ideally, you should be given all the data and allowed to make your own judgment. But this would mean giving you what thousands of scientists throughout the world are hardly able to master collectively. Then, should not their names at least be mentioned, as is Einstein's, partly to give credit where credit is due, and especially to identify individual responsibilities? Even that, however, can be done only exceptionally, because of the way in which individual contributions have become interwoven into the complex fabric of modern science. For this reason, very few scientists will be mentioned by name in the course of our tour. But they are there anonymously—often, it must be added, overshadowed by the shortcomings and biases of the guide—behind every object that we will be given to contemplate, behind every process that we will try to understand. This must be kept in mind.

The Itineraries

A Look at the Map

Before we set off on our tour, a brief look at a cell map is in order, so we can get our bearings and have some idea of where we are going and how the visit is to be organized. In this perfunctory survey, we will do no more than identify by name the main parts of the cell and their functions. Clarification will come in due time, in the course of our tour.

The most obvious feature of any eukaryotic cell, perceived even by the early cytologists, is the distinction between the centrally located nucleus and the surrounding cytoplasm. These two parts of the cell are related to each other somewhat like the stone—nucleus is derived from the Latin word for nut—and the pulp in a cherry. Like a cherry, the cell is covered by a skin, or membrane, called the plasmalemma, or plasma membrane.

Map of the cell.

The nucleus is the repository of the cell's genetic library, which is filed in chemically coded form in distinct units, the chromosomes. As a rule, these are so intermingled as to form what appears to be a single mass of chromatin, irregularly subdivided into denser (heterochromatin) and less dense (euchromatin) parts, and including one or more specialized structures called nucleoli. A membranous envelope completely encloses this mass, which is impregnated by a fluid named nucleoplasm.

The main functions of the nucleus are directly related to information processing. They include conservation and, if need be, restoration of the genetic library and, especially, transcription, a complex and highly selective process by which certain specific instructions are read from the information store and sent out to the cytoplasm for expression. Genes exert their commanding influence over the cell by these mechanisms. When a cell prepares to divide, the nucleus has to perform an additional duty, which consists in the complete and accurate duplication of the genetic library. Subsequently, the nucleus goes through a complex reorganization called mitosis, in which the chromosomes become temporarily visible as discrete rods, and which results in the formation of two nuclei.

The cytoplasm consists of a formless jelly, the cytosol, bolstered by a cytoskeleton and containing a number of embedded organelles. These subserve a variety of functions, which can be roughly classified under "internal affairs" and "external affairs."

The internal affairs of a living cell are mainly concerned with biosynthesis and energy production. Biosynthesis is an ongoing activity, even in a nongrowing cell, because cells are not chemically static. They continuously break down and rebuild most of their constituents in a highly dynamic fashion. This activity consumes a great deal of energy. So do the other forms of work that cells accomplish in relation to motion, molecular transport, generation of electricity, information transfer, and sometimes light emission. Cells cover these requirements by breaking down energy-rich foodstuffs supplied either from the outside or from the cell's own reserves and, in green plants and photosynthetic bacteria, by absorbing and utilizing sunlight. The sum total of these reactions makes up what is generally referred to as metabolism, itself subdivided into anabolism (biosynthetic processes) and catabolism (breakdown reactions).

The systems that carry out these various activities are situated in the cytosol and in a number of organelles that maintain close relationships with it. Among these are the mitochondria, often designated as the cell's power plants, sites of the major oxidative reactions and of the mechanisms whereby the energy released by these reactions is retrieved and made available to the cell in utilizable form; the chloroplasts, which house the photosynthetic machinery of green plant cells; the microbodies, a heterogeneous family of metabolic organelles, of which peroxisomes are the most important members; a number of different motor units concerned with cell movement; the ribosomes, which are the centers of protein synthesis and, as such, the main executors of the genetic commands issued by the nucleus; and, finally, a variety of cytomembranes, which are primarily involved with the cell's external affairs (below), but which, in addition, house a number of important metabolic systems.

Under the heading of external affairs we group the various activities involved in communication and in exchanges of matter between the cell and the world around it. These activities are shared between the plasma membrane, which is the actual cell boundary, and an elaborate network of intracellular membranes related to the plasma membrane and organized into a large number of closed, pouchlike structures. Capable, directly or indirectly, of establishing transient connections with each other or with the plasma membrane, these structures serve in the storage, processing, and intracellular transport of materials that are either taken in from the outside and broken down intracellularly or made inside the cell for extracellular discharge. A key feature of these exchanges is that they occur without the membranes involved ever suffering the slightest gap or tear. There is thus always a membrane boundary between the contents of the pouches and the cytosol. For this reason, these membranes act like the plasma membrane to the extent that they also mediate exchanges between the cell and its surroundings by means of the segregated spaces that they delimit.

Designated vacuome by the early cytologists—the term is largely abandoned, but deserves to be resurrected—this cytomembrane system consists of two distinct, though closely interconnected, sections, dedicated respectively to import and to export, and themselves subdivided into functionally distinct subsections. The import department includes specialized areas of the plasma membrane concerned with uptake of extracellular materials by a mechanism called endocytosis; a storage compartment made up of endosomes, concerned mainly with the sorting and routing of the materials taken up; and a complex of digestive vacuoles or lysosomes, in which these materials are broken down. The export system starts with the rough-surfaced endoplasmic reticulum, which collects and processes newly made export proteins manufactured by ribosomes bound to its limiting membranes. This structure communicates, by means of the smooth-surfaced endoplasmic reticulum (without attached ribosomes), with a complex system known as the Golgi apparatus. Further processing, as well as sorting, of the export materials goes on in these two subdivisions. From the Golgi apparatus, the materials are then eventually directed, after storage and concentration in secretion granules, toward the cell periphery, where they are discharged by exocytosis. A special set of vesicles conveys materials from the Golgi apparatus to the lysosomes. Others are involved in recycling of the membranes that participate in these various processes.

Based on this blueprint, our tour will be divided into three distinct itineraries.

Itinerary 1: The Outskirts and Surface of the Cell, the Vacuome

We will start by approaching the cell progressively, coming from a blood vessel, so as to have an opportunity to see some of the extracellular structures. Then, after a good look at the cell surface, we will explore the vacuome, entering by endocytosis and exiting by exocytosis. In between, we will take advantage of the local mass-transportation system to visit all the different parts of the complex and to watch firsthand, and practically share, the intracellular travels and experiences of such materials as either are imported by the cell from the outside or are manufactured by it for export. This first part of our tour will also provide us with convenient opportunities for becoming acquainted with the main cell constituents—the proteins, polysaccharides, and lipids—and with some of the structures that they build together.

Itinerary 2: The Cytosol and Cytoplasmic Organelles

Our second itinerary will take us directly into the cytosol, from which we will be able to call successively at all the major organelles that are connected with it. In doing so, we will be able to learn something about metabolism and the major principles that govern energy transformations and biosynthetic mechanisms. By the time we reach the ribosomes, we will receive our first introduction to biological information transfer and to the molecules that mediate it, the nucleic acids.

Itinerary 3: The Nucleus

Finally, our third itinerary will cover the nucleus, where we will see the genes in action, as well as the chromosomes going through the complex transformations that are associated with mitosis and with a related process called meiosis. Briefed as we will be by then, we will be able to address some key biological problems, including the origin of life and the mechanism of evolution, as well as such important issues as the mechanism of cancer and the future of biotechnology. We will take advantage of the turmoil that accompanies cell division to effect our final exit.

Underwater Equipment Needed

One last point before we set out: cells live in a watery world, even when the organisms of which they are part do not. Take our own cells, for instance. Except for the outer layers of our skin, which consist of mummified dead cells, all cells are immersed in liquid, either blood or some fluid derived from blood. Similarly, plant cells are bathed by sap. Even the hardiest bacteria need some moisture around them. They may survive complete dryness, but only in a dormant state, with all their processes arrested until they are reawakened by water. The same may happen with more complex cells, such as those of molds and of seeds.

Thus, cell tourists really resemble those underwater explorers popularized by the films of Jacques Cousteau. They carry out their visits swimming. Indeed, much of what they see would shrivel away but for the surrounding water, much like the delicate creatures that cover the surface of submerged rocks. So strap on your aqualung and plunge. We are off.

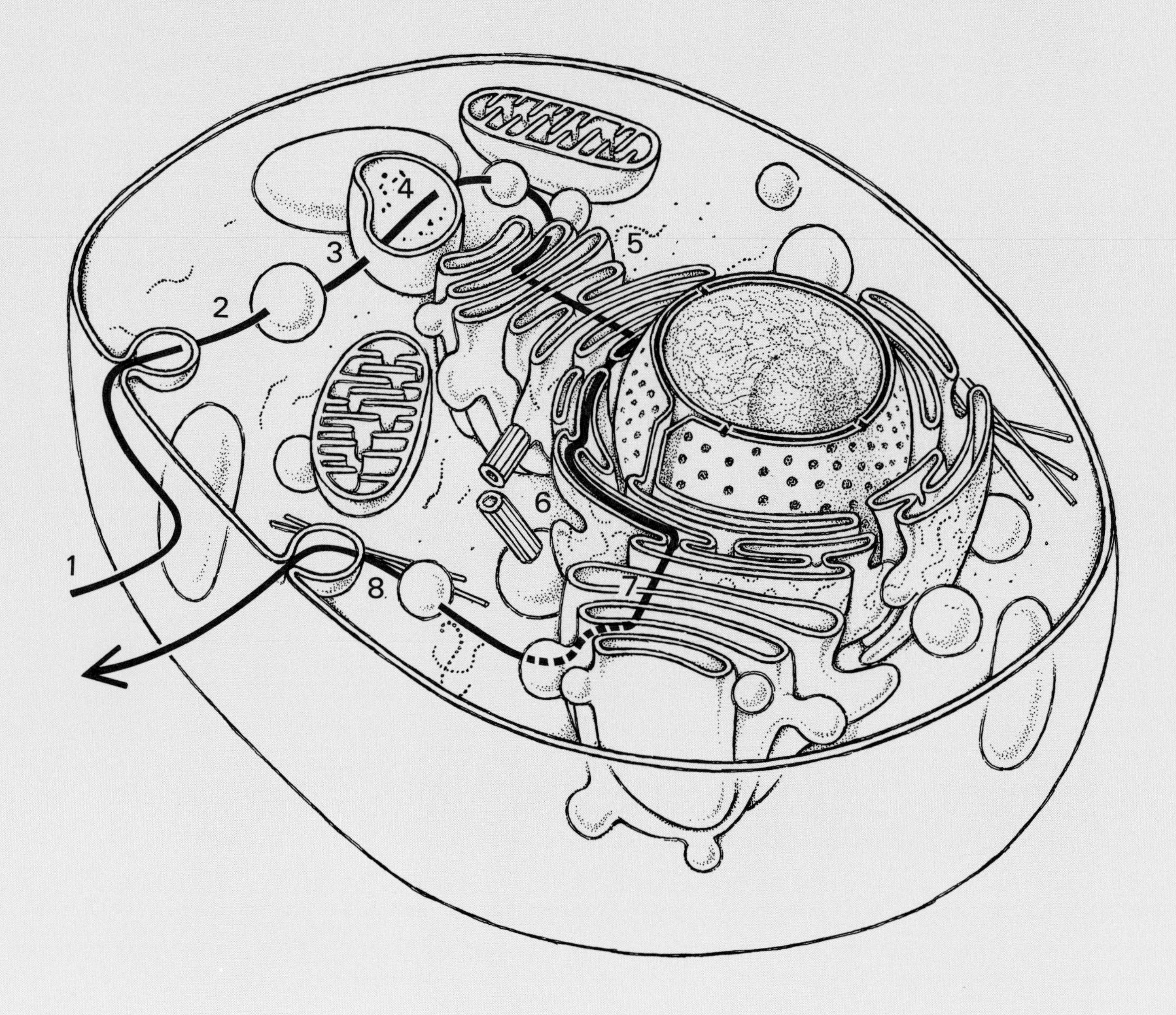
1
2
3
4
5
6
7
8

ITINERARY I

The Outskirts and Surface of the Cell, the Vacuome

1 Approach the cell through extracellular structures and explore its surface.

2 Enter by way of endocytosis and watch sorting process in endosome.

3 Reach lysosome by endosome-lysosome fusion.

4 Visit lysosomal space.

5 Board lysosome-Golgi shuttle and proceed immediately upstream into endoplasmic reticulum to reach a rough-surfaced recess.

6 Watch synthesis of secretory proteins and follow their processing down endoplasmic reticulum and back into Golgi apparatus.

7 Visit Golgi apparatus.

8 Board secretion granule and leave by exocytosis.

2 Extracellular Structures, With an Introduction to Polysaccharides and Proteins

Before we actually enter a living cell, we should take a look at its surface, which has many interesting features. But it is rarely a simple matter to find a cell, for most cells are surrounded by a more or less elaborate network of outer defenses and scaffoldings that often completely hide them from view and greatly hamper approach to them. Such structures are not parts of the cells but are built from precursor materials that are secreted by the cells and that subsequently join together into a variety of combinations of almost every possible shape and consistency, from the softest of jellies to the toughest of shells and woods. Sometimes they become impregnated with minerals, which may provide their organic matrix with the durability of stone or the hardness of enamel. These extracellular constructions act as props for the cells; they provide every sort of visible form that life creates on our planet. Without them, there would be no trees, no flowers, no animals; nothing but an amorphous covering of oozy slime made of a myriad of naked cells crawling over each other.

An Architect's Dream

Extracellular structures are particularly extensive in plants, where they form a continuous rigid scaffolding within which each individual cell is provided with a small chamber of its own—the original "cells" first seen by Robert Hooke as microscopic holes in a slice of cork. The main component of this framework is appropriately called cellulose. Most abundant of all organic substances on earth, cellulose is described by chemists as a

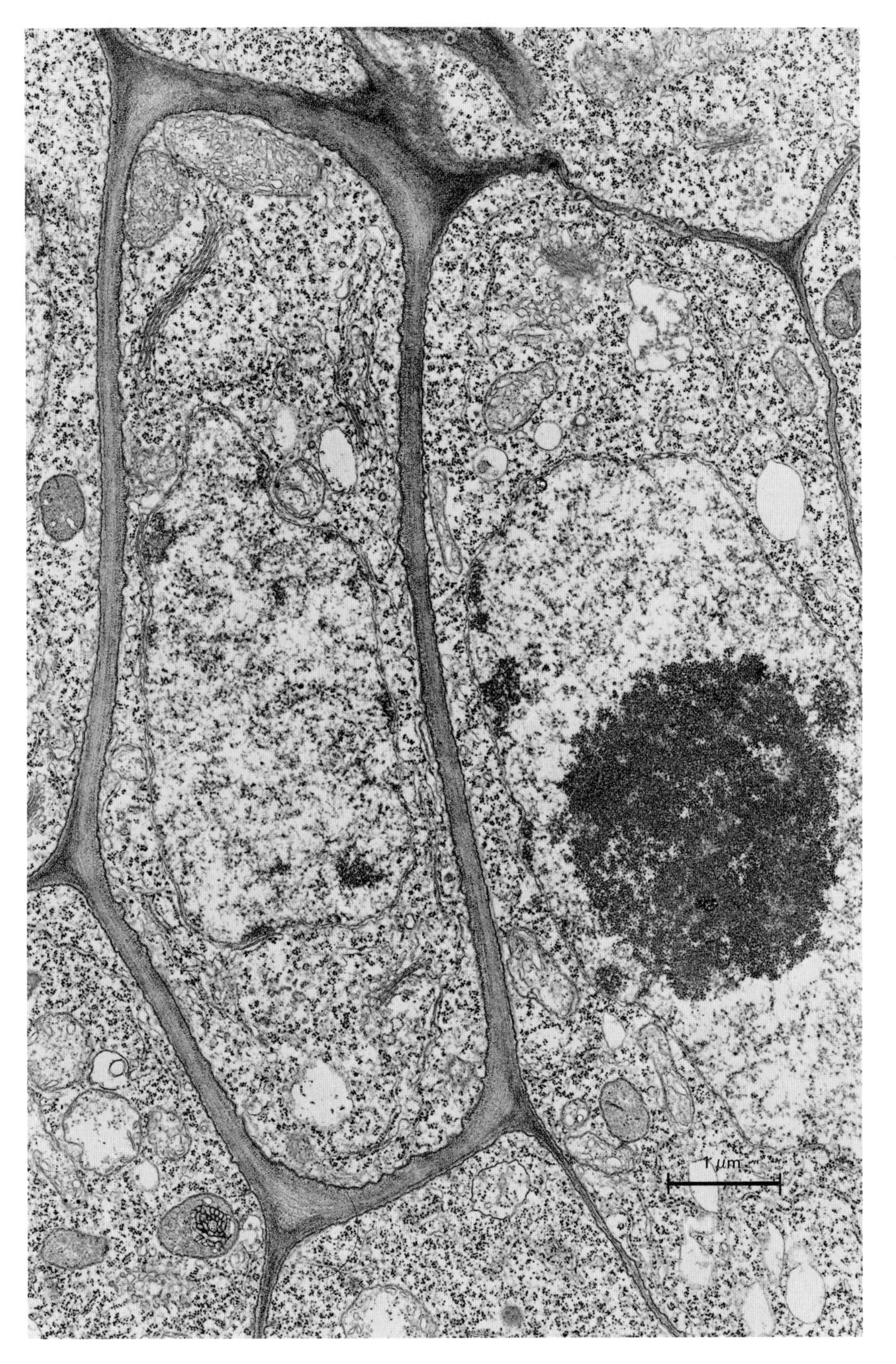

Electron micrograph of a transverse section of a root tip of the bean *Phaseolus vulgaris* showing how each cell occupies a closed chamber. Only the walls of the chambers were seen by Robert Hooke (see the illustrations on p. 6).

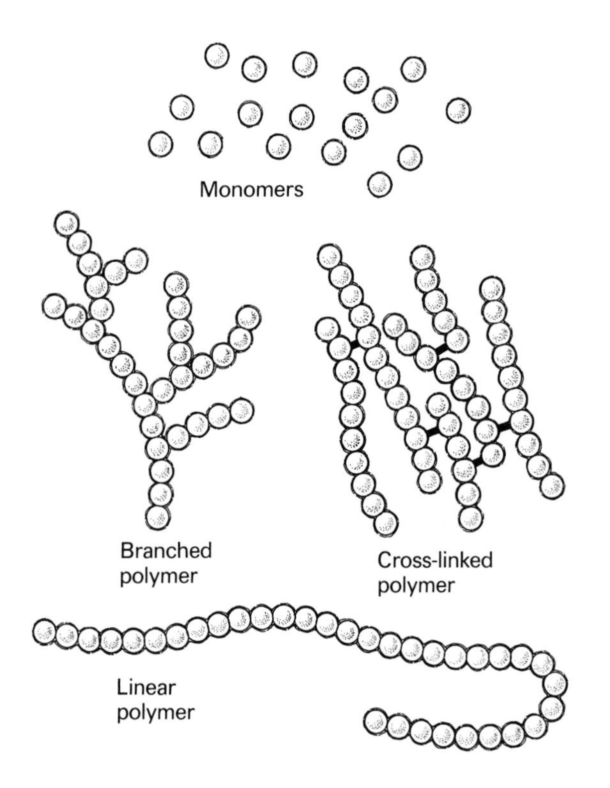

Biological polymers are made of monomers, which are variously shaped molecules ranging between about 0.3 and 1 nm in size. Magnified a millionfold, they would be just visible as tiny flecks.

polymer of the simple sugar glucose. Polymers are giant molecules made of many (Greek *polys*) parts (Greek *meros*) called monomers (Greek *monos*, single). All our plastics and artificial fibers are polymers. So are the main constituents of living organisms, including the polysaccharides (Greek *sakkharon*, sugar), to which cellulose belongs, and the proteins, which we will consider in a moment. Unlike most man-made polymers, which are usually of an indefinite size, limited essentially by the mold in which they are cast at the time of polymerization, natural polymers have definite molecular sizes. They consist of macromolecules (Greek *makros*, large), in which the building blocks are assembled in a specific fashion. Their molecular masses range mostly from a few thousands to a few hundreds of thousands of daltons. (The dalton is the mass of the hydrogen atom.) Exceptionally, they may reach or exceed one million daltons. The corresponding molecular sizes are of the order of a few nanometers, if the molecules are globular. Fibrous molecules may be as much as several hundred nanometers long. Most structural macromolecules are fibrous and are naturally endowed with the property of combining into characteristic multimacromolecular assemblages. Quite often these assemblages are reinforced by chemical cross-links.

Cellulose is a particularly simple polymer, made entirely of a single type of monomer, glucose. It is thus identical in gross composition with starch, the main carbohydrate reserve substance of plant cells, and with glycogen, its animal counterpart, which are also simple glucose polymers. But, through a quirk of its chemical structure, cellulose is extremely resistant to degradation, both biological and chemical. After purification, it serves as the almost indecomposable fibrous component of cotton, linen, and paper, and of such chemical derivatives as cellulose acetate or cellophane. Only certain microorganisms can degrade cellulose. Their favorite ecological niche is the digestive tract of herbivores, much to the advantage of the animals, who owe their ability to utilize cellulose to these friendly guests. Most other mammals—including ourselves—pass out cellulose essentially intact. It is the "fiber" in our food and forms much of the solid part of stools. Plants use a number of other structural molecules besides cellulose. Trees, in particular, depend largely for their exceptional resilience on a phenolic polymer called lignin (Latin *lignus*, wood).

Bacteria also are completely surrounded by a rigid wall. It includes a number of substances that are peculiar to the bacterial world. This wall acts as a protective shell that allows the bacterial cell to withstand very harsh environmental conditions. It is also important from the medical point of view. Infectious microbes depend largely on

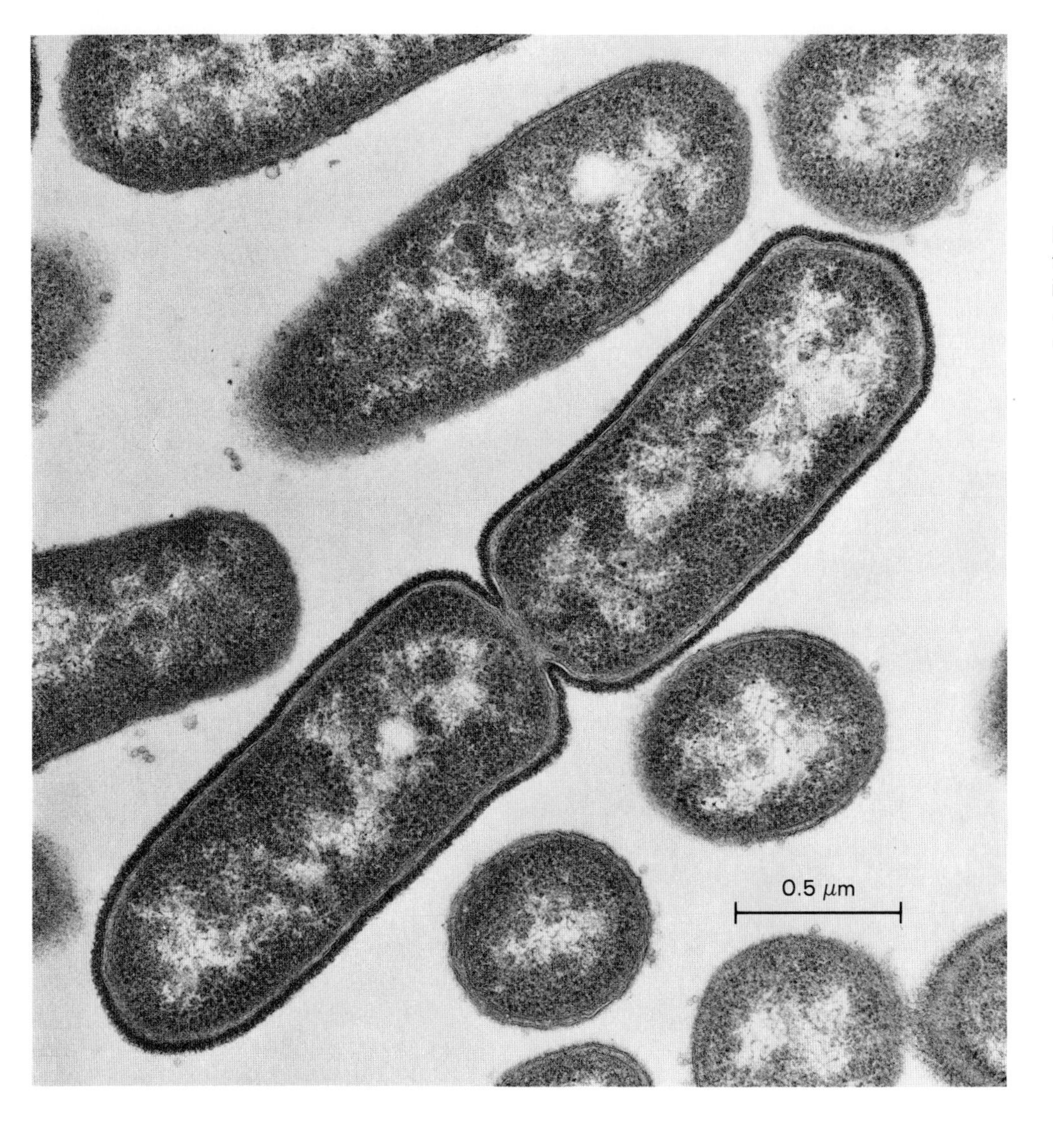

Electron micrograph of a section through a group of *Bacillus fragilis* illustrating the bacterial cell wall. The cells in the middle are in the process of separating.

their outer covering to resist or elude our defenses. Some even use it posthumously to launch a toxic attack. In turn, our own cells, especially the white blood cells, which form our main defense corps, "see" only the wall of the invader. They are alerted to it, and they ultimately kill the bacteria by a complex series of motions triggered off by that chemical recognition. (Some of the mechanisms are considered in Chapters 4 and 5.) Interestingly, penicillin and related antibiotics actually kill microbes by interfering with the construction of the cell wall, thus blocking bacterial growth. Lysozyme, a natural antibacterial agent occurring in white cells and in tears, kills bacteria by breaking down their cell walls.

In the animal world, cells are rarely encased individually. Most often they huddle into intimately clustered groups of characteristic shape, which themselves are then bolstered and buttressed by extracellular structures of various sorts. This kind of arrangement is largely responsible for the richness of organization and evolutionary capability of animals. For one thing, it has allowed the development of nervous systems. Unfortunately, time does not permit us to explore the admirable manner in which each particular tissue or organ is architecturally adapted to perform its function. But we should at least take a look at the main components that are used in this remarkable construction work. Several of these components belong to the all-important class of proteins, of which we will encounter many representatives all along our tour. Here is obviously a case where some chemical briefing is mandatory.

A Look at Proteins

In 1838, the Dutch chemist Gerardus Johannes Mulder, a pioneer in the analysis of "albuminoid" substances (Latin *albus*, white; *albumen*, egg white), adopted the name protein (Greek *proteios*, primary) to designate what he thought was the basic constituent of heat-coagulable nitrogenous substances such as blood fibrin, milk casein, and egg albumin, which were beginning to be recognized at that time as belonging to a common class. The term protein was suggested to Mulder in a letter by the Swedish scientist Jöns Jacob Berzelius, one of the founders of chemistry and the father of the concept of "catalysis."

Mulder and Berzelius were guilty of vast oversimplification, inevitable in their day, but the proposed terminology was prophetic. Not only have the proteins turned out to be the primary agents of all living processes, but they also share with the Greek sea-god Proteus the ability to take on innumerable different shapes. Proteins are also protean.

We can readily appreciate this by adjusting our molecular eyeglasses to a high degree of resolution, so that we can examine in detail objects only a few nanometers in size. At our adopted millionfold magnification, such objects would still be only about the size of a gnat. Under an appropriate magnifying glass, however, gnats and other small creatures of similar size reveal a wealth of different shapes and structures. So do proteins if we increase the magnification further, except that their forms are more abstract, resembling in their sleek curvatures the biomorphic sculptures of Hans Arp, rather than the articulated angularities of insects. Some are globoid, almost spherical; others have more tormented shapes, bent, twisted, or bulging with tuberosities; yet others thin out to slender threads, often helically coiled. These shapes are not immutable. They swell, pulsate, elongate, contract, or uncoil, sometimes with dramatic suddenness.

Most proteins are situated inside cells, where their main function is concerned with catalysis, a fact which, had he but known it, would have delighted Berzelius. The Swedish chemist coined this word (Greek *kata*, down; *lysis*, loosening) to designate the property of substances (called catalysts) that facilitate the occurrence of chemical transformations without themselves being consumed in the reactions. Inorganic catalysts are known and are used in the chemical industry, but the real magicians of catalysis are to be found in living cells. Life, even that of a humble microbe, depends on the performance of thousands of chemical reactions, most of which either cannot be reproduced under artificial conditions or can be so only under conditions—such as high pressure, temperature, or acidity—that are incompatible with life. In living cells, these reactions take place with remarkable ease and rapidity, thanks to the mediation of specific catalysts called enzymes. This word, which is derived from the Greek *zyme*, yeast, recalls the historical role of the study of alcoholic fermentation in the characterization of the first intracellular catalysts (see Chapter 7). The term ferment has also been used, for the same reason, but is now abandoned.

Virtually every chemical reaction that occurs in living organisms is catalyzed by a specific enzyme, often assisted, as we shall see in the second part of our tour, by one or more cofactors, or coenzymes. All known enzymes are proteins. Thus the number of different protein varieties is considerable—several thousand, at least, in any given cell, where they generally account for some 20 per cent of the wet weight of the cell, or more than half its dry weight. In addition to their functions as catalysts,

Top: *Serpent,* by Hans Arp.

Bottom: Plasticine model of a molecule of myoglobin, the oxygen carrier of red muscle cells. The white wormlike structure represents a coiled polypeptide chain (see the illustration on p. 33), the dark disk an iron-containing heme group, which binds oxygen (see Chapter 9). Hemoglobin, the oxygen carrier of red blood cells, consists of four subunits shaped like the myoglobin molecule.

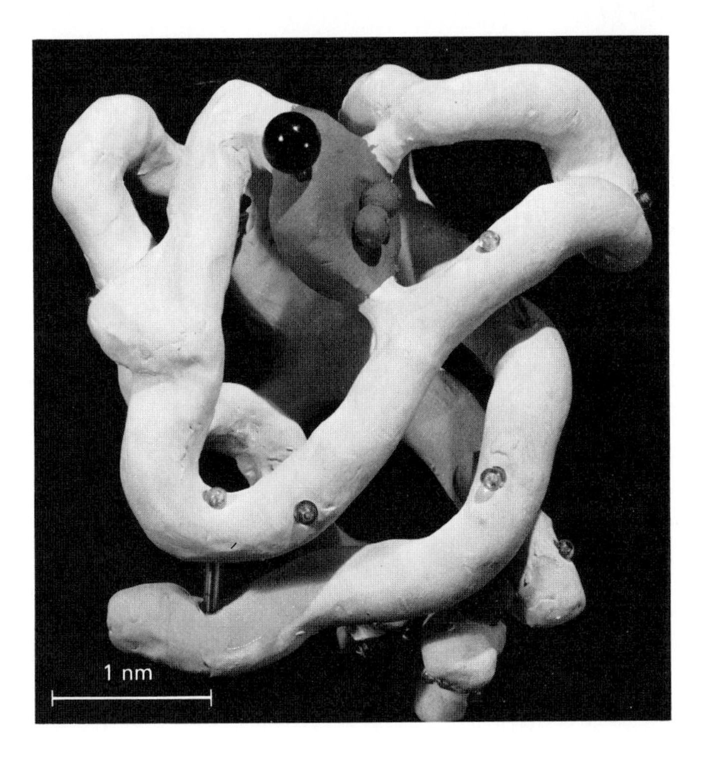

proteins also serve in regulation, transport, locomotion, and many other activities. They also play an important role as structural components, both intracellularly and extracellularly. The proteins are indeed the primary constituents of living cells; they deserve their name.

Proteins readily lose their unique shape, as well as their catalytic power, when subjected to heat or other physical or chemical agents. This process is called denaturation. Long believed to be irreversible, protein denaturation has often been used as an argument in support of the need for a special force or principle to guide the assembly of living structures. This view is no longer held, and denaturation has, in fact, been shown to be reversible in a number of cases. There is nothing exceptional about the shape of a protein molecule. It is no other than the most probable conformation, the one the molecule is most likely to assume if given the opportunity to do so.

It is instructive to look with our molecular eyeglasses at a mixture of denatured protein molecules. Except for differences in size, they now have the same general appearance. All we see, in lieu of the beautiful variety of shapes and structures, is a tangle of long, exceedingly thin threads. Now we understand better why denaturation often behaves as an irreversible phenomenon. Unless special conditions are provided, each thread will remain caught in the tangle, unable to wind back into its characteristic shape. Even more important, denaturation tells us something of major significance about the structure of protein molecules, which otherwise might well have escaped our scrutiny. All proteins are filamentous molecules. The innumerable shapes with which they present us are all generated by coils and twists of a threadlike arrangement of atoms.

As we know from chemical studies, the common structure behind all these different shapes is a backbone made of a simple, six-atom repetitive unit recurring as many as several hundred times:

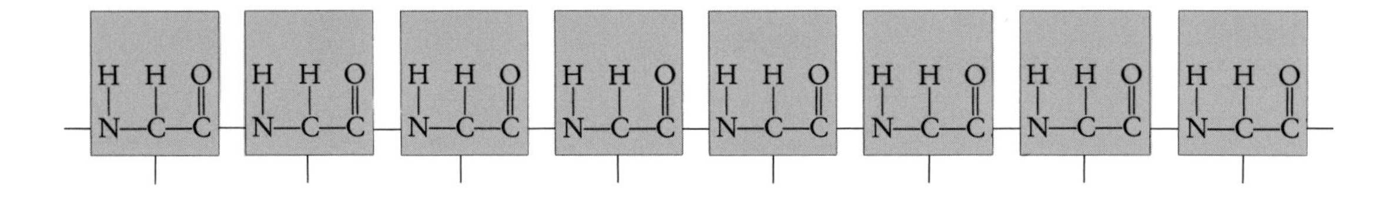

The central carbon of each unit has one valence that is not engaged in the backbone. It bears one of twenty distinct chemical groups, which are responsible for the specific properties that distinguish different protein molecules from each other. The link between the units is called the peptide bond, a name that goes back to the discovery that this bond is hydrolyzed by the digestive enzyme pepsin, a component of gastric juice (Greek *pepsis*, digestion):

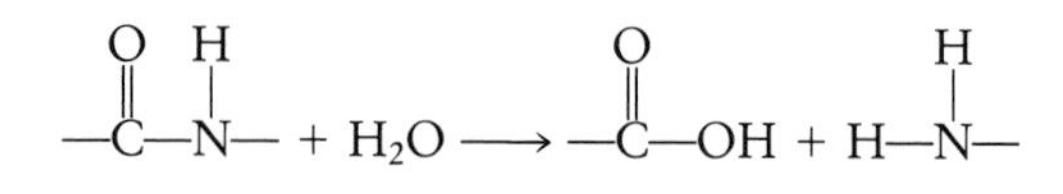

When all such bonds in a given backbone have been split, the resulting products all have the structure:

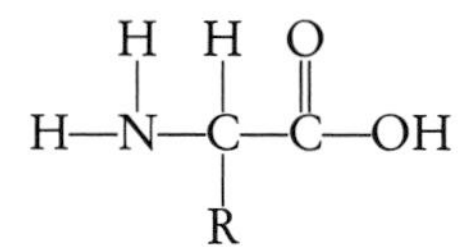

They are designated amino acids, because they contain both the amino group (—NH_2) and the carboxyl acid group (—COOH); more specifically, α-amino acids, because the two groups are attached to the α-carbon; and, even more specifically, L-α-amino acids, because they all (except for the optically inactive glycine) exhibit the L type of configuration around the asymmetric α-carbon. Their stereoisomers, the D-amino acids, occur in certain bacterial constituents, including some antibiotics, but not in proteins.

Twenty different amino acids take part in the formation of proteins. Their structures (see Appendix 1) conform (with one minor variant, proline) to the structure near the top of this page, differing from each other by the nature of the side group, R. The names of the amino acids are listed in the table on the facing page, along with the abbreviations and symbols used to represent them. As we shall see later, some of the symbols also represent another important group of substances, the nucleosides. Within the appropriate context, there is no risk of confusion. If there is, the three-letter abbreviations are employed.

Amino acid	Abbreviation	Symbol
Alanine	Ala	A
Arginine	Arg	R
Asparagine	Asn	N
Aspartic acid	Asp	D
Cysteine	Cys	C
Glutamic acid	Glu	E
Glutamine	Gln	Q
Glycine	Gly	G
Histidine	His	H
Isoleucine	Ile	I
Leucine	Leu	L
Lysine	Lys	K
Methionine	Met	M
Phenylalanine	Phe	F
Proline	Pro	P
Serine	Ser	S
Threonine	Thr	T
Tryptophan	Trp	W
Tyrosine	Tyr	Y
Valine	Val	V

Amino acids attached to each other by peptide bonds are designated residues. The resulting chains are called peptides. The term peptide is often preceded by a Greek prefix indicating the number of amino acid residues in the chain—for instance, *di*, two; *tri*, three; *tetra*, four; *penta*, five; *oligo*, a few; *poly*, many.

Peptide chains conform to the general structure:

They have a free amino end-group (N-terminal at the left), a free carboxyl end-group (C-terminal at the right), and $n - 1$ peptide bonds.

Peptides differ from each other by the number (n), nature, and ordering, or sequence, of their amino acid residues. They may be likened to words of variable length written with an alphabet of twenty letters. The simplified one-letter symbols make this particularly clear. For instance, the tetrapeptide Cys-Glu-Leu-Leu becomes CELL, and the nonapeptide Ala-Arg-Cys-His-Glu-Thr-Tyr-Pro-Glu becomes ARCHETYPE. In fact, this whole book could be written in peptide language, were it not for the absence of B, J, O, U, X, and Z in that alphabet.

Peptides are flexible structures, owing to the possibility of rotation around the N—C and C—C axes in the backbone, though not around the peptide bond itself. As a result, they adopt more or less contorted shapes, depending on the attractions and repulsions existing between their parts and on the ability of these parts to either bind or exclude water molecules. In addition, the folding of a peptide chain is often influenced by the presence of substances with which it is able to associate. Many of these interactions are physical and do not involve the formation of true (covalent) chemical bonds. They depend on two types of forces, which are often designated by the names of the scientists who discovered them: the French physicist Coulomb and the Dutch chemist van der Waals.

Coulomb forces are electrostatic. They govern the attraction between electric charges of opposite sign and the repulsion between charges of the same sign. Some amino acids have R groups that are charged, either positively or negatively, under physiological conditions. Others, without being charged, are polarized—that is, they show a local charge displacement that creates a positive and a negative pole. All such groups can interact electrostatically.

Polar groups bearing a hydrogen atom can bind electrostatically to negative or negatively polarized groups by means of a special bond, called the hydrogen bond, which involves some sort of sharing of the hydrogen atom. A very important such bond in biology is the following:

$$\overset{\ominus}{N}-\overset{\oplus}{H}\cdots\overset{\ominus}{O}=C$$

This bond (the dotted line) can join two peptide linkages. It thereby plays an important role in the conformation of proteins, and consequently also in all the structural, catalytic, and other functional properties that depend on this conformation. As we shall see in the later part of our tour, the phenomenon of base-pairing in nucleic acids, which governs all the transfers of genetic information in the living world, depends on the same bond, and on another, similar, hydrogen bond:

$$\overset{\ominus}{N}-\overset{\oplus}{H}\cdots\overset{\ominus}{N}$$

Van der Waals forces are responsible for the attraction that exists between nonpolar groups made only of carbon and hydrogen, such as constitute the hydrocarbons found in gasoline and other petroleum products. A number of amino acids have R groups that can establish connections by means of van der Waals forces.

A point of key importance with respect to these interactions is that the water molecule has an asymmetric structure, which makes it polar.

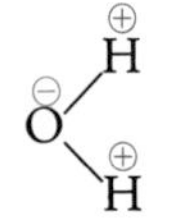

Hence, it has the ability to bind electrostatically to all electrically charged or polarized groups. These are called hydrophilic for this reason. In contrast, nonpolar groups have no affinity for water; they are hydrophobic, water repellent. We are all familiar with this phenomenon: oil does not mix with water, certain plastic surfaces don't get wet; they are hydrophobic. Actually, this term is a little misleading. Hydrophobic groups do not really repel water; they are excluded by it, as a result of the strong hydrophilicity of the water molecules themselves, which keeps them huddled together by means of hydrogen bonds.

Most of the cellular milieu being aqueous, interactions with water have much to do with the kind of conformation that a polypeptide chain will tend to adopt. In very schematic language, we may say that whenever this is structurally feasible, a polypeptide chain will fold in such a way as to expose the greatest number of hydrophilic groups on its surface, or in clefts accessible to water, and to bury most of its hydrophobic groups together in internal regions from which water is excluded and in which the groups can interact with each other. Proteins that lend themselves to this kind of group segregation on the basis of water affinity are generally soluble in water. Those that are unable to bundle themselves inside a hydrophilic shell tend to home to hydrophobic regions, where their hydrophobic parts are made welcome by van der Waals types of interactions. As we shall soon find out, membranes provide the main resting place of this kind.

An important element in determining the conformation of proteins is represented by the tendency of polypeptide chains to twist into a corkscrew kind of arrangement stabilized by hydrogen bonds between nearby peptide linkages. The most common such arrangement is the α-helix, which has a pitch of 0.54 nm and contains 3.6 residues per turn. It forms a relatively rigid rod about 1 nm thick, with a knobby surface shaped by the R groups of the residues. Some amino acids cannot fit into this arrangement, and break it. Consequently, α-helical parts are usually relatively short segments. Whole protein molecules often contain several such rods, joined, usually at an angle, by less regular parts. Exceptionally, the sequence of a polypeptide may be such as to allow the α-helical disposition to be maintained over a very long distance. Such molecules are naturally filamentous. Other arrangements besides the α-helix, sometimes involving two or three polypeptide chains twisting together (coiled coils), may also provide the molecules with a threadlike struc-

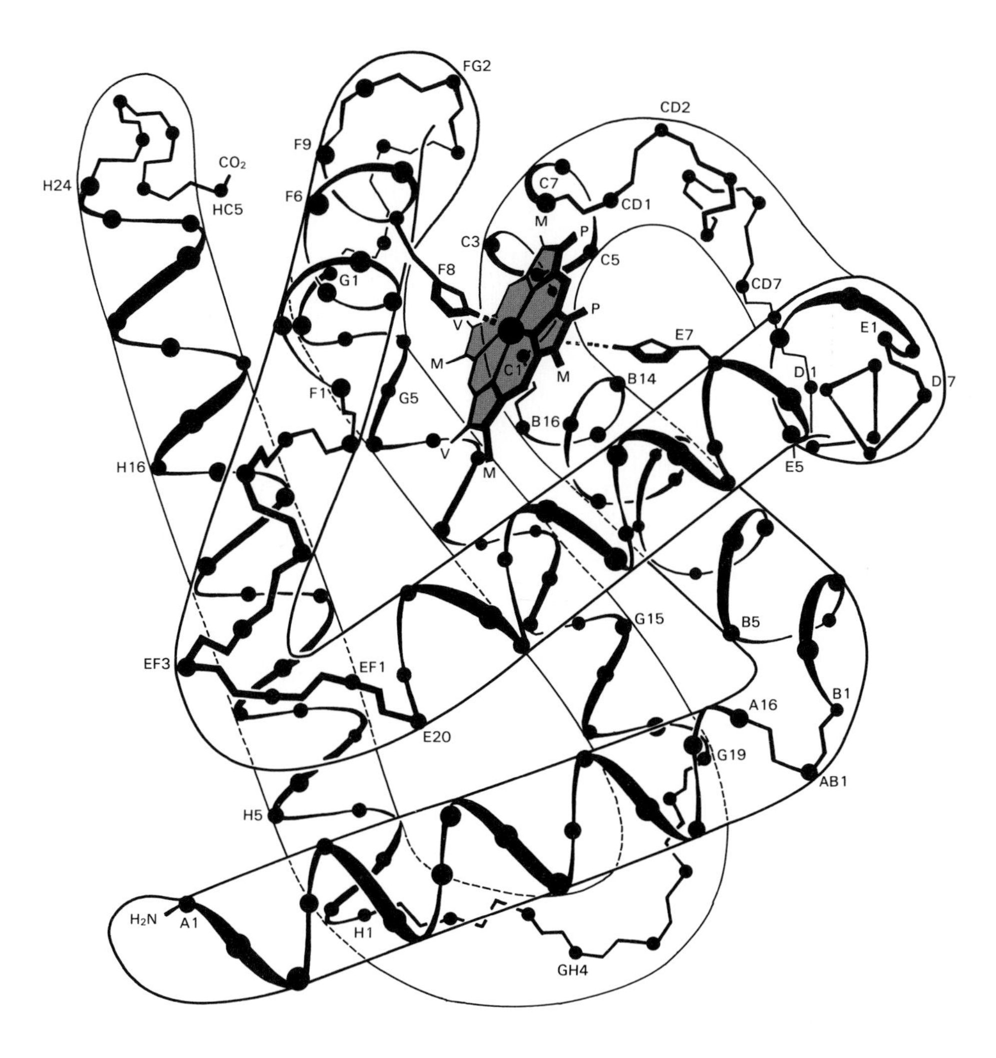

Structure of a myoglobin molecule. The protein includes eight α-helical segments folded around the inserted heme (shown in color). (Compare model on p. 29.) The letters A through H refer to α-helical segments, numbers to amino acid residues.

ture. The functions of such fibrous proteins are mainly concerned with structural support and locomotion.

To these various intrinsic factors that determine the shape that a given polypeptide chain will adopt must be added the interactions between the chain and other molecules of either protein or nonprotein nature. Many of the most important functions of proteins depend on such interactions, which often have a crucial molding effect on the conformation of the molecules involved. Sometimes a given structure is sealed by a true covalent bond between two parts of the same polypeptide chain or between two chains. The most common such bond is the disulfide linkage —S—S—, which arises by the oxidative joining of the thiol (—SH) groups of two cysteine residues:

$$2\ \text{Cys—SH} \longrightarrow \text{Cys—S—S—Cys} + 2\ \text{H}$$

Many proteins consist of a single polypeptide chain; others contain two or more distinct chains or subunits, linked together by noncovalent attractive forces, sometimes also by covalent bonds, such as disulfide bridges.

So much for this preliminary briefing. More will be said about proteins when we meet some particularly characteristic member of this immensely important, as well as varied, group of natural molecules. But now we should

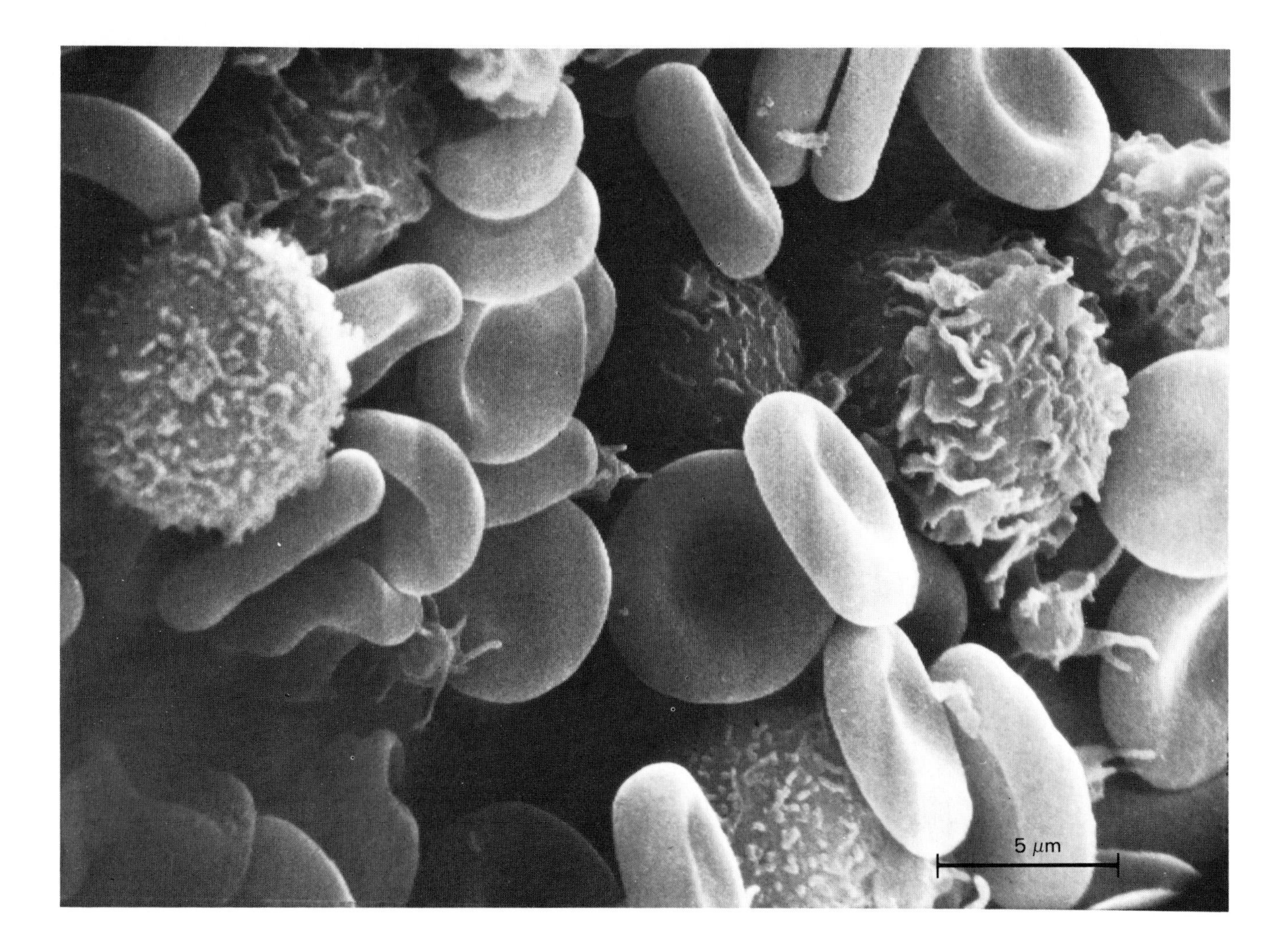

start our tour without further ado. A convenient approach to our goal is by way of the circulatory system, which carries most of the traffic to and from cells.

Scanning electron micrograph of blood cells from a patient suffering from Hodgkin's disease. The biconcave disks are erythrocytes; the other cells are leukocytes.

Blood Cells, Our First Companions

Blood vessels are turbulent waterways, especially the larger arteries where we find ourselves at the start of our journey, rhythmically propelled by a powerful swell at every heartbeat. All around us are cells that, like ourselves, are carried along and tossed about by the swiftly moving current of plasma, the fluid part of the blood.

Different types of junctions.

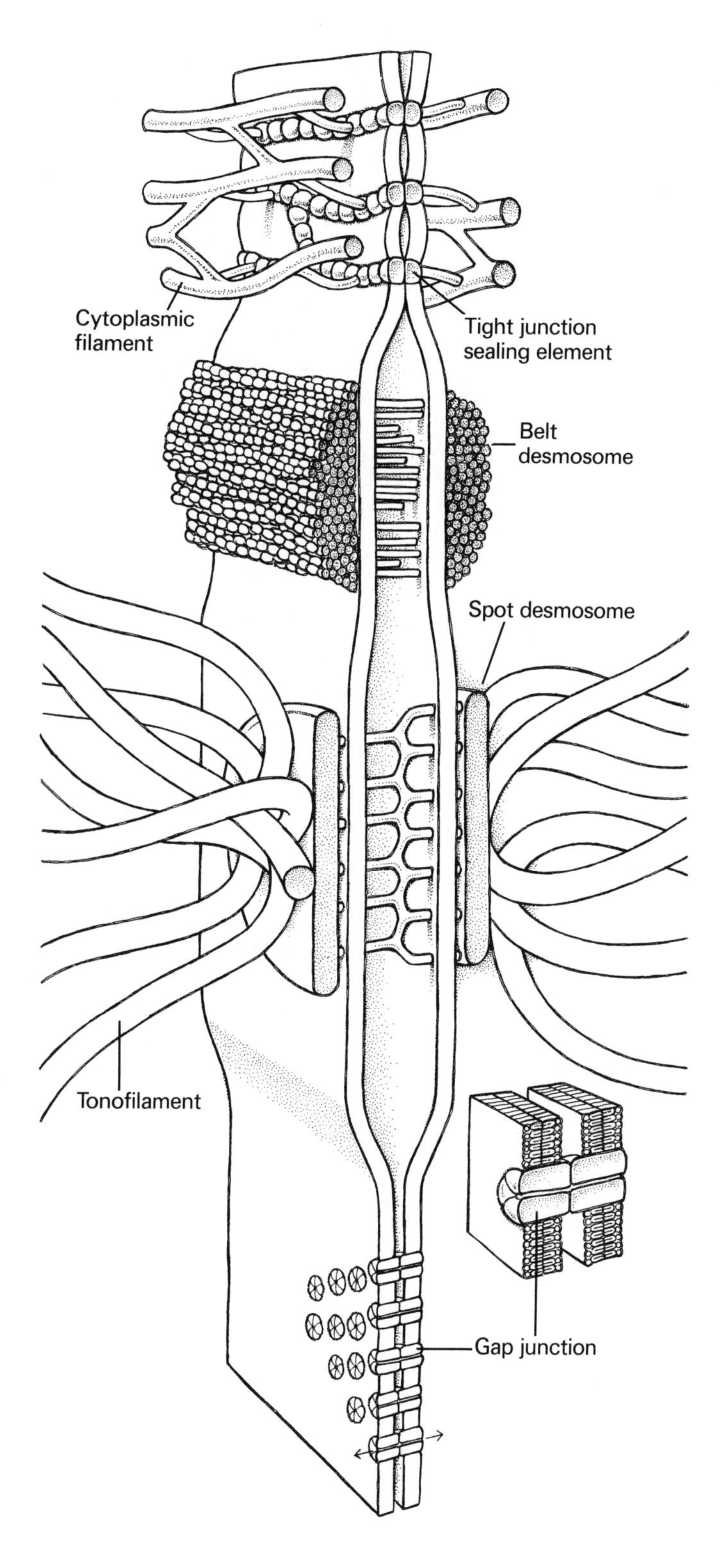

Most of these blood cells are small biconcave disks filled with a ruby-red material. They are the red blood cells, or erythrocytes (Greek *erythros*, red), whose main functions are to bring oxygen from the lungs to the tissues and to help bring the waste-product carbon dioxide back to the lungs. They are extremely valuable to the economy of the organism and contain a very important protein, the red oxygen-carrier hemoglobin. But as cells they are hardly worth a second glance, since they are completely degenerate and virtually moribund. Red blood cells have no mitochondria, no ribosomes, no intracellular membrane system, and, in mammals, no nucleus.

On rare occasions, we may bump into a white cell, or leukocyte (Greek *leukos*, white). White cells occur in different sizes and shapes. But these fleeting encounters hardly afford us a satisfactory view. We will meet some of them again later, under more restful conditions.

As the arteries become narrower and the force of the current begins to spend itself, we start catching glimpses of the cells that line the walls of the blood vessels. They are called endothelial cells (Greek *endon*, inside; *thele*, nipple) and form a single layer looking somewhat like an irregular tiling or pavement. They are exceedingly flat over most of their surface, so that their nuclei bulge out into the lumen of the vessel, thus giving the wall a knobby appearance. Endothelial cells are cemented together by different types of junctions to form a thin, continuous, tubular sheath that serves as an important filter and regulator of exchanges between blood and tissues. Such junctions are not peculiar to endothelial cells. Most cells are attached in this manner and organized into sheets, columns, clusters, sacs, or any other of the various multicellular arrangements of which the tissues and organs are composed. The strongest intercellular junctions are the desmosomes (Greek *desmos*, bond; *soma*, body), in which the two cells are connected by a dense joint and riveted together by bundles of transverse fibers extending deeply into their cytoplasms; these are known as tonofilaments (Greek *tonos*, tension). Several other types of junctions exist, including one, called gap junction, in which adjacent cells are connected by closely packed rows of hollow, cylindrical pegs, about 15 nm long, and 8 nm in diameter, with a 1.5- to 2-nm bore. Plugged at each end into the membranes of adjacent cells, these structures keep the two cells separated by a narrow gap but allow the intercellular passage of electric currents and of small ions and molecules through their inner channel.

Without additional supporting structures, the multicellular connections established by these junctions would be quite unable to stand even the weakest of stresses. The junctions would resist, but not so the membranes they serve to join, which, as we shall see, are very flimsy films, hardly stronger than soap bubbles. Blood vessels are particularly in need of such reinforcement to withstand the pressure imposed upon them by the contractions of the heart. Thus the larger arteries are enveloped in a thick, resilient casing, almost impassable by even the smallest molecule. But as the vessel becomes narrower and the blood pressure decreases, the wall becomes thinner. By the time we reach the capillaries, which are very narrow, hairlike ducts (Latin *capillus*, hair), only a thin sheet, called basement membrane (see p. 40), remains to back up the endothelial lining. Here is where most exchanges between blood and tissues take place—and where we ourselves should leave if we don't want to get caught in the venous return traffic. But to do so, we must pry open a junction between two endothelial cells. As we wonder how to do this, unexpected help comes from a passing leukocyte that, before our very eyes, is undergoing an odd change in behavior.

A moment ago, this cell was just drifting along. Now it clings to the endothelial lining and forcibly worms its way between two cells in what seems to be a most frantic activity. It is, in fact, displaying chemotaxis (Greek *taxis*, arrangement), in response to exposure to some chemical. Prompted somehow by the encounter, the cell starts to follow the chemical's trail back to its source—that is, it moves up the chemical's concentration gradient, cutting its way through intervening structures by means of specially secreted enzymes. This remarkable, and still poorly understood, phenomenon is elicited by substances that are commonly released at a site of some injury or bacterial infection. It is most useful to the organism, therefore, as it automatically brings the white cells to where their defensive properties are needed. As an unexpected fringe benefit to our party, it has opened a gap for us through the capillary wall.

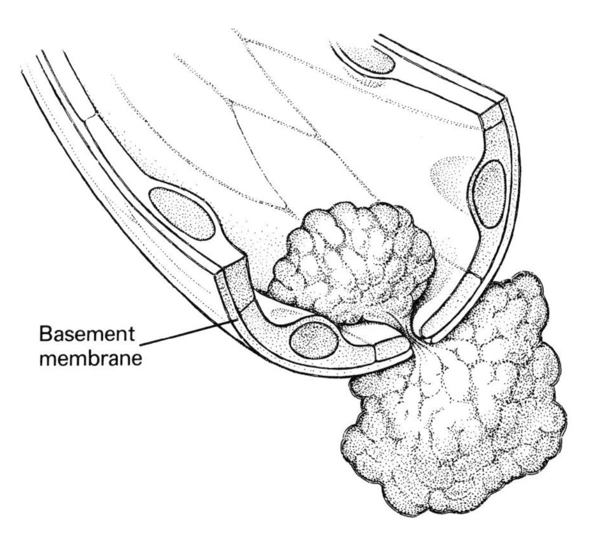

Leukocyte, attracted by chemotactic signal, squeezes through capillary endothelium.

Breaking through the Extracellular Structures

As we enter the extracellular spaces, we come upon an awesome sight, reminiscent, except for its submarine character, of those wild jungle scenes dear to the illustrators of early travel books. A thick tangle of filaments stretches like lianas between towering trunks tilted in every direction. In some areas, the mesh is almost impenetrable. In others, it is only a tenuous filigree. Such, at least, is the view that we get at the molecular level with our millionfold magnifying glasses. If we magnify our surroundings a few hundredfold less, we find that microscopic order arises out of this apparent molecular disorder in the form of fibers, sheets, plates, and other structural elements.

An awesome sight. The interior of a primeval forest on the Amazon.

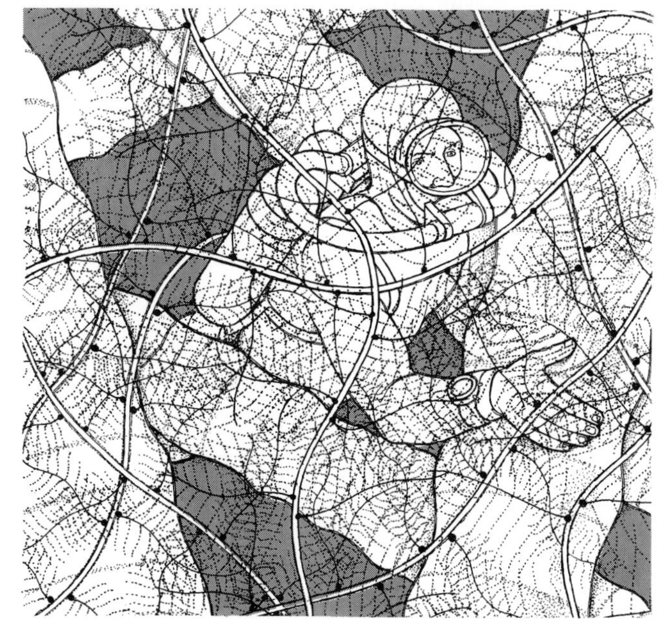

Cytonaut works his way through connective-tissue jungle. Note thick collagen trunks and tangle of proteo-glycan lianas.

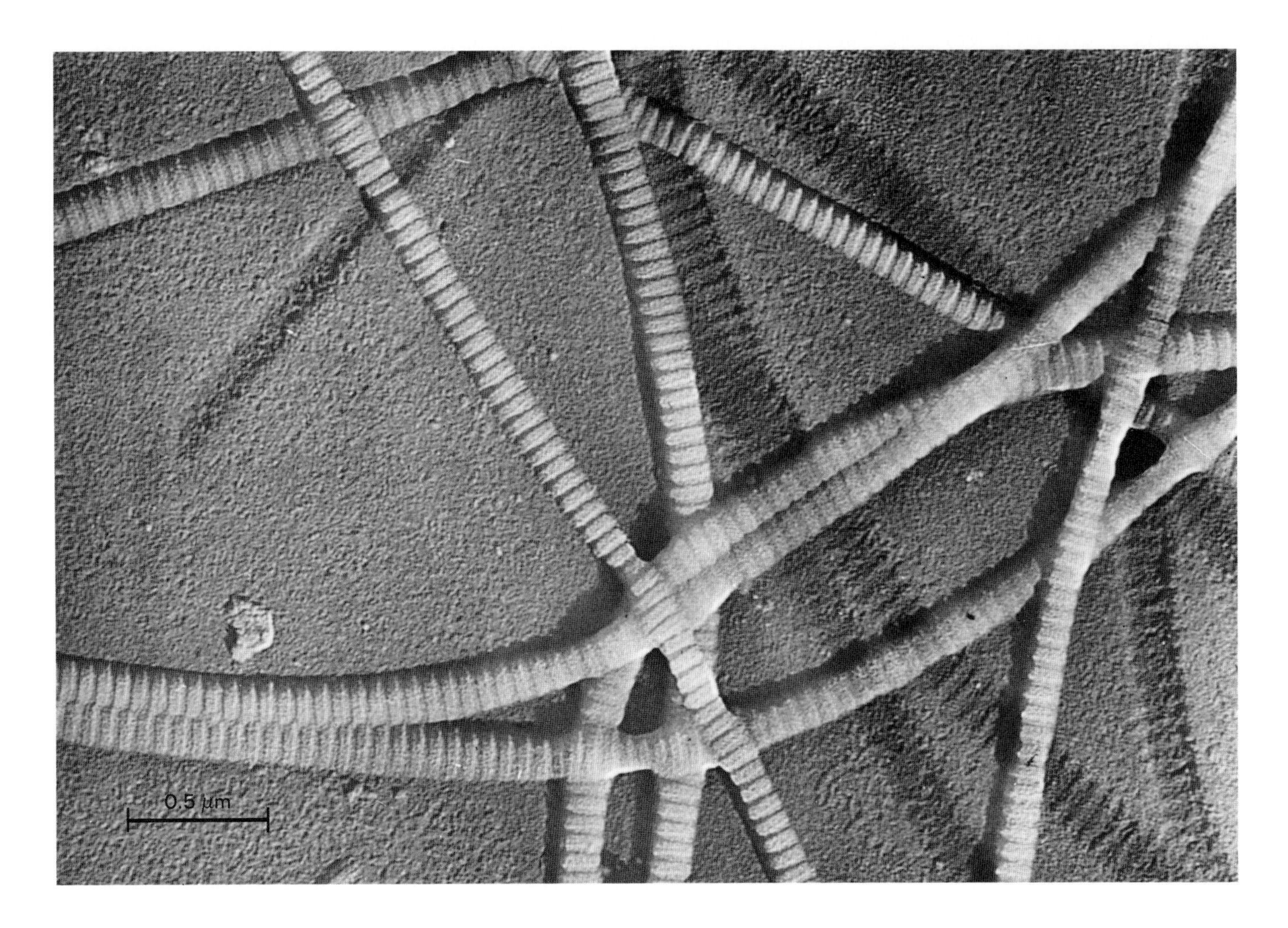

Electron micrograph of metal-shadowed replica of collagen fibers from human skin.

The main molecular component of these extracellular structures is a substance called collagen, which is extracted industrially from bones to make gelatin and animal glue; hence its name (Greek *kolla*, glue). Collagen is a protein. Its basic unit is a long polypeptide chain of 1,055 amino acids, exceptionally rich in the amino acids glycine, proline, and hydroxyproline. Note that the last one is not on the list of twenty given on page 31. It is made from proline residues after the polypeptide has been synthesized, and it is peculiar to collagen. Also peculiar to this polypeptide is its twisting into a left-handed helix of about 1-nm pitch, quite different from the α-helix mentioned earlier. Three such chains join into a right-handed helical thread about 1.5 nm thick and 300 nm long, with a 3-nm repeat (or 9-nm pitch for each individual chain).

Called tropocollagen (Greek *tropê*, turn), the triple-stranded molecule is the building block of collagen. With our molecular eyeglasses, we would see it as a short length

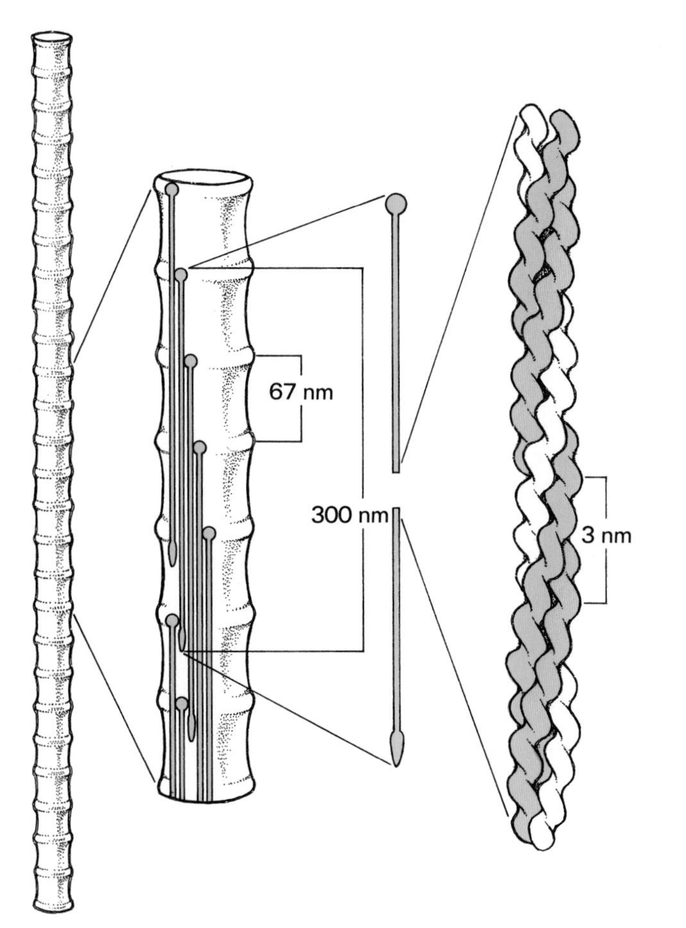

Structure of collagen fiber showing its staggered arrangement of triple-stranded tropocollagen molecules.

of string, a little over one-twentieth of an inch thick and close to a foot long. How such frail material can build the massive trunks that surround us is explained by the natural propensity of tropocollagen molecules to join side by side in staggered fashion. This property allows them to assemble spontaneously into fibers of virtually any thickness or length. The arrangement is such that the C-terminal heads of adjacent tropocollagen molecules are staggered by exactly 67 nm. Five such repeats (335 nm) are the minimum needed to accommodate the full length of a tropocollagen molecule (300 nm), leaving a gap of 35 nm between the heads and tails of consecutive molecules. Because these gaps are in register with each other across the whole thickness of the fiber, they create a characteristic cross-striation by bands 67 nm apart. Each of these bands corresponds to a region where one-fifth of the fiber thickness is free. This free space is important. In bone, it is the site of mineralization.

Tropocollagen is made as a precursor molecule, called procollagen, that possesses extra N-terminal and C-terminal pieces that prevent self-assembly from taking place prematurely. Only after the procollagen has been discharged extracellularly is it processed by enzymes that remove the interfering parts. After the fibers have formed, they undergo further changes, including strengthening by cross-links. Such modifications are believed to continue throughout life and to contribute to the progressive stiffening of connective-tissue structures with age.

Collagen fibers are the main components of the rigid framework whereby supporting structures are reinforced. They act, so to speak, like the steel rods in reinforced concrete or the fibers in fiberglass. Arranged in parallel, they serve to make all sorts of longitudinal parts, up to tendons and ligaments. Their three-dimensional intertwining forms the scaffolding of tissues and organs, including the authentic skeletal parts, such as cartilage and bone.

Several different types of collagen contribute to the building of these various structures. A special kind, called type IV, tends to make flat arrangements. Together with other proteins, such as laminin (Latin *lamina*, thin plate) and fibronectin (Latin *nectere*, to bind), and with proteoglycans (see below), it forms a resilient sheetlike material, between 50 and 100 nm thick, out of which structures known as basement membranes are made. This appellation, which goes back to the early histological literature, is unfortunate. Basement membranes are in no way constructed like the cellular membranes we are about to see; they are really walls. Sometimes they surround individual cells, as do the walls of bacterial or plant cells. More frequently, they envelop or support multicellular arrangements of various sizes and shapes. One of their functions is to provide cells with the kind of carpeting they need to creep or stick. Another is to act as molecular filters. The basement-membrane casing of capillaries plays a particularly important role in screening the substances that are allowed to gain access to the tissues from the blood. The most exacting such filter occurs in brain capillaries (blood-brain barrier).

In certain areas, as in the arterial wall, use is made of another filamentous protein called elastin. Like collagen, elastin assembles into fibrils, which themselves form more elaborate arrangements, mostly fibers and flat plates, or lamellae. Unlike similar structures made of collagen, elastin fibers and plates tend to adopt a sinuous or scalloped shape, such that they can be stretched to 1.5 times their length and spontaneously shorten again when released. As their name indicates, they provide elasticity.

As in reinforced concrete or fiberglass, the fibrous network of the supporting structures is embedded in an amorphous matrix or filler, called ground substance, which, at the molecular level, is seen as a mesh of tenuous polymeric molecules. These include a number of important polysaccharides, many of them attached by one end to a protein stalk (proteoglycans). Depending on the nature of these molecules and on their density, the resulting matrix may be no more than a viscous fluid, may behave like a jelly, or may reach the hardness and resilience of cartilage or a lobster shell. Sometimes, as in bones, teeth, coral, mollusk shells, and other biomineral structures, the matrix spaces are largely occupied by crystalline deposits of mineral salts.

As a rule, one more obstacle stands in the way of the intruder trying to approach a cell. It is a carbohydrate-rich covering called surface coat, or glycocalyx (Greek *kalyx*, husk). Its thickness and appearance vary according to cell type from a hardly detectable fuzz to a bona fide basement membrane. Bound to the cell membrane by relatively weak forces, it serves to maintain a special microenvironment around the cells, protects them against certain physical or chemical attacks, and helps them to recognize kindred neighbors and to establish connections with them and with connective-tissue structures.

3 The Cell Surface, With an Introduction to Membranes and Lipids

Most striking to the traveller who first sets eyes on the cell surface are its unevenness and changeability. With rare exceptions, cells display surface patterns of great complexity that are characteristic for each type. Some cells are ravined by deep clefts or pitted by craterlike depressions. Others are deformed by protrusions, called pseudopods (Greek *pous, podos*, foot), or studded with fingerlike projections identified as microvilli (Latin *villus*, shaggy hair) or as cilia (Latin *cilium*, eyelash). Yet others have their surface pleated by filmy veils. The variety is endless.

A Very Flexible Skin

The scanning electron microscope has revealed the rare beauty of these patterns but fails to show their constant movements and kaleidoscopic changes. Cilia beat, membranous veils wave and undulate, microvilli bend and twist, pseudopods bulge, craters erupt suddenly in the midst of a quiet surface, spewing secretory products, or, in contrast, invaginations yawn open, to be sucked deep into the cell. Sometimes, as we recently witnessed, the whole cell appears convulsed or starts crawling away, set in motion by some elusive chemical signal.

In spite of these incessant, sometimes tumultuous contortions, the cell always remains completely clothed by a tight-fitting membrane that adapts to every change in shape with what looks like effortless plasticity. This membrane is called the plasma membrane, or plasmalemma (Greek *plasma*, form; *lemma*, husk). It is a thin film, about 10 nm thick. Magnified a millionfold, this

Scanning electron micrograph of intestinal mucosal cell illustrating microvilli.

amounts to no more than 1 cm, or 0.4 inch—an uncommonly tenuous casing for a structure the size of a large auditorium (which, we have seen, is the size of a cell similarly magnified). Actually, we will find out when we explore the internal aspect of this membrane that it does not alone bear the brunt of containing the cell; it is helped by a number of supporting structures that bolster it on its inside face.

Containment is only one of the many functions of the plasma membrane, which is really a major cell organ. It is a highly complex and dynamic structure that regulates virtually every interaction between the cell and its environment, including other cells. It does so thanks to a unique mode of construction that involves two main types of molecular constituents—phospholipids and proteins.

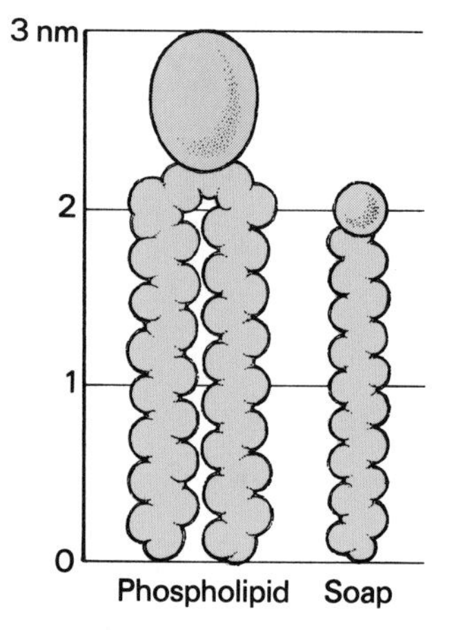

Models of phospholipid and soap molecules.

Lessons of a Soap Bubble

The secret of membrane construction—not just the plasma membrane, but all other biomembranes—is the lipid bilayer (Greek *lipos*, fat). The simplest example of a lipid bilayer is a soap bubble. Soaps, the salts of fatty acids, are linear molecules made up of only carbon and hydrogen, except for a terminal carboxyl group carrying a negative charge, COO^-. In terms of the classification that we have seen in the preceding chapter in relation to the amino acids, they have a long, hydrophobic tail and a hydrophilic head. Such molecules are called amphipathic (or amphiphilic), which is the Greek way of saying that they have two loves. The lipids of membranes are more complex than the simple soaps, but they, too, are amphipathic. They have a forked hydrophobic tail that consists of two fatty acid chains and a bulky hydrophilic head that contains a negatively charged phosphoric acid group, linked to another group that is often positively charged; they are called phospholipids.

When amphipathic substances are mixed with water, they spontaneously adopt a molecular configuration that satisfies their two conflicting requirements simultaneously. They associate in such a way as to dip their hydro-

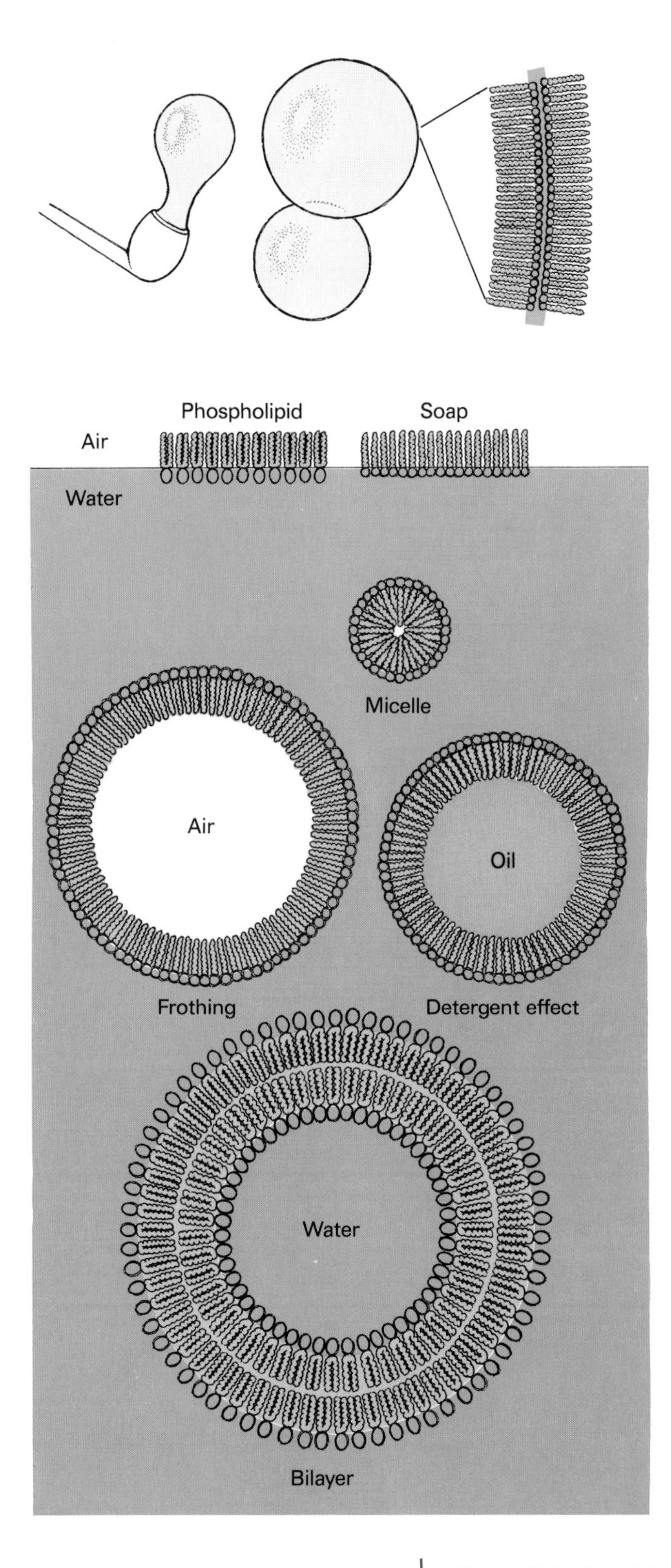

Different kinds of structures formed by amphipathic substances and air, water, water and air, or water and oil.

philic heads in the water, while keeping their hydrophobic tails out of it and in contact only with each other and with any other hydrophobic materials, such as oil, plastic, or air, that may be around. Essentially three structures can be generated in this way: monolayers, micelles, and bilayers. Molecular monolayers form at interfaces between water and air or some hydrophobic fluid. Micelles form in water, as spherical clusters of intertwined tails surrounded by heads. Stir a micellar solution with air, and the micelles will join and spread as monolayers around imprisoned air bubbles (frothing, surfactant effect). Stir it with oil or use it to rub a greasy surface, and the micelles will similarly change to monolayers that coat dispersed droplets of oil or grease (emulsifying or detergent effect). As to bilayers, they develop under certain conditions as partitions between two phases of similar nature. Two arrangements are possible. If the two phases are hydrophobic—air, for instance—the tails will stick outside on both faces of the bilayer and the heads will be inside, joined by a film of water. This is the structure of soap bubbles. If the two phases are aqueous, the tails will be inside the bilayer and the heads outside.

Phospholipids have a great tendency to form bilayers when they are mixed with water. Their forked hydrophobic tails are too bulky to fit comfortably inside micelles. In contrast, they are readily accommodated in bilayers. These close spontaneously into vesicles, which may consist of a single bilayer or of several concentric bilayers, depending on how they are prepared. Called liposomes, such artificial phospholipid vesicles have become an im-

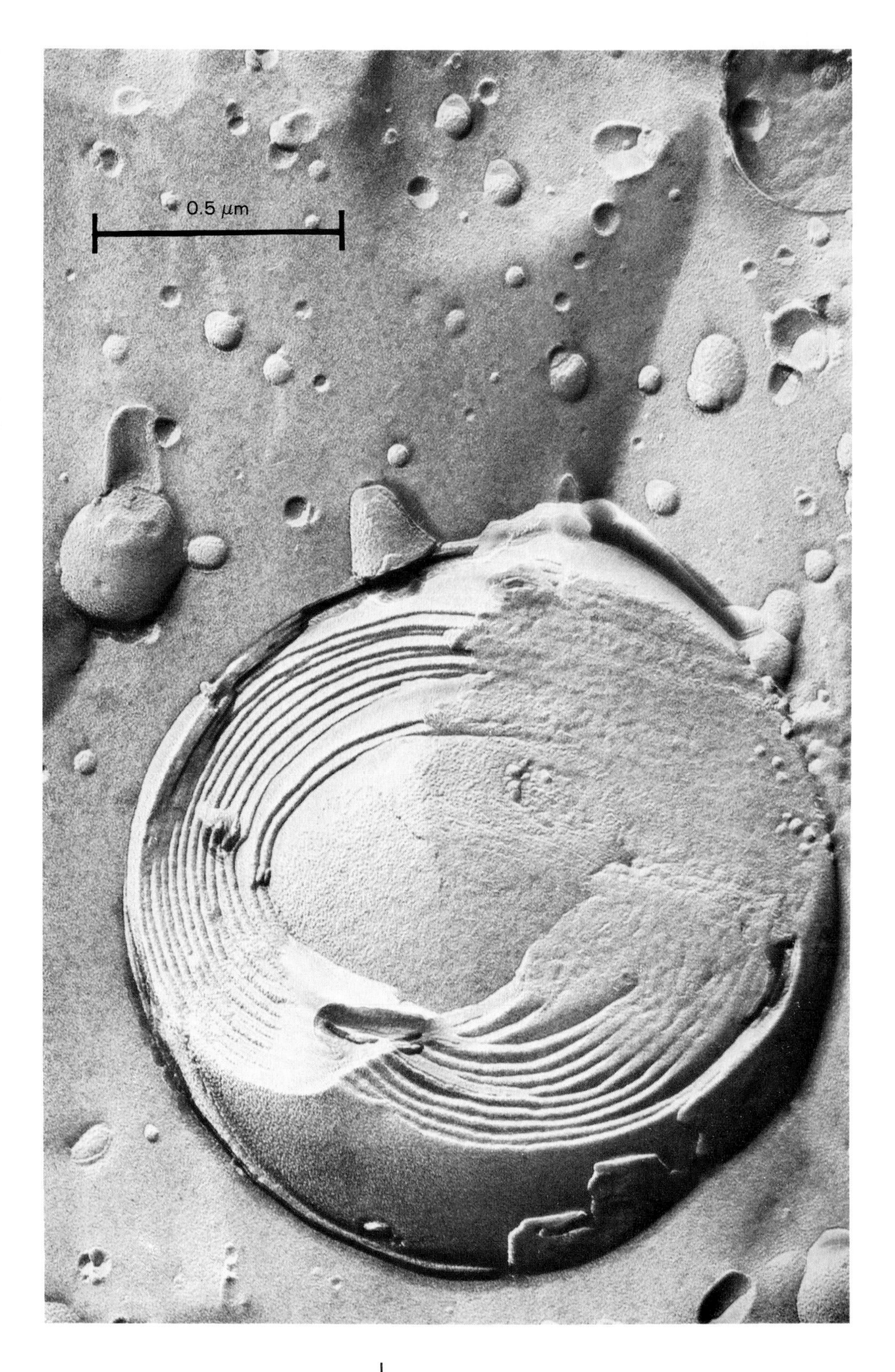

Electron micrograph of a metal-shadowed replica of a freeze-fractured multilayered liposome.

portant tool in cell research. The natural propensity of phospholipids to assemble into bilayers in an aqueous milieu lies at the heart of the structure of normal membranes, which all have as their basic fabric a phospholipid bilayer, from 5 to 6 nm thick. In addition, they contain various protein molecules, of which more will be said later.

Several important properties of biological membranes, as well as of soap bubbles, are explained by their lipid bilayer structure. One property is their flexibility. A lipid bilayer is an essentially fluid arrangement within which the molecules can freely move about in the plane of the bilayer and reorganize themselves into almost any sort of shape without losing the contacts that satisfy their mutual attraction.

This kind of flexibility is characteristic of soap bubbles, except that they are more rigid than cell membranes owing to their higher surface tension. They can suffer distortion, when blown upon, for instance, but have a strong tendency to adopt the spherical shape, which minimizes surface tension. The cell membrane, with one face supported by cytoplasmic structures and the other in contact with a watery milieu, has a lower surface tension and is therefore more flexible than a soap bubble.

The fluidity of lipid bilayers requires that the hydrophobic tails be able to slide freely past each other. This ability, in turn, is influenced by temperature. Below a certain critical temperature, called the transition temperature, which depends on the nature of the lipids concerned, the hydrophobic chains congeal into an ordered, rigid structure, no longer compatible with the functions that a membrane has to fulfill. Indeed, cells adapted to different temperatures have membrane lipids of correspondingly different transition temperatures. Bilayer fluidity is also influenced by inserted substances that interfere either with the sliding of the hydrophobic chains or with their congealing. Cholesterol, which is a key component of the plasma membrane in all eukaryotic cells, owes its importance to such interactions.

A second important property of lipid bilayers is that they are self-sealing. Again look at a soap bubble. If you cut it in half, you do not get two half-bubbles, but rather

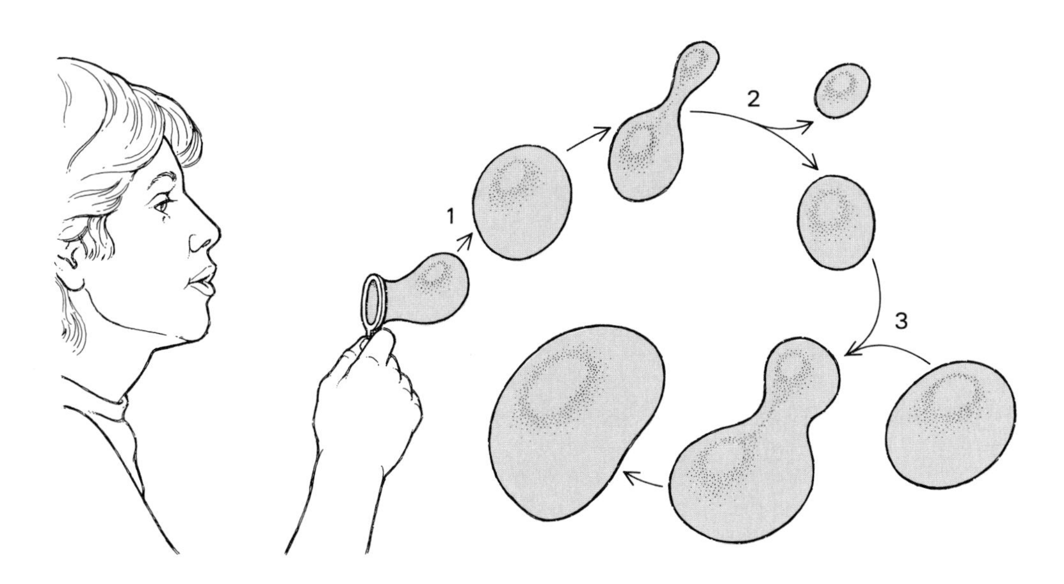

Soap bubbles (1) detach and seal; (2) divide; and (3) fuse.

two bubbles, smaller but whole. If two soap bubbles collide, they fuse, and the whole structure reorganizes to make a single and bigger soap bubble. So it is possible to have fusion and fission of lipid bilayers, always on the basis of the self-sealing properties inherent in this type of structure.

Coming back to the cell, it is possible to push a needle through a cell membrane and take it out again: the puncture site will close automatically. It is even possible to cut a cell with a microknife and obtain two pieces, each completely surrounded by a sealed plasma membrane. Conversely, cells can be made to fuse, like soap bubbles, by merger of their plasma membranes. This phenomenon happens physiologically—for instance, in muscle development—and it can be induced artificially. It is exploited for the preparation of monoclonal antibodies (see Chapter 19). Membrane fusion also plays an important role in many intracellular processes, as we will see later in the tour.

A third property of lipid bilayers (of the membrane type) is their impermeability to molecules that are soluble in water. Such molecules cannot traverse the bilayer, because in order to do so they would have to cross the oily film made by the hydrophobic tails of phospholipid molecules. To get physically through such a film, a substance must itself be hydrophobic or it must take advantage of thermal agitation and squeeze through the occasional gaps that open up in the bilayer as a result of molecular movements. This is how water and other very small molecules are exchanged between the cell and the environment. As a rule, however, the bilayer serves as a highly effective barrier, allowing the cells to retain their own constituents and ward off extracellular substances with a minimum of leakage. This is sound economy, resembling that of medieval cities, which protected themselves with a wall and a moat. But bridges and gates are needed to permit, and at the same time to regulate, the two-way traffic that necessarily must take place to support life within the city. This the lipid bilayer, which is little more than a passive moat, cannot do by itself; it requires the participation of membrane proteins, molecules endowed with a higher degree of specificity.

From Bilayers to Membranes: the Role of Proteins

As mentioned in the preceding chapter, proteins that are unable to bury their hydrophobic groups inside a hydrophilic shell find their natural residence in membranes. There they can make all their parts happy by inserting their hydrophobic domains into the lipid bilayer and allowing the hydrophilic ones to emerge into the watery outside world. Depending on the organization of these domains, the molecules will either float iceberglike on the bilayer, with their hydrophilic parts all facing the same side, or straddle the bilayer, with hydrophilic parts protruding on both faces (transmembrane proteins). Some of these arrangements are remarkably complex. Bacteriorhodopsin, the major component of the photochemical apparatus of the microbe *Halobium* (see Chapter 10), spans the bilayer seven times. The acetylcholine receptor (see Chapter 13) consists of two units, each of which is made of five transmembrane polypeptides clustered around a central channel. In these and other known examples, the protein parts that cross the bilayer consist of α-helical rods containing from 21 to 27 amino acid residues, most of them hydrophobic.

Such proteins, which have one or more hydrophobic parts embedded in the lipid bilayer of a membrane, are called integral, or intrinsic, membrane proteins (whether they traverse the bilayer or not). Those that are attached to either side of the membrane but do not penetrate the lipid bilayer are called peripheral, or extrinsic.

Thanks to the essential fluidity of the lipid bilayer, which was commented upon earlier, membrane proteins are free to move about in the plane of the membrane by lateral diffusion. This freedom is essential to a number of functions—we will soon turn it to our own use—but there are limits to it, imposed by the interactions between the membrane proteins and other components lying within and outside the membrane. Some of these patterns are subject to modification by specific molecules that bind to membrane proteins, with consequences that may be of great functional importance.

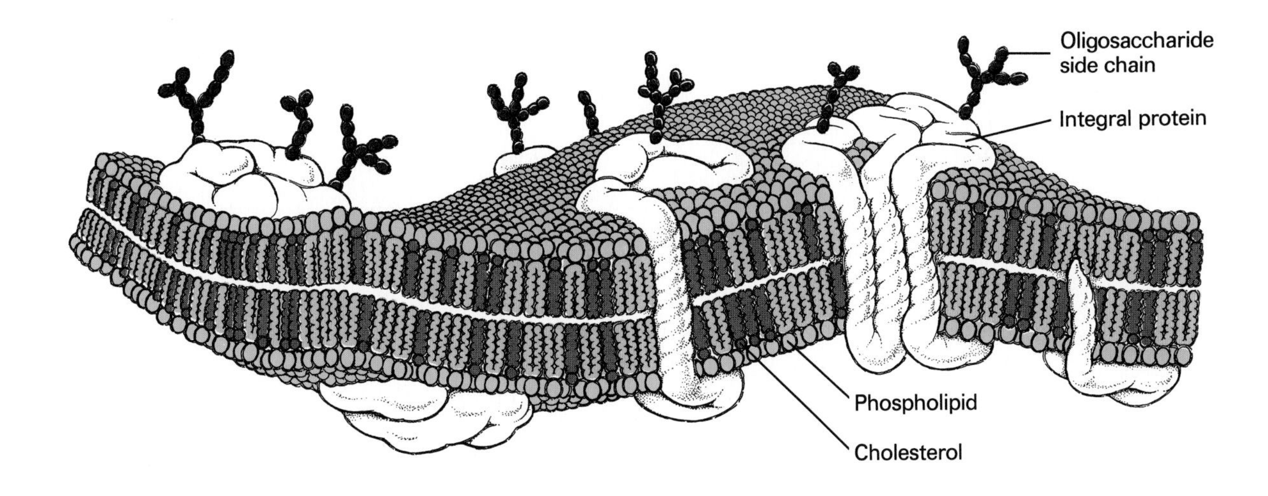

A molecular view of the plasma membrane. Integral proteins either float iceberglike on the lipid bilayer or they straddle it; they move about by lateral diffusion. Oligosaccharide side chains of glycoproteins form molecular down—or a forest of antennae—on the cell surface.

A characteristic of the plasma membrane is that many of its outside protein components are glycoproteins—that is, they bear oligosaccharidic side chains made of sugar molecules. It also contains some glycolipids. As many as a dozen or more sugar molecules of several different kinds may participate in the formation of these carbohydrate components, which often have a branched structure. They cover the cell surface with a fine molecular down. In actual fact, they are not down; they are feelers. Magnify them another hundredfold, and you will readily appreciate this. You are now reminded of one of those underwater scenes where shoals of fishes and other creatures browse peacefully among animals that look like plants, until suddenly a drooping limb turns into a deadly tentacle that snaps up an unsuspecting victim and drags it inside. In our cellular seascape, the potential prey are molecules present around the cells; the tentacles are the oligosaccharide side chains (and other parts) of the membrane glycoproteins, many of which function as receptors.

Surface Receptors

The concept of a receptor was formulated in the beginning of this century by the German scientist Paul Ehrlich, known for his contributions to immunology and chemotherapy. He used the lock-and-key analogy, introduced by his contemporary, the chemist Emil Fischer, to explain how there may be on the surface of cells certain strategically located chemical groupings—the receptors, or binding sites—that specifically bind a certain kind of molecule—an antibody, for instance (see pp. 50–51), or a drug, in general terms a ligand (Latin *ligare*, to bind). Just as there are many locks and many keys, there are many receptors and many ligands.

Modern chemistry has added to this concept that of conformational change: when occupied by its ligand, the receptor has a shape different from that when it is free—that is, the folding pattern of the polypeptide is altered. When this kind of change happens to a transmembrane protein or to a molecule capable in turn of affecting the conformation of a transmembrane protein, some sort of communication is established between the outside and the inside of the cell. Many of the most important interactions between a cell and its environment, including other cells, occur by way of receptors.

There are many kinds of receptors on the surface of any given cell, each represented by up to hundreds of thousands of molecules. Many, but possibly not all, are glycoproteins. They cover the cell with a forest of molec-

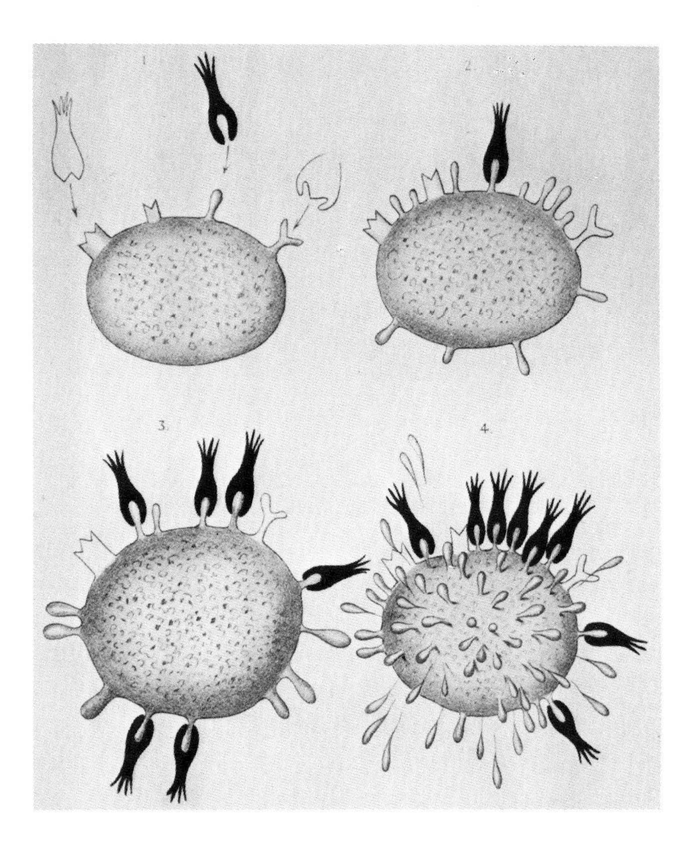

The picture at the left was drawn by Paul Ehrlich in 1900 to illustrate his receptor concept.

A receptor, when occupied by a ligand on the cell surface, changes conformation and activates an intracellular trigger, as shown below.

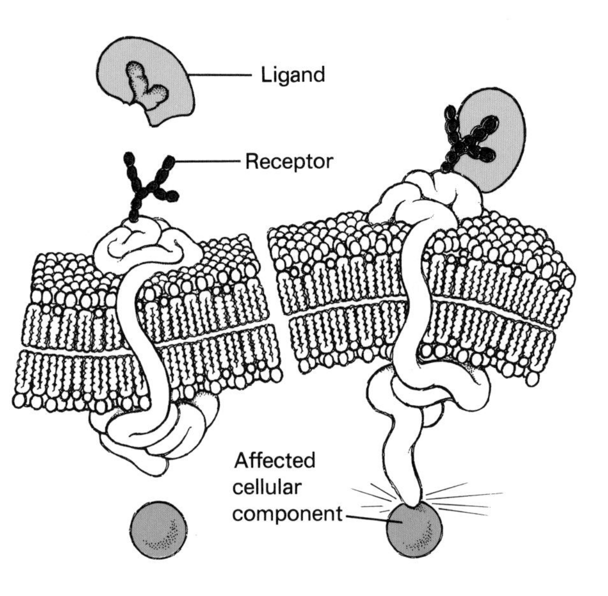

ular antennae and offer the observer a most entertaining display. As molecules of all kinds drift by, carried by the eddies and currents the cell maintains around itself, no moment passes without some spectacular catch occurring somewhere. Sometimes, more than one receptor molecule participates in the binding process. It may even happen that hundreds of receptors become involved collectively in the immobilization of a large object, such as a bacterium.

An event that often follows capture, and is readily seen from our vantage point, is intake, just as in our underwater analogy. The processes involved make up some of the most dramatic parts of the receptor show; they result in the formation of pitlike depressions into which receptors and prey disappear from our view, drawn in by an invisible force. We will take advantage of this mechanism to effect our first entry into the cell and will consider it in detail in the next chapter.

Among the substances that get caught by receptors, many are specific messengers, called hormones, that are manufactured elsewhere in the body and arrive at the cell by way of the bloodstream. Hormones that bind to a cellular receptor generally end up being taken in. Before that, however, their binding triggers some characteristic cellular response. From where we are, we can observe only the quiver of the receptor as it becomes occupied.

Not until we visit the cytosol will we be able to see the devices to which the receptors are connected on the other face of the membrane and the mechanisms whereby those devices cause the cell to respond in some specific way (see Chapter 13).

A subtle, but very important, change mediated by some receptors when they are occupied is gating—the temporary opening of a channel through which given substances or ions are allowed to enter or leave the cell. To understand the significance of such a process, we need to look more closely at the mechanisms whereby molecules cross the plasma membrane.

A second condition is the occurrence of appropriate conduits across the lipid bilayer. These are usually provided by transmembrane proteins acting as specific carriers for the translocated substances or, in some special cases, as controllable gates or channels, operated by the binding of a specific ligand or by an electric perturbation.

The third condition is the availability of energy. The decisive factor here is whether the direction of transport is down a concentration gradient (from where the substance is more concentrated to where it is less) or up a gradient.

Molecular Traffic across the Plasma Membrane

The maintenance of cellular life depends on the continuous passage of many different substances—most of them highly hydrophilic—into and out of cells across the plasma membrane. Sugars, amino acids, and other nutrients must get in to satisfy growth and energy needs, while waste and breakdown products must get out to avoid piling up inside. In addition, ions must be moved in or out in order to maintain the ionic composition of the intracellular milieu, which is very different from that of the surrounding medium. For one thing, it is much richer in potassium ions and much poorer in sodium ions. These inequalities cause leakages, which must be compensated by transport in the reverse direction. All this adds up to considerable two-way traffic across a boundary that, it must be remembered, is made largely of a continuous phospholipid bilayer, almost impermeable to most hydrophilic molecules.

The first condition to support such traffic is a large enough surface area. In fact, the main function of microvilli, those fingerlike projections found densely crowded on the surface of some cells (p. 42), is to increase the area available for exchanges between the cells and their environment. They are characteristically found on cells, such as those that line the intestinal mucosa or the kidney tubules, that are exceptionally active in such exchanges.

Different kinds of molecular transport.

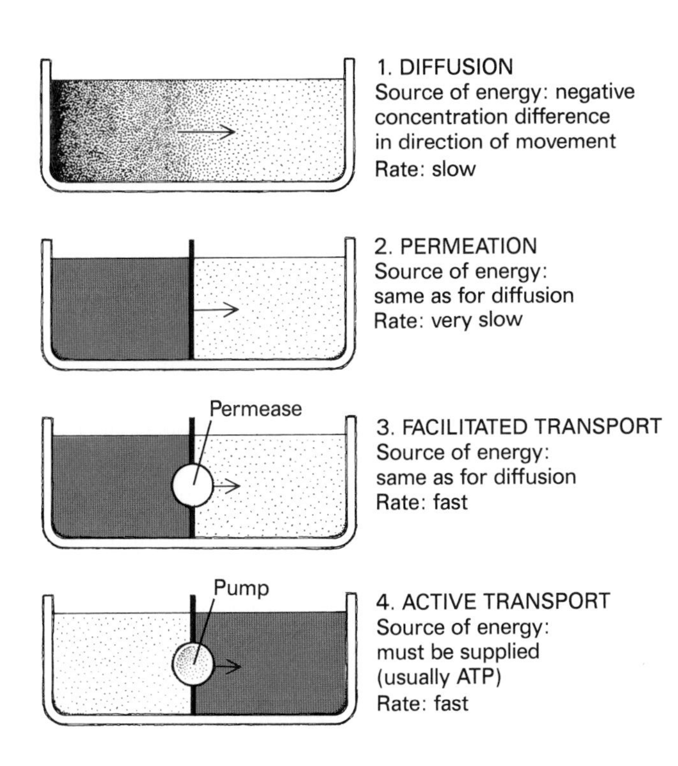

Downhill transport can occur spontaneously (see Appendix 2), but it tends to level the concentration gradient that supports it and thereby to exhaust its source of energy. For transport to continue, the gradient needs to be maintained through the production or supply of the substance on one side of the membrane and its consumption or removal on the other. Downhill transport is called diffusion if it is unhampered and permeation if it is restricted by a permeability barrier. The transport is said to be facilitated when it is helped by some sort of carrier or translocator system acting catalytically. Translocators that catalyze facilitated transport are called permeases. The term ionophore (Greek *pherein*, to bear) is used to designate a special group of carriers that mediate the transport of certain ions across membranes.

When transport occurs against a concentration gradient, the process must be directly supplied with energy, and the machinery involved is correspondingly more complex. The systems that carry out such active transport are generally referred to as pumps. The most important among them translocate positively charged ions, especially of hydrogen (H^+), sodium (Na^+), potassium (K^+), or calcium (Ca^{2+}). Their activity often results in the creation of an electric imbalance across the membrane (membrane potential). Energy transduction in mitochondria (see Chapter 9) and chloroplasts (see Chapter 10) is crucially dependent on this kind of charge displacement, as are all the manifestations of bioelectricity. Pumps, aided by controllable ion gates that allow very rapid perturbations of the generated membrane potentials, are behind nerve conduction, brain function, muscle excitation, cardiac rhythm, gland stimulation, and many other phenomena. They feed those hidden currents that modern medicine has learned to tap for diagnostic purposes. They support the discharges of up to several hundred volts whereby the torpedo fish and the electric eel stun their prey.

At present, the battery of transport machines that stud the cell surface reveals itself to us only by streams of ingoing and outgoing molecules and ions. But we will have an opportunity to see something of their inner works later in the tour, when we get to the other side of the plasma membrane (see Chapter 13).

Identity and Immune Recognition

One last important function of the plasma membrane is to provide cells with an identity card. This consists of a number of specific chemical groupings, known as transplantation, or histocompatibility, antigens. The first such antigens to be discovered were those that determine the blood groups A and B. As is known, some of us are blood type A, others are B, and still others are AB or O. In other words, we fall into one of four classes that represent the four possible combinations (2^2) that can be achieved with two characters, depending on whether they are present or absent.

Today, many transplantation antigens have been recognized in the human species. Their number and polymorphism are such as to make it highly improbable that any two individuals should have exactly the same combination. In practice, it happens only in identical twins. Transplantation antigens are displayed more or less completely on the surface of every cell of a given individual and are characteristic of that individual. As a means of identification, they are as reliable as fingerprints.

In the organism, these chemical identification marks are continuously being inspected by a special cellular defense corps, the lymphocytes, the agents of the immune system, which have the ability to recognize by its surface identification marks any invader that might have infiltrated our defenses and to destroy or help destroy it. They circulate through blood and lymph from a number of bases, which include the spleen, the thymus, the lymph nodes, the tonsils, and various so-called lymphoid patches.

There are two types of lymphocytes, known as T and B for their main source of origin, thymus and bone marrow (originally bursa of Fabricius, a lymphoid organ of birds). Within each type, there are several subclasses. T lymphocytes—at least the main subclass designated cytotoxic—are the infantry of the immune system; they are specially equipped to kill other cells by direct contact through a "kiss of death" mechanism. The B lymphocytes are the artillery, or rather give rise to the artillery, in the form of cells (plasma cells) that discharge missiles known as antibodies, which are endowed with the ability to combine specifically with their target. These antibodies, or immunoglobulins, which are of protein nature, do not kill by themselves, but they serve as recognition devices for a number of killing mechanisms. In particular, when bound to their target, they cause it to become attached to a receptor present on the surface of white blood cells, which then engulf and destroy the enemy thus branded. This mechanism serves as a major defense against microbes and viruses (see Chapter 4). Antibodies also serve to alert a soluble killing system carried by the bloodstream and known as complement.

Obviously, it is extremely useful for us to have such a defense corps. In fact, we could not survive without it. But this advantage carries with it a danger of mistaken identity and summary execution of friends. Here is where our transplantation antigens come into play. Early in fetal life, our lymphocytes learn to recognize the specific pattern displayed on the surfaces of our cells and to treat the cells thus labeled as friendly. But their inspection is very thorough, and they will spot even a small deviation from the pattern defined as "self." For instance, it is believed that they can detect and kill some cancer cells, which have almost the same identity card as normal cells. Lymphocytes certainly have no trouble recognizing cells originating from another individual, and they thereby tend to oppose the successful surgical transplantation of a tissue or organ. When transplants are made, a detailed typing of the transplantation antigens of the recipient and of the potential donors is done to allow the best possible matching. After the operation, the patient is treated with immunodepressant drugs, which weaken the rejection mechanism but which also, unfortunately, reduce the patient's ability to withstand infectious attacks and perhaps to reject cancer cells. There is a better way of eluding immune rejection, if only we knew how it works. Fetuses use it to entice their mothers into overlooking the foreign identification marks that they carry as part of their fatherly inheritance. Some women have immune systems that refuse to be subverted in this manner; they suffer repeated spontaneous abortions caused by immune rejection.

Lymphocytes are organized like no other defense force in the world, in that each individual lymphocyte can recognize only a single type of foreign grouping, as though each member of the corps were able to grapple with only a single type of invader. Since there are millions, if not billions, of such distinct groupings, most of our lymphocytes are never called to combat, and those that are are necessarily few. Quite often, in fact, they are much less numerous than the invaders.

This may look like an extremely inefficient way of doing business, but it seems to be the only way in which immense versatility can be combined with perfect reliability of recognition. And this, of course, is of paramount importance. Just imagine the havoc that would be created by a trigger-happy corps of the kind that fired first and asked questions afterward. Apparently, efficiency and safety can be achieved simultaneously only on the basis of the one-lymphocyte-one-target principle. Such being the case, Nature has provided a neat solution to the problem of numerical weakness. When a lymphocyte encounters and recognizes its specific target, it starts multiplying. This is yet another, and particularly important, example of a receptor-mediated response. The recognition occurs through binding of the target to specific surface receptors of the lymphocyte, and the binding results in a mitogenic response (one that stimulates mitosis). Thanks to this mechanism, a whole army, or clone (Greek *klôn*, twig), of identical lymphocytes directed against the target is generated by successive cell division. The organism becomes immunized.

This powerful mechanism has only one drawback. It takes time to be established and may arrive too late when the invading force is a strong one. Hence we use vaccination, a way of developing an appropriate defense corps by introducing a "dummy" enemy, such as attenuated viruses or killed bacteria, which can no longer cause a serious disease but still carries the identification marks that set off recognition, lymphocyte multiplication, and antibody formation. Recently, specific surface proteins or peptides extracted from the pathogen, or reproduced artificially either by genetic engineering methods (see Chapter 18) or by organic synthesis, have been used for the same purpose (synthetic vaccines).

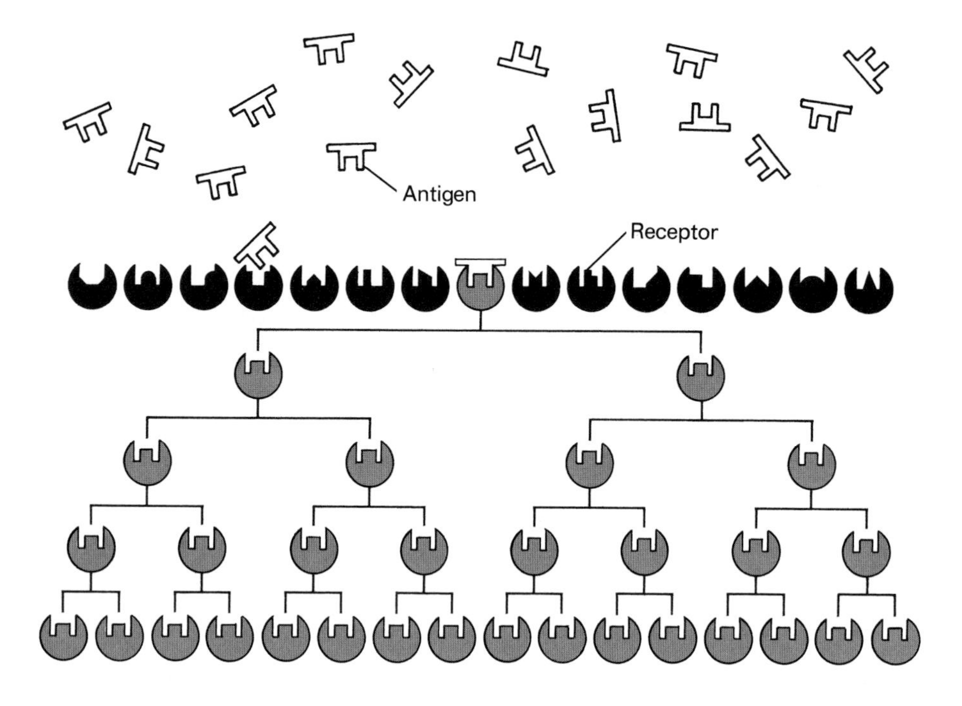

Functioning of the immune system by clonal selection.

We will return in some of the following chapters to the remarkable mechanisms whereby lymphocytes acquire their specific surface sensors in the course of differentiation (see Chapter 18) and are later stimulated to divide upon contact with the appropriate antigen (see Chapter 13). Like all living wonders, unfortunately, immune defense sometimes goes wrong. It may be congenitally defective, to the extent that the affected children must be kept constantly in a sterile environment. Or it may fail later in life, as in the recently described acquired immune deficiency syndrome (AIDS). Conversely, it may start striking friends as well as foes and launch treacherous autoimmune attacks against a patient's own liver, kidneys, joints, or other organs.

4 Entering the Cell: Endocytosis and Vesicular Transport

The various ways into a cell, whether by permeation, facilitated transport, or active transport, that were briefly described in Chapter 3 are restricted to small molecules. With rare exceptions, even single macromolecules do not cross the plasma membrane; members of our party are clearly excluded. Fortunately, there is a more accommodating means of entry, which is open not only to macromolecules, but also to much larger objects, including viruses, bacteria, and fragments of other cells. It depends on a membrane-mediated process of bulk uptake, called endocytosis.

The Endocytic Route

To understand the term endocytosis, we must go back one century. The scene is Messina in Sicily. Ilya Metchnikoff, a self-exiled Russian zoologist, is observing a transparent starfish larva through his microscope while, as he recalls in his memoirs, his "family has gone to a circus to see some extraordinary performing apes." As his eyes follow a wandering cell moving like some free amoeba through the larva's tissues, "a new thought suddenly flashes across my brain" that perhaps every organism contains a corps of mobile cells whose task it is to detect, pursue, engulf, and destroy unwelcome intruders such as microbes or viruses. This intuition made Metchnikoff one of the founders of immunology. With the help of a Hellenist friend, he coined the term "phagocyte" (Greek *phagein*, to eat) to designate the "eating cell." The act of engulfing a solid particle such as a

Left: Phagocytosis of a red blood cell by a mouse macrophage. Notice the thin rim of cytoplasm that completely envelops the engulfed erythrocyte.

Below: Two views of pinocytosis by a human monocyte. Invaginations of the plasma membrane are just about to close and detach as vesicles. Note the fuzzy "coat" around budding vesicle in right-hand picture. This coat is lost very rapidly after closure of the vesicle. It is not seen around the pinocytic vesicles in the cytoplasm.

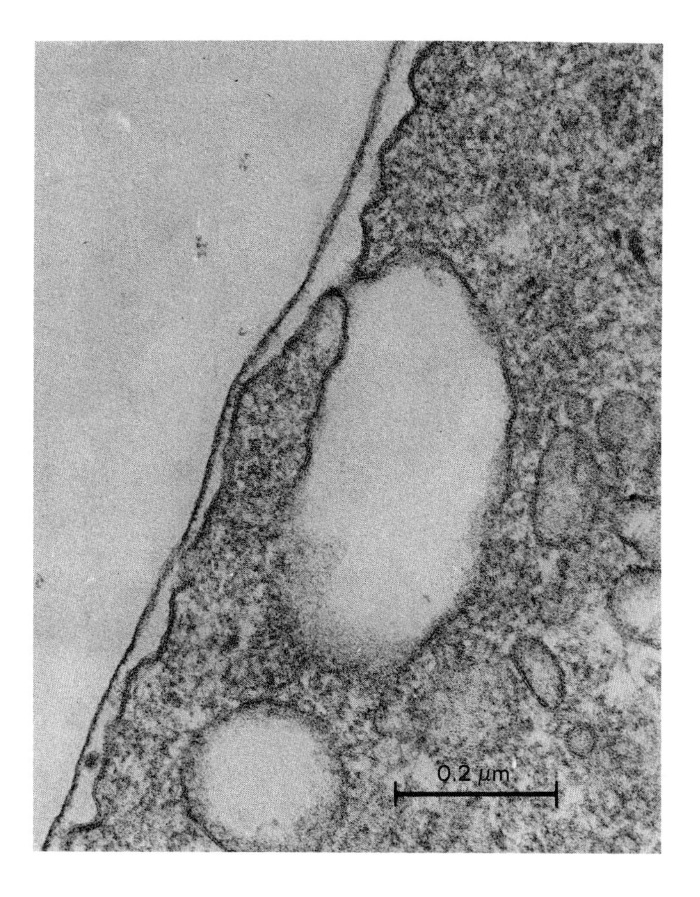

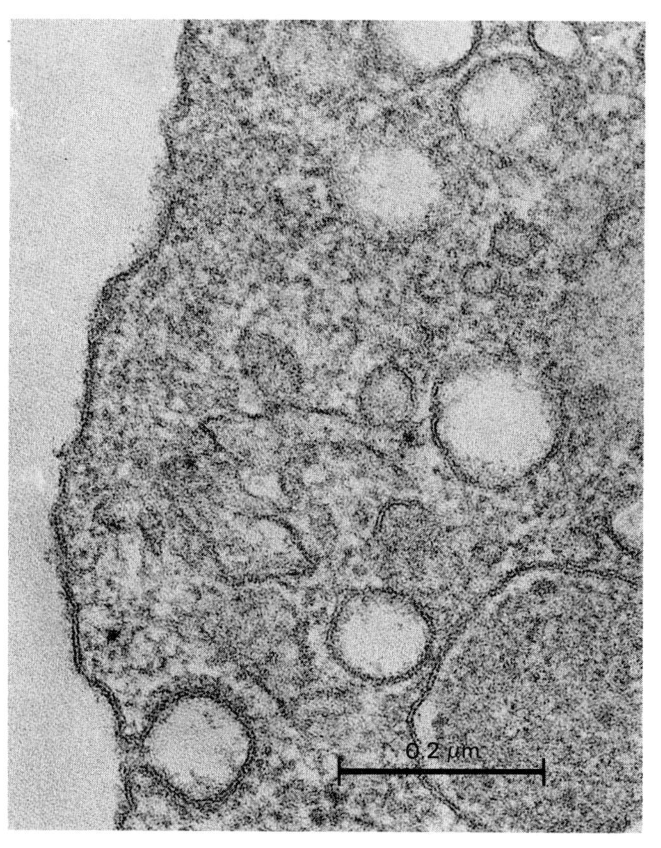

bacterium thus came to be called phagocytosis, literally the cellular act of eating. Then, in the early 1930s, the American biologist Warren Lewis discovered that cells could also engulf droplets of fluid and called this phenomenon pinocytosis (Greek *pinein*, to drink). Eventually, phagocytosis and pinocytosis were found to be manifestations of a more general mechanism of uptake, to which the name endocytosis was given.

Endocytosis can occur in many forms but depends invariably on the plasma membrane to provide the vehicle of entry. Whatever object is taken in, it always enters the cell wrapped in a sealed membranous sac that originates from an invagination of the plasma membrane. What we have learned of the fluidity and self-sealing properties of biomembranes helps us understand the more physical aspects of this phenomenon. Imagine a small patch being pulled inward from the surface of a soap bubble and eventually detaching as a miniature bubble imprisoned within the larger one. What does the pulling and how it is engineered are much more complex questions. We must wait until we can watch the process from inside the cell before we can address them.

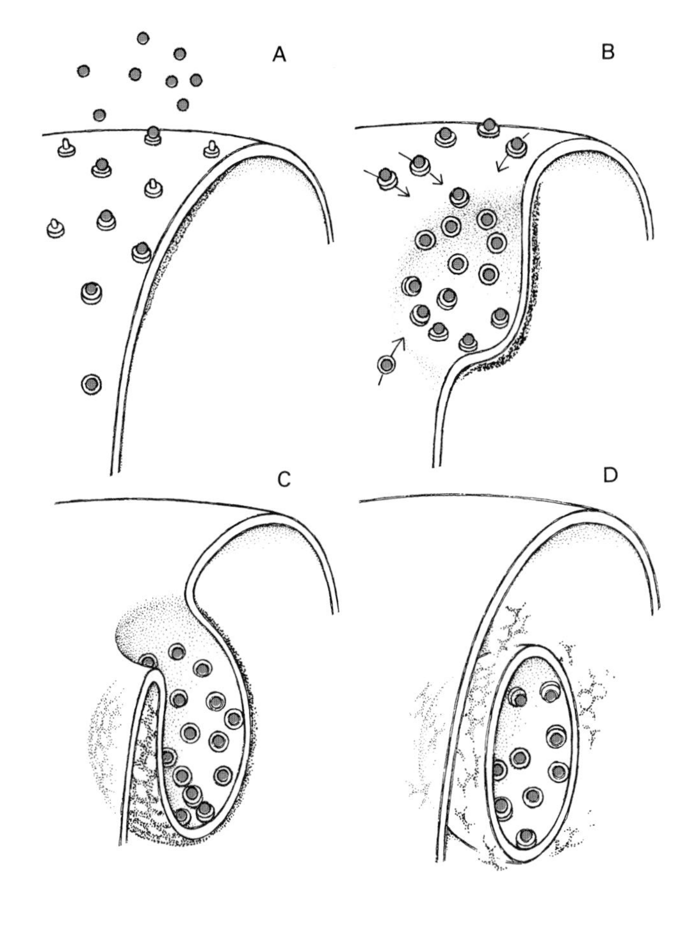

Receptor-mediated endocytosis: (A) ligands bind to receptors; (B) ligand-loaded receptors cluster in a coated pit; (C) the coated pit becomes a deep invagination; and (D) the invagination has detached as a coated vesicle, which is in the process of losing its coat.

In the meantime, some very interesting things about endocytosis can be observed from the outside. Most arresting is the participation of surface receptors. As mentioned in Chapter 3, binding of a ligand to a receptor is often followed by intake. In fact, many receptors have no function other than to select extracellular materials for intracellular intake. The process involved is called receptor-mediated endocytosis. The best way to observe this process is by introducing into the pericellular medium molecules of a ligand that is specifically bound by a given type of receptor. As ligand molecules are picked up, the occupied receptors can be seen to converge swiftly with their catch toward pitlike depressions of the membrane, and then to disappear progressively from view as the pits become deeper and their rims narrower. Eventually, only a pinhole opening remains, soon to vanish completely while the membrane regains its smooth, undisturbed appearance. Actually, a whole patch of membrane has been excised in this process and is now moving freely through the cytoplasm below in the form of a closed vesicle containing the engulfed material. Terms such as phagosome, pinosome, endosome, and phagocytic, pinocytic, or endocytic vacuole have been used to designate such a vesicle. The word endosome has acquired a special meaning as the name of the acidic sorting station in which endocytized materials undergo their first intracellular processing (see p. 58).

The clustering of occupied receptors in endocytic pits strikingly illustrates the ability of membrane proteins to move freely along the plane of the lipid bilayer by lateral diffusion. What actually brings them together is not clear. Cross-linking by divalent or multivalent ligands may play a role in some cases. But, in others, the receptors seem to

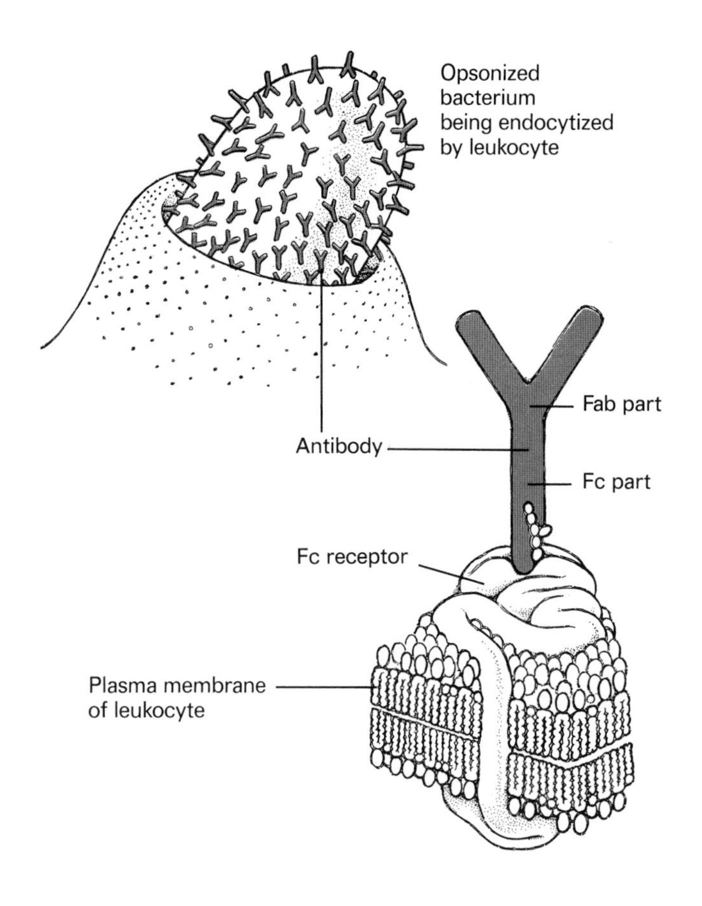

Phagocytosis of antibody-covered (opsonized) microbe mediated by Fc receptor (zippering mechanism).

be continuously on the move. Whether occupied or not, they disappear into endocytic pits and reappear on the cell surface in an uninterrupted stream, somewhat in the mode of an escalator or moving belt. As in these examples, the machinery that does the moving is hidden. We will take a look at it when we visit the cytosol. All we can distinguish from the outside is the shadow of a cagelike structure at the back of the endocytic pit; it seems to pull the membrane patch inward and to strangle it into a separate vesicle. Made of clathrin (see Chapter 12), this structure appears in cross section as a bristly coat. Hence the name "coated" given to the pits and vesicles that are thus lined.

A different kind of endocytic uptake, one that is highly dependent on cross-linking of receptors, occurs when the ligand molecules are attached to the surface of a large object, such as a bacterium. Once contact between the object and the cell membrane is established by means of a few receptor-ligand linkages, additional receptors are recruited from neighboring parts of the cell surface, and their binding causes a progressive envelopment of the ligand-coated object by the receptor-covered piece of plasma membrane. Thus, this molecular "zippering" effect actually induces the engulfment process. The name phagocytosis is generally reserved to designate this kind of uptake.

The mechanism whereby antibodies brand targets for destruction by leukocytes (Chapter 3) depends on this kind of process. Antibodies are Y-shaped protein molecules that attach to the corresponding antigens by means of their forked end, known as the Fab part (see Chapter 18). This part of the antibody carries the immunological specificity of the molecule and is therefore different for every individual type of antibody. The stem of the Y, called the Fc part, is the same for all antibodies. Now imagine a microbe entering an immunized organism and there encountering antibodies directed against some of its cell-wall components. It becomes covered with antibody molecules that all bind by their Fab forks and therefore have their Fc stems sticking out. This Fc pelt will induce zippering of the microbe by a white blood cell, thanks to the occurrence of specific Fc receptors on that cell's membrane. The early immunologists gave the suggestive name opsonin (Greek *opson*, seasoning) to such phagocytosis-inducing antibodies. They are the "dressing" that makes the objects they cover palatable to the eating cells.

We will understand the significance of endocytic receptors much better once we find out what happens to the materials that are engulfed with their help. An important point to remember, in the meantime, is that the nature and density of receptors occurring on the cell surface vary from one cell type to another, and even in the same cell, according to its functional state. This means that each cell selects its particular "menu" out of the environment by means of its surface receptors.

Endocytic uptake is not restricted to molecules and objects that are selected by receptors. Extracellular components are also taken in randomly as the dissolved contents of engulfed droplets of fluid. This is pinocytosis in the etymological sense of the term. Called fluid-phase endocytosis, it can lead to sizeable uptakes, since cells easily "drink" the equivalent of their own volume every day. Just imagine a human being guzzling from 15 to 20 gallons a day.

Even more impressive is the quantity of membrane that is being translocated by the endocytic activity. It may amount to the interiorization of as much as twice the surface area of the plasma membrane, with all its infoldings and digitations, every hour. Cells hardly make new membrane at this rate. They continuously recycle back to the surface the membrane patches used in endocytosis.

A Hazardous Start

Our visit to the cell's interior is about to start. But first we should get properly dressed. For reasons that will soon become clear, we need an acid-resistant and lysosome-proof coat if we want to avoid an abrupt—and fatal—termination to our tour. A waxy substance extracted from the cell wall of leprosy bacilli, which owe to this substance their remarkable resistance to intracellular destruction, will provide us with the appropriate material. On top of that, we still have to don an overcoat carrying a suitable ligand, so that we can be recognized and admitted by the cell we plan to visit. Should it be a white blood cell, our garment would be made of antibodies, so as to satisfy the cell's Fc receptors. But other cells may require a different overcoat. Remember that surface receptors vary from one cell to another. This means that we can, to some extent, choose the target of our visit by covering ourselves with the appropriate ligand.

With due regard to our size—still comparable to that of a large bacterium, even after shrinking a millionfold—our proper mode of entry should be phagocytosis. However, there will be more to learn if we ask the cell to stretch it a little and to accommodate us within a coated pit. There, together with a variety of receptor-bound molecules, we are about to experience endocytosis from the vantage point of its objects or victims. Even to the forewarned, it can be an unsettling experience to feel oneself being gradually enshrouded by creeping membrane folds. But worse is yet to come, after our last connection with the outside world seals shut and our flimsy craft casts off into the deepening cytoplasmic gloom, past bulky shadows that can just be discerned through the half-transparent membrane.

The first sign that our engulfment has been completed is a progressive acidification of the medium inside our membranous vessel. The hydrogen ions, H^+, or protons, responsible for this phenomenon come from the cytosol outside, delivered by an energy-dependent proton pump (Chapter 3). This pump presumably already exists in the pits on the cell surface, but the protons it ejects are diluted and washed away by the extracellular fluid. When the pits close into vesicles, the protons remain trapped inside, where they accumulate rapidly until they reach a concentration of about 20 microequivalents per liter (pH 4.7), as opposed to a cytosolic concentration near neutrality (0.1 microequivalents per liter, or pH 7.0). An important consequence of this acidification is that it decreases considerably the affinity of many ligands for their receptors, thus causing them to fall off the membrane. As it happens, our overcoat is made of ligands of this type, and we are now swimming freely in the stinging fluid that fills the endocytic vacuole, thankful for the acid-resistant coat that we borrowed from leprosy bacilli. Many other ligands have joined us, but some remain attached to their receptors.

While these events are going on, the membrane around us presents us with a spectacle strangely reminiscent of what we watched on the cell surface. Again we see receptors—some occupied, but many others free—clustering in deepening clefts and finally vanishing into the cytoplasm in the form of small, flattened vesicles that either are returned to the cell surface or are routed to some other destination that we cannot discern at present. The main function of this activity is membrane retrieval, without which, as we have seen, endocytosis could not continue

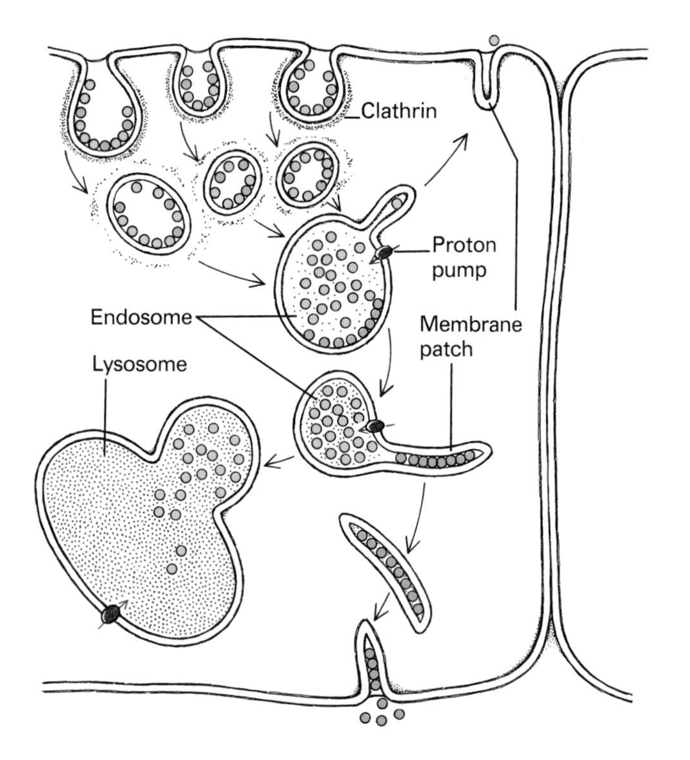

Sorting in an endosome after endocytosis. In this example, two types of ligands (represented by yellow and blue dots) are taken up together by receptor-mediated endocytosis. As a result of the acidification (shown in red) of the endosome content by a proton pump, the yellow ligands fall off their receptors and are subsequently conveyed to a lysosome by endosome-lysosome fusion. Before that, membrane patches detach from the endosome and are recycled back to the cell surface. One returns to the site of entry with some endosome contents (regurgitation). Another moves to the opposite side of the cell with a cluster of blue ligands that have remained receptor-bound and discharges them outside the cell (diacytosis, see p. 61).

very long. But the process is in no way random; it involves a lot of highly discriminate sorting and reorganization. Defined receptors and other membrane components are selectively concentrated on the excised patches and removed with them, accompanied by such engulfed materials as remain or become attached to the receptors in the acid medium created by the proton pump. At the same time, patches are also added to our surrounding membrane, brought in by newly arrived endocytic vacuoles that fuse with our own. We are in fact in an authentic organelle, the first station on the cellular import line. In it, receptors unload their catch and are recycled for a new round of duty, with the exception of a few that take their load further into the cell. Names such as intermediary compartment, receptosome, or endosome have been given to this organelle. The last is a generic term meaning endocytic vacuole (see p. 55) that now tends to be reserved for the endocytic sorting station.

All this membrane reshuffling does not go on without a certain amount of "slobbering" or regurgitation of fluid contents. By and large, however, dissolved materials and detached ligands remain inside endosomes (or are returned to this compartment by a new endocytic gulp). With them, we continue our adventurous cruise through the cytosol, somehow pulled and guided by invisible strings, parts of the cytoskeleton, as we will see in Chapter 12. Suddenly, right in front of us, looms a dark, ungainly blimp of a dirty brown color. Collision is inevitable.

The crash is mild, but its outcome is fateful, since it has caused the membrane of our vessel to merge with that of the blimp, hurling us into what turns out to be a thoroughly unpleasant environment. Everywhere we look are

Our endosome has collided and fused with a lysosome, projecting its contents into highly corrosive, digestive fluid. Scenes of destruction are everywhere.

scenes of destruction: maimed molecules of various kinds, shapeless debris, half-recognizable pieces of bacteria and viruses, fragments of mitochondria, membrane whorls, damaged ribosomes, all in the process of dissolving away before our very eyes. Obviously, we have been projected into some frightfully corrosive fluid. For any but the hardiest of beings, this would be the end of the trail. It would have been for us, but for our protective coat. In chemical terms, the kind of massive and utterly indiscriminate process of dismantlement we are witnessing is called digestion. The blimp is a digestive pocket, or lysosome (lytic body). We will spend all of Chapter 5 exploring this inhospitable, but very necessary, part of the cell. But first, let us reconstruct the events that led us into it.

The Sealed-Room Trick

In just a few minutes' time, we have been spirited through several hermetically sealed walls: first from the extracellular spaces into an intracellular endosome, and then from the endosome into a lysosome. At the same time, we have seen patches being either added to the tenuous fabric that surrounds us or removed from it. At no time, however, has this fabric suffered the slightest tear. Even Sherlock Holmes might have found this one a poser. Or, more likely, he would have regaled Dr. Watson with a discourse on the elementary properties of lipid bilayers: for the secret of our translocation lies in membrane coalescence. This phenomenon forms the physical basis of vesicular transport, the main means of mass transportation through the multiple membrane-bounded compartments that we have started exploring in this first itinerary of our tour.

Vesicular transport depends on two distinct types of events, which, in terms of containers, may be described as fusion and fission; in terms of contents, as mixing and separation; and in terms of mechanisms, as *cis* and *trans* membrane mergers. To clarify these definitions, let us look first at the second leg of our journey, when our endosome joined with a lysosome. This, clearly, was a fusion event, leading to mixing of contents. It was deter-

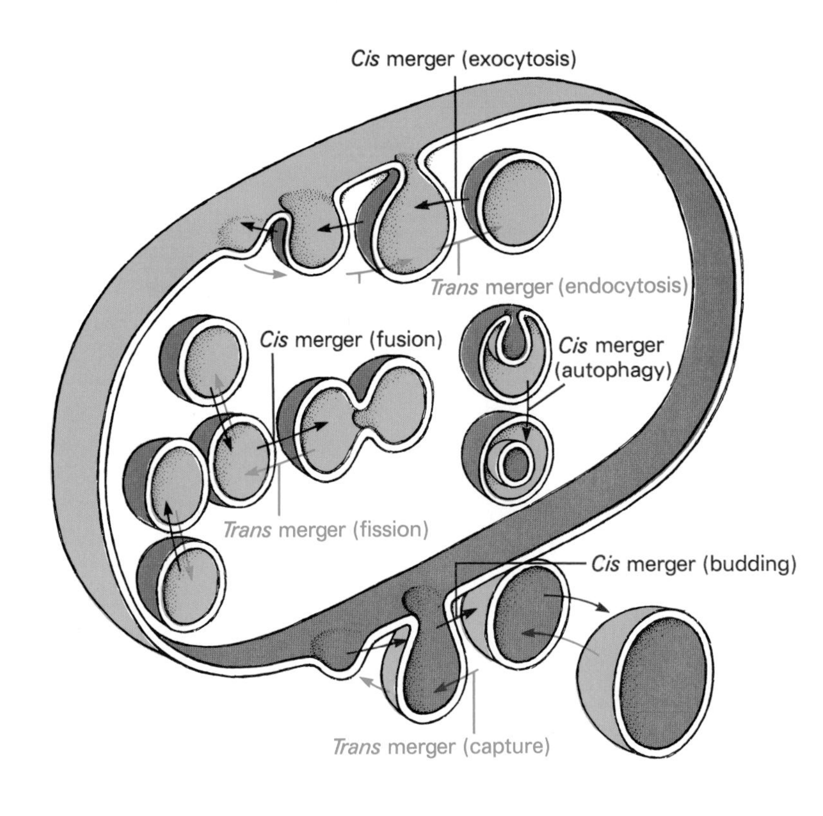

Different forms of vesicular transport mediated by *cis* or *trans* membrane mergers.

mined by the merger and reorganization of the membranes of the two bodies, approaching each other by way of their cytoplasmic faces (the faces of the membranes that are in direct contact with the cytoplasm). This, by convention, is what we will refer to as a *cis* membrane merger. Besides being alliterative with *cyt*, the definition makes sense: *cis* means on the near side; *trans* means on the far side. It is only fair that we should look at the membranes from the point of view of the cell.

It is easy to see that this type of *cis* merger is a mechanism of general significance: any two vesicles that interact in this way mix their contents. A special case of *cis* merger obtains when one of the partners is a cytoplasmic vesicle and the other is the plasma membrane. Mixing then occurs between the vesicle's contents and the outside world: the vesicle's contents are discharged from the cell. This phenomenon, called exocytosis, plays a major role in secretion. It will be examined in greater detail in Chapter 6.

Now consider the earlier event that first got us into the cell—namely, the closure of the endocytic invagination. It represents a fission event, in the sense that a small volume is cut off from the outside world and becomes isolated as the content of an endosome. If we look at its mechanism, we find that this phenomenon also depends on a merger of membranes, in the form of narrowing lips that in this case are led to approach each other by their outside, or *trans*, faces. Again, the mechanism is of general significance. Any vesicle undergoing it divides into two. In particular, the final unmooring of the vesicles we saw detaching from our endosome took place by a *trans* type of membrane merger.

There goes the mystery of the sealed room. All that we need in order to move bulk materials through any number of sealed walls, whether in import, as just experienced, or in export (see Chapter 6), are alternating fusion and fission events determined by *cis* and *trans* membrane mergers, respectively. "Elementary, my dear Watson." But is it? Only in a purely descriptive sense. When it comes to the forces that maneuver the membranes into position so that their lipid bilayers can merge and reorganize and to the factors that control the remarkable specificity of the events, we are still largely in the dark. For it must be pointed out that these mergers obey strict rules. A cytoplasmic vesicle does not simply fuse with the first membrane-bounded object it happens to bump into.

Vesicular transport also represents the main mechanism for moving membrane material through the cell. As far as is known, membrane patches always travel as closed vesicles, although these may be very flat and virtually devoid of content. This is how the piece of membrane that brought us in, or its equivalent, is returned to the cell surface. Such recycling, as we have seen, is an imperative requirement enforced by the enormous membrane consumption associated with endocytosis. It takes place largely from endosomes, but continues also to some extent after endosome-lysosome fusion. Fragile receptors are thereby rescued from lysosomal destruction. Exceptionally, materials that bind tightly to the excised membrane patch may be similarly saved. The vesicles that form in this way may travel directly to the plasma membrane and be reinserted in it by exocytosis. Or they may follow a more circuitous pathway (see Chapter 6).

The Role of Endocytosis

Finding a lysosome as the first reception room of the cell is hardly what one would expect from a friendly host. But cells do not care to be visited. They need to be fed. Endocytosis is first and foremost a feeding mechanism. For many single-celled organisms, such as protozoa, and for lower invertebrates, it is *the* feeding mechanism. Food, as we all know, must be conveyed to a stomach for digestion. And this is exactly what endosome-lysosome fusion accomplishes.

It is true that as we go up the animal scale the need for such a feeding mechanism decreases. Our cells find plenty of nourishing small molecules in the blood and extracellular fluids, which together form what the French physiologist Claude Bernard has called the "milieu intérieur." Our cells thus can afford to feed by molecular transport (Chapter 3) rather than by bulk uptake. However, the endocytic route has not been closed by evolution. It has, instead, become increasingly selective and refined and has adapted to a wide variety of different functions. Most of these involve digestion in the lysosomes and will be considered in the next chapter. But there are exceptions to this rule. For instance, most endosomes that form on the blood side of the flat endothelial cells that line the surface of blood vessels are not intercepted by a lysosome. Instead, they migrate to the tissue side of the cell, where they unload their contents by fusing with the plasma membrane. In this case, endocytosis is followed directly by exocytosis and serves to transport certain blood constituents across a cellular sheet. This process is called diacytosis (Greek *dia*, through), or sometimes transcytosis.

In some cells, endocytic feeding and diacytosis take place simultaneously. Liver cells, for example, direct some of the materials they engulf to their lysosomes, others to the bile canaliculi. Both kinds of materials enter by the same route, as the mixed contents of the same vacuole. Only after their uptake are they separated and sent to their respective destinations by means of the sorting process that we have described.

It may also happen that endosomes are altered in some manner that delays their fusion with lysosomes. They can then serve as temporary storage vacuoles. There is evidence that this occurs in the egg cells of certain insects that use endocytized exogenous proteins for the formation of yolk. Upon fertilization, fusion of the yolk granules with lysosomes is activated, and digestion of the yolk proceeds, to feed the growing embryo. Exactly how endosome-lysosome fusion is delayed in such cases is not known, but there is evidence that the process can be controlled from inside the endocytic vacuole. Some endocytized substances that bind to the membrane—for instance, concanavalin A, a glycoprotein extracted from jack beans—have this ability. So have certain microorganisms; the tuberculosis bacillus is one. Thanks to this property, it is able to launch a successful intracellular invasion, remaining and proliferating within an endosome. Its cousin, the leprosy bacillus, has evolved a different mechanism. It makes a digestion-resistant coat—we have taken advantage of it—and actually resides and proliferates within lysosomes.

5 The Meals of a Cell: Lysosomes and Intracellular Digestion

A major consequence of the kind of fusion-fission events described in Chapter 4 is to allow intermittent communication and sharing of contents among all membrane-bounded pockets that can engage in this sort of activity with each other. The term space or compartment designates a set of pockets related in this manner.

The Lysosomal Space

The lysosomes, of which there may be several hundred per cell, make up a typical space. This kinship would, however, be far from obvious to a casual observer. Indeed, diversity and polymorphism are the most characteristic features of the lysosomal compartment. Lysosomes come in all shapes and sizes, and their inner structures, especially, are extraordinarily variable. This heterogeneity is reflected in the morphological nomenclature. Many different words have been coined to describe particular entities that we now know to be lysosomes. Among them are dense bodies, residual bodies, myelin bodies, multivesicular bodies, cytosomes, and cytosegresomes, to mention only a few. The pathological vocabulary is even richer, since many of the abnormal cellular inclusions seen in diseased tissues also are lysosomes.

As soon as we look at biochemical function, however, the kinship of all these different structures becomes obvious: all are sites of digestion. At the same time, their structural heterogeneity becomes understandable. The contents of lysosomes consist for the most part of materi-

The concept of space or compartment. This diagram shows how three vesicles containing different components (A, B, and C) can be converted into three identical-looking vesicles in which the components are now mixed together, thanks to the ability of the structures to fuse with each other by *cis* membrane mergers and to divide by *trans* mergers.

als in the process of being digested and of indigestible residues. Inspect a few hundred stomachs chosen at random, and you will find that their contents, too, vary greatly, depending on when the owner had his last meal and what he ate. But, having once realized this fact, you would have no difficulty in recognizing all these organs as stomachs. The very diversity of their contents would be diagnostic. This has happened in cell biology. Nowadays, when an electron microscopist sees a particle surrounded by a single membrane and with a "messy" internal structure, he knows from previous experience that he is most likely looking at a lysosome.

In chemical terms, to digest means to hydrolyze—that is, to split with the help of water the various bonds

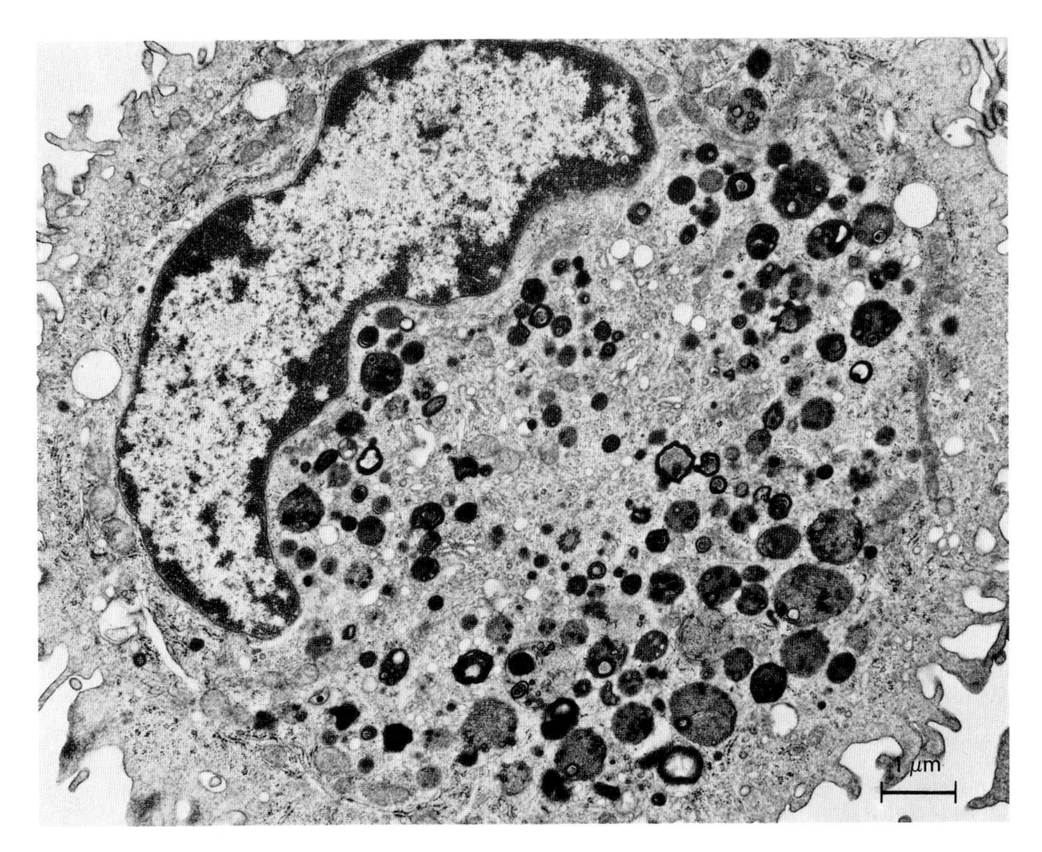

Morphological heterogeneity of lysosomes. In this electron micrograph of a rabbit alveolar macrophage, most cytoplasmic structures are lysosomes. Note the diversity of sizes, shapes, and inner structures. The crescent-shaped nucleus in the upper left-hand corner seems to envelop the lysosome-rich region, sometimes referred to by German scientists as the *Hof*, the court.

whereby the building blocks of natural macromolecules are linked. Examples are the peptide bonds, which join amino acids in proteins; the glycoside bonds, which link sugars in polysaccharides; and the ester bonds between acids and alcohols. Most of these bonds are quite stable and are broken only under harsh conditions of temperature and of acidity or alkalinity. Living organisms could neither realize nor tolerate such conditions, yet they readily digest their food. They do so with the help of special catalysts, called hydrolytic enzymes, or hydrolases, which are secreted into their digestive cavities.

Hydrolases are specific catalysts. Each splits only a well-defined type of bond. Since food is usually made of many different constituents that contain many kinds of chemical bonds, it follows that the act of digestion requires the concerted or sequential participation of many different enzymes. Indeed, the digestive juices secreted into the gastrointestinal tract contain many different hydrolases, and that is why the human organism can utilize a variety of complex foods of vegetal and animal origin. This ability is, however, limited. For instance, it does not include cellulose, as was mentioned in Chapter 2.

These basic notions are essentially valid for lysosomes also. In each lysosome, we find a large collection of different hydrolases—more than fifty have been identified—that together are capable of digesting completely or almost completely many of the major natural constituents, including proteins, polysaccharides, lipids, nucleic acids, and their combinations and derivatives. Like that of the human intestine, however, the digestive ability of the lysosomes is subject to certain limitations.

While they act on many different chemical bonds and substances, the lysosomal digestive enzymes have one property in common: they act best in a slightly acid medium. In technical terms, they have an acid pH optimum, usually between 3.5 and 5.0. In keeping with this property, lysosomes maintain an acid interior with the help of a proton pump, as do the endosomes from which they derive (Chapter 4).

The lysosomal space occupies a central position in the economy of the cell. Incoming roads bring materials to be digested not only from the outside, but also from inside the cell. The products of digestion make up the main outgoing stream. In addition, the lysosomal space is connected by a special line to a subsection of the cell's export machinery that supplies it with the necessary enzymes (see Chapter 6). As is to be expected for a membrane-limited space, this commerce uses two means of transport: permeation for all materials capable of traversing the membrane, eventually with the assistance of special pumps or permeases, and vesicular transport for all the others.

A major problem, which so far has received no satisfactory explanation, concerns the manner in which cells succeed in containing their lysosomal space. The lysosomal membrane is made, like other biomembranes, of proteins and phospholipids that are intrinsically digestible by lysosomal enzymes. Indeed, when pieces of the membrane are excised by autophagic segregation (see p. 69), they are completely broken down. Yet, they resist digestion when in the lysosomal lining. In fact, they make up an astonishingly safe shield that effectively protects the surrounding cytoplasm against the corrosive contents of the lysosomal compartment. The key to this mystery is not known. We can only surmise that the surface components, probably glycoproteins, that line the inner face of the lysosomal membrane are folded in such a way as to put their chemical bonds beyond the reach of the hydrolases to which they are susceptible. We must assume further that this special conformation is maintained with the help of metabolic systems present in the adjoining cytoplasm and is lost when the connection with the cytoplasm is severed.

Of course, this kind of explanation is really no explanation at all. It recalls the answer given in Molière's *Malade imaginaire* by a medical candidate who is asked in a mock examination why opium puts people to sleep. "Because opium has sleep-inducing virtues whose nature it is to numb the senses," the aspiring physician answers solemnly (in Latin, to sound more learned). We are no better when we state with equal profundity that the lysosomal membrane is resistant to digestion "because its surface lining has an enzyme-resistant conformation." Unfortunately, that is where we stand at the present time.

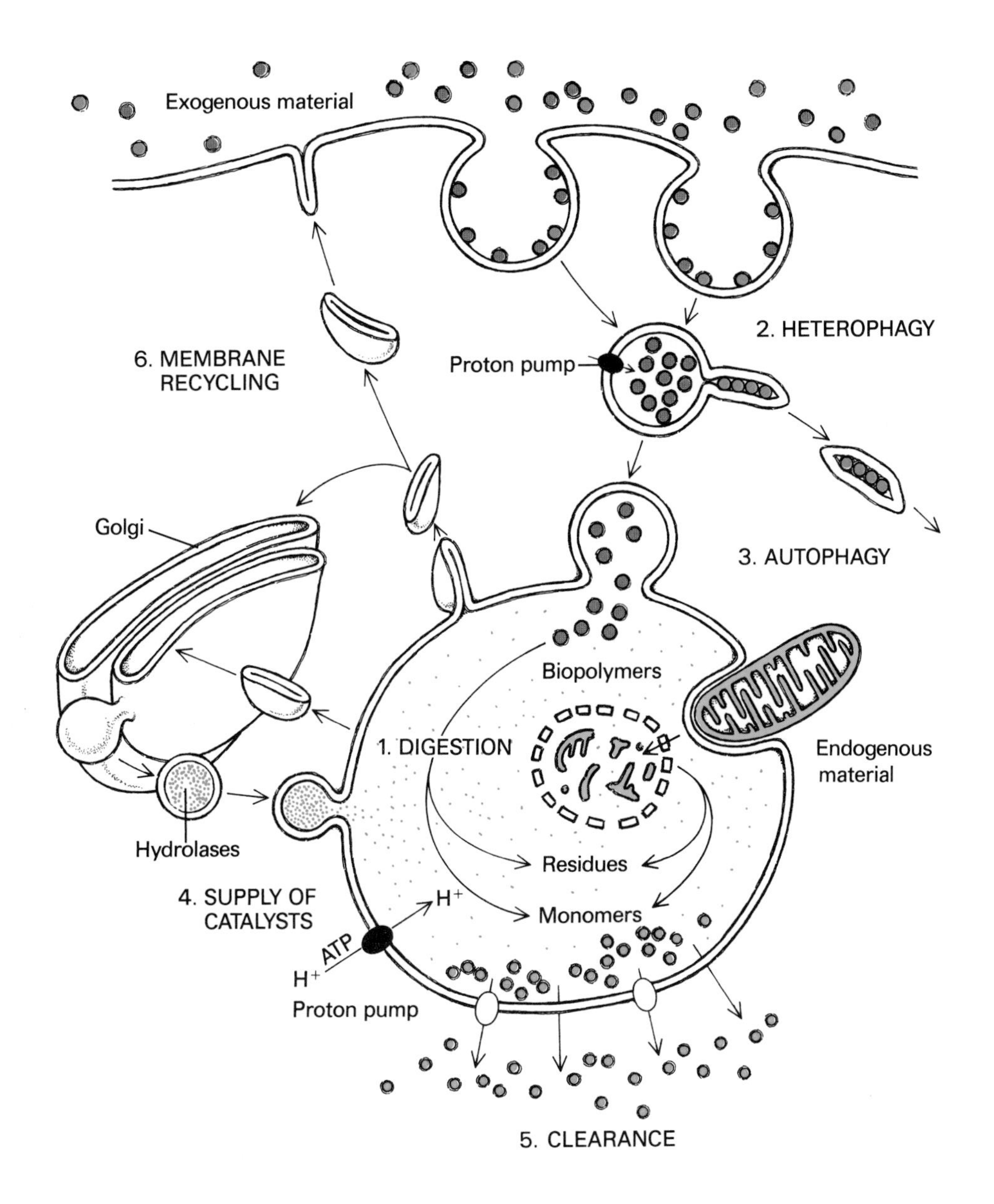

Composite view of lysosome function. The large circular body represents the whole lysosome space.

1. The main event occurring in the lysosome space is the digestion (hydrolysis) of biological polymers to monomers, catalyzed by hydrolytic enzymes (hydrolases) requiring an acid medium (protons, or H^+ ions) for optimal activity.

2. Substrates of digestion are introduced into the lysosome space from outside the cell by endocytosis (*trans* merger) followed by endosome-lysosome fusion (*cis* merger), often after preliminary sorting of contents and membrane in the endosome.

3. Other substrates enter from inside the cell by intralysosomal budding followed by severance (*cis* merger) of the bud.

4. Newly made enzymes are conveyed to the lysosomes by Golgi vesicles (*cis* merger). Protons are actively transported into the lysosomes by a proton pump, with the help of energy provided by ATP (see Chapter 7).

5. Products of digestion diffuse or are transported across the lysosomal membrane into the cytosol. Indigestible materials unable to traverse the membrane in this way remain in lysosomes as residues.

6. Much of the membrane material added to lysosomal membranes is removed by *trans* merger and recycled back to the cell surface, either directly or by way of the Golgi apparatus. A small amount of membrane is interiorized by autophagic segregation and digested. Note that the *trans* face of the lysosomal membrane is resistant to digestion when its *cis* face is in contact with the cytoplasm but not after segregation.

Physiological adaptations of heterophagy.

A. Defense: a leukocyte captures and destroys a pathogenic bacterium.

B. Cleaning of lung alveolus: a macrophage sweeps up dust particles, microbes, viruses, and other foreign objects.

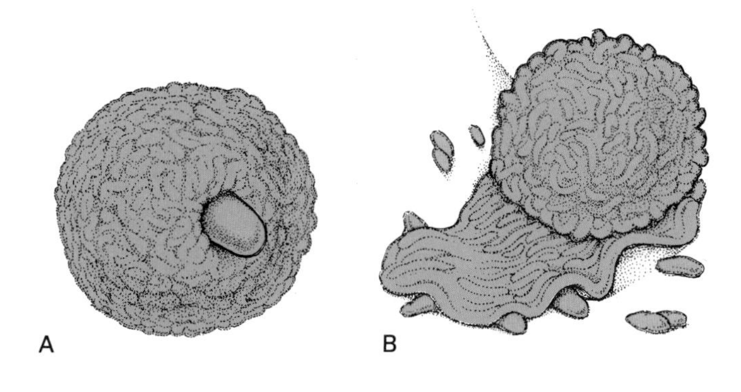

Heterophagy

Uptake and digestion of extracellular materials is called heterophagy (Greek *heteros*, other). It is a faculty that comes down to us from those early single-celled ancestors that first learned to chase after living prey and to kill and digest them intracellularly. It represents the main feeding mechanism of protozoa and lower invertebrates and persists almost unaltered in the leukocytes, whose function it is to pursue and destroy invaders.

There is, however, an important difference between a free-living protozoon and a leukocyte. For the former, endocytosis is a matter of life and death. Its survival depends on a daily catch of bacteria. The leukocyte, on the other hand, lives in an unctuously rich fluid full of sugars, amino acids, and other small molecules that can be used directly, without prior digestion. It has no need for heterophagy. On the contrary, heterophagy kills it. Leukocytes are constructed so as to consume only one big meal in their lives. They are made in the bone marrow, loaded with lysosomal hydrolases and other deadly weapons, and then sent out to search for the enemy. When they encounter them, they gobble up as many as they can. Soon after, they die, tricked by natural selection into committing mortal gluttony for the higher good of the organism. Until suffering this fate, they feed mostly on small molecules that they take in from the environment through the various transport systems with which their plasma membrane is fitted. They are essentially osmotrophic (Greek *osmos*, push; *trophê*, nourishment). Free-living protozoa, on the other hand, are phagotrophic.

Human cells, and in general the cells of most multicellular organisms, are essentially osmotrophic. To them, endocytosis and lysosomal digestion do not have the same survival significance as they have in the lower organisms, where these processes represent the sole means of food collection. Even in higher organisms, however, heterophagy still serves some nutritive function, because certain foodstuffs, such as iron or cholesterol, are carried by proteins that have to be taken up and processed for the nutrients they carry to become available.

In addition, in multicellular organisms, heterophagy is adapted to a wide variety of functions that serve the organism as a whole, rather than the individual cells. One such adaptation—to immune defense—has already been encountered in leukocytes. A related activity is carried out by macrophages, a group of cells that are scattered throughout the tissues and are characterized by a particularly avid and indiscriminate phagocytic activity. The macrophages also act in defense but, in addition, have a general role as "cleaners" and "garbage collectors." We find them in the lungs, for instance, endlessly scrubbing the surface of the alveoli, sweeping up dust, soot, tar, bacteria, viruses, any particulate material brought in by the inspired air. Whatever they cannot digest they deposit in their lysosomes. When these are full, the cells curl up and die, mission accomplished, and are ejected with the sputum.

Other cells are more selective and remove specific molecules from their environment by means of their endocytic receptors. As a rule, the function subserved by this activity is regulation by destruction. For instance, many hormones are endocytized and destroyed in lysosomes after binding to surface receptors and eliciting their intracellular effects. Thus, by being at the same time a sensor and an endocytic catcher, the receptor both mediates and limits hormonal activity.

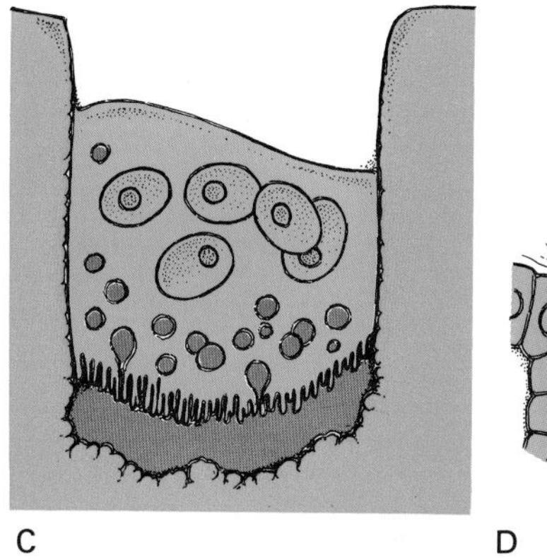

C. Bone remodeling: an osteoclast breaks the bone matrix down into fragments by means of secreted acid and lysosomal hydrolases; it completes the breakdown by endocytosis and lysosomal digestion of the fragments.

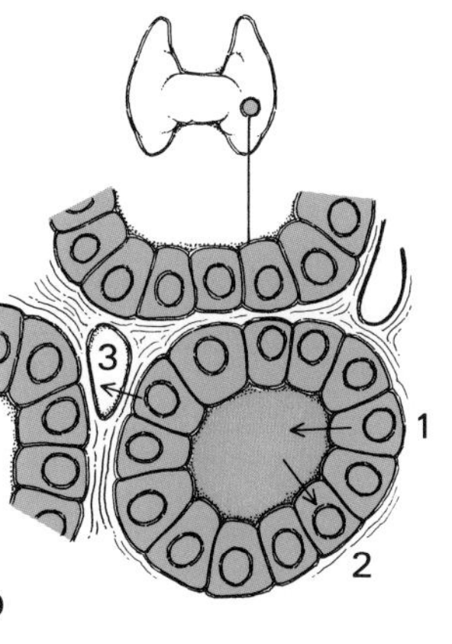

D. Thyroxin production in the thyroid gland: (1) thyroglobulin is secreted into the lumen of a follicle; (2) thyroglobulin is endocytized by follicular cells and digested in lysosomes; (3) one product of digestion is thyroxin, which diffuses into pericellular spaces to be picked up by blood and lymph.

E

E. Renal reabsorption: proteins and other macromolecules that have filtered through renal glomerulus are endocytized by proximal tubule cells and digested in lysosomes.

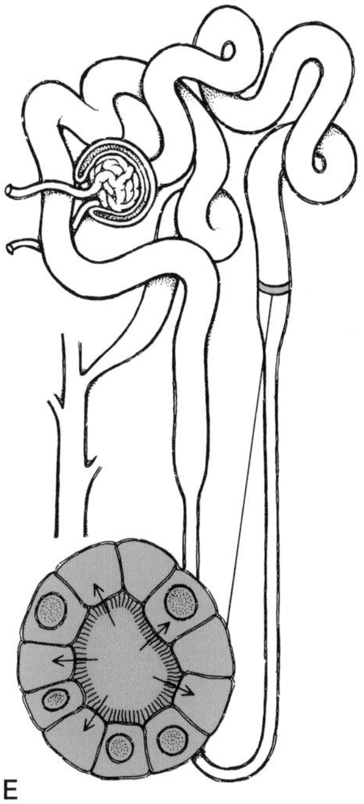

Heterophagy is also involved in the turnover (see p. 68) and remodeling of insoluble extracellular structures. Bones, for example, contain cells, called osteoclasts (Greek *klastos*, broken), that spend their time like moles, digging galleries and tunnels through the bone matrix. To do this, they secrete acid and lysosomal enzymes into the blind end of the tunnel; they thereby dissolve the crystals of hydroxyapatite (a complex of calcium phosphate and hydroxide) that make up most of the bone mineral and dismantle the scaffolding of collagen fibers that form the organic matrix of the tissue. They then complete the job by phagocytosis and intracellular digestion of the fragments. The damage is repaired by builder cells, called osteoblasts (Greek *blastos*, germ), that construct new bone elements, or trabeculae (diminutive of Latin *trabs*, beam).

Sometimes endocytosis is used for retrieval. This happens to the proteins that leak through the kidneys. While passing through the renal tubules, they are reabsorbed, digested, and recovered as amino acids. A remarkable case of retrieval is seen in certain cells—in particular the fibroblasts, which manufacture a good part of the collagen and other components of connective tissue. These cells display on their surface a special receptor for lysosomal enzymes, which are thereby both removed from the extracellular spaces, where they can do harm, and returned to the lysosomes, where they can do good. The same receptor also occurs intracellularly, as a key piece of the machinery whereby newly made lysosomal enzymes are conveyed to their destination (see Chapter 6).

It may even happen that heterophagy is geared to a synthetic activity. This occurs in the thyroid gland, where the hormone thyroxin arises as the product of the lysosomal digestion of thyroglobulin. This is an iodinated protein made by the thyroid cells and discharged by them into a common extracellular reservoir, subsequently to be taken up again and digested by the same cells upon appropriate stimulation.

These few examples illustrate the remarkable functional diversity of heterophagy. Starting as powerful nutritional aids, lysosomes have thus provided evolution with a particularly rich theme on which it has composed countless variations. And this is only one side of the coin.

Autophagy

Cells also practice autophagy; that is, they "eat" and digest little pieces of their own substance (Greek *autos*, self). They do this not only when they are deprived of food and forced to live on their own resources, but even when they are abundantly provided with nutrients. The rate at which this self-destruction

Schematic representation of autophagy. Four stages are shown from left to right: (1) interiorization by intralysosomal budding of the cytoplasm, (2) segregation by *cis* merger, (3) digestion, and (4) clearance. The size of the segregated material distinguishes macroautophagy from microautophagy. Multiple microautophagic events lead to the formation of multivesicular bodies.

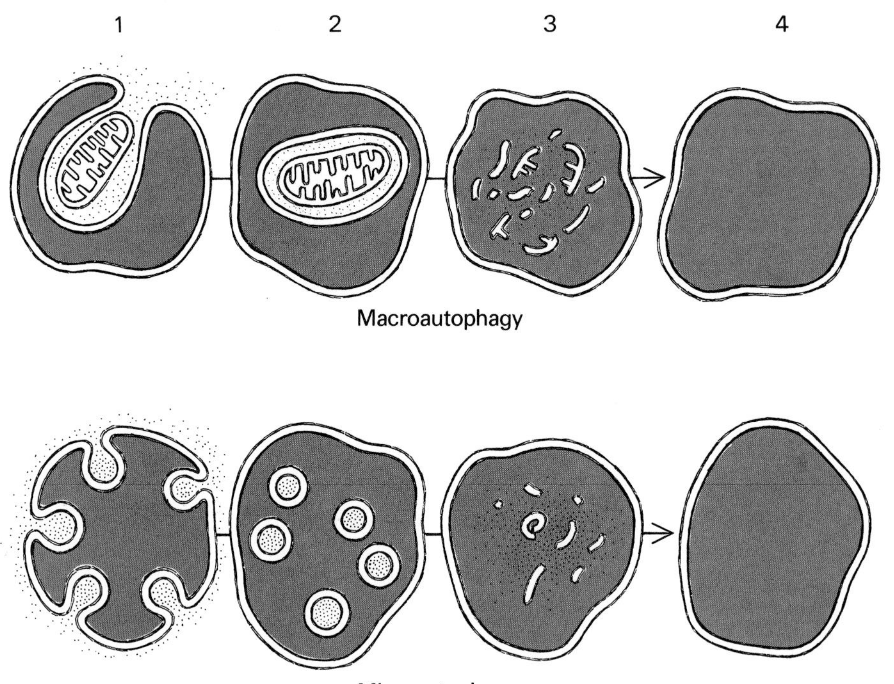

goes on is astonishing. An average liver cell, for example, destroys most of its contents in less than a week.

This you would hardly suspect from examining the cell, even with the sharpest of morphological or chemical tools. A hepatic cell may live for many years, and during all that time its structure and its chemical composition hardly change. The intense molecular turmoil that goes on behind this façade came to light only when isotopes became available, so that chemically identical molecules could be distinguished from each other by their atoms. It was then found that cells continually destroy and rebuild their constituents at a remarkably rapid rate. They are like those old houses that look exactly as they did when they were first built but, owing to multiple repairs, are left with few of their original window panes, or tiles, or even bricks or boards. But what can take centuries in a house is completed in a matter of days in a living cell.

Turnover, as this molecular renewal is called, is a finely regulated activity. Each cell constituent has its own characteristic average lifespan. Protein molecules, for instance, generally live between a couple of hours and a few days, depending on their nature. This implies that each type of protein has its own characteristic rates of synthesis and destruction and that these two rates must balance each other perfectly if the cell is to remain unaltered. The same is true for most other cell constituents, with the exception of DNA, which does not turn over as such, although it does undergo local injuries and repairs.

To destroy their own constituents, cells need digestion just as much as they do when they break down exogenous materials of similarly complex nature. Several different systems participate in this digestive activity. Among them

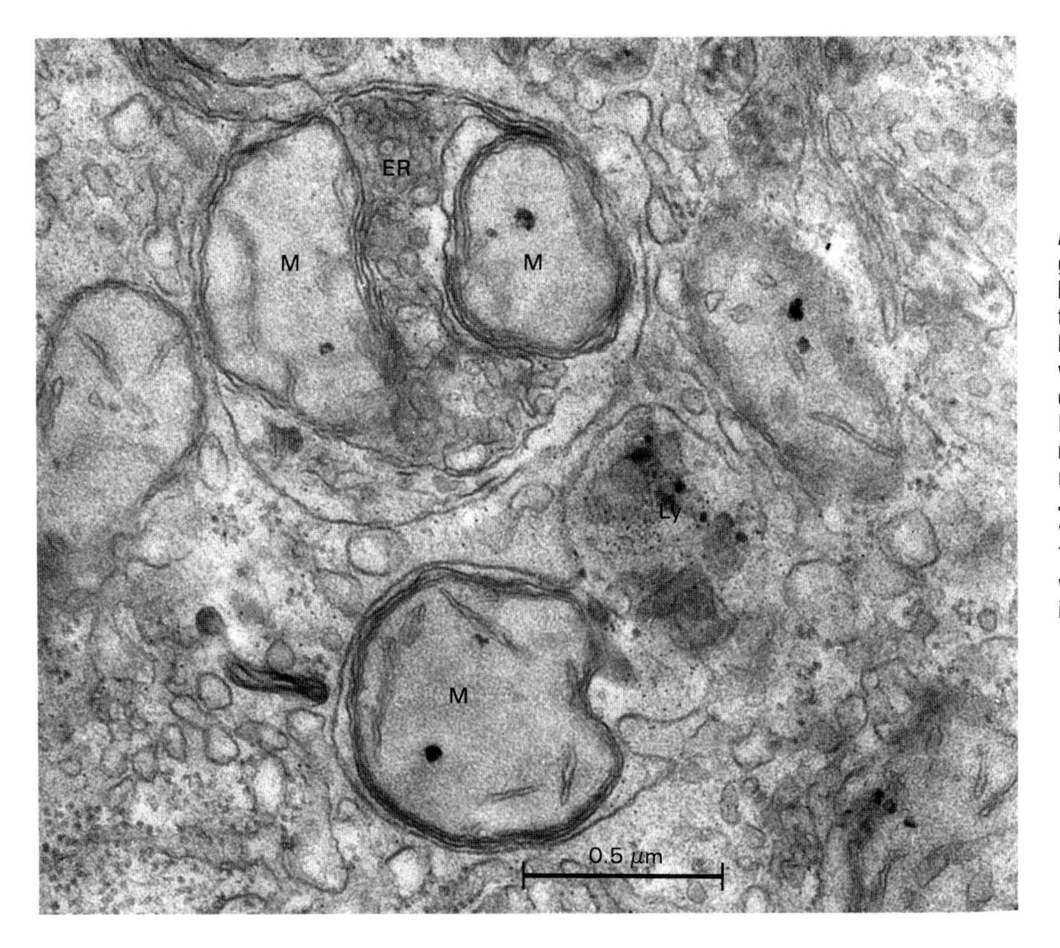

Autophagic vacuoles. Electron micrograph of a thin section through the liver of a rat that was injected with the hormone glucagon, which stimulates autophagy. The autophagic vacuole at the top contains two mitochondria (M) and endoplasmic reticulum (ER). Remnants of a second membrane are seen near the outer membrane of this autophagic vacuole. Just below it is another autophagic vacuole containing a mitochondrion. This vacuole seems to have fused with a lysosome (Ly), recognizable by its dense, polymorphous contents.

are the lysosomes, which serve as dumping ground and disposal site for a variety of intracellular materials, especially bulky objects such as whole chunks of cytoplasm, mitochondria, clusters of ribosomes, and membrane fragments abstracted from the endoplasmic reticulum. These objects enter the lysosomal compartment by a variant of the sealed-room trick (Chapter 4) that is essentially similar to the mechanism whereby cytoplasmic buds detach from cells. In budding, a piece of cytoplasm jutting out of the cell is amputated as a result of the progressive narrowing and final severance—by a *cis* type of membrane merger—of its connecting link. Let the same mechanism operate on a blob of cytoplasm that protrudes into the lumen of a lysosome and the bud ends up segregated inside the lysosomal compartment, a prey to cellular autophagy. Multiple events of this kind lead to the formation of characteristic structures described by morphologists as multivesicular bodies.

In the early stages that follow the act of autophagic segregation, the snared cytoplasmic objects can still be recognized inside their membranous trap. At first, they are surrounded by two membranes, as required by the budding mechanism just described. Later, the piece of membrane surrounding the sequestered bud degenerates and is no longer visible. Eventually, the contents of the bud become themselves unrecognizable, as their digestion progresses. The term autophagic vacuole designates all the structures that can be identified by any of the above characteristics as arising by autophagy. It has been claimed that other membranous structures besides lysosomes may serve in the initial segregation mechanism. Autophagic vacuoles arising in this manner would contain no digestive

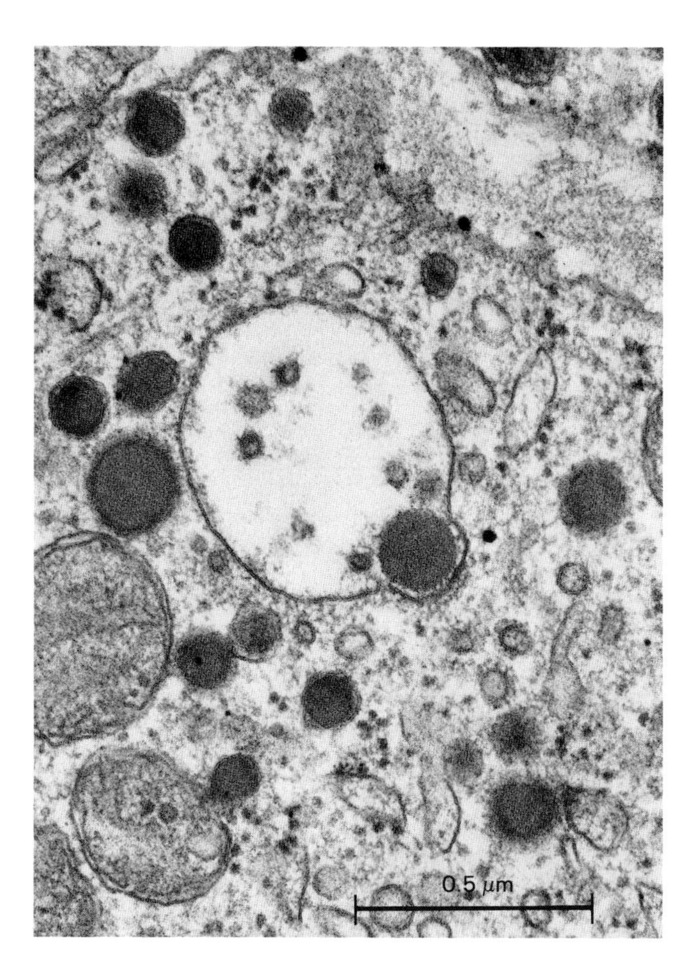

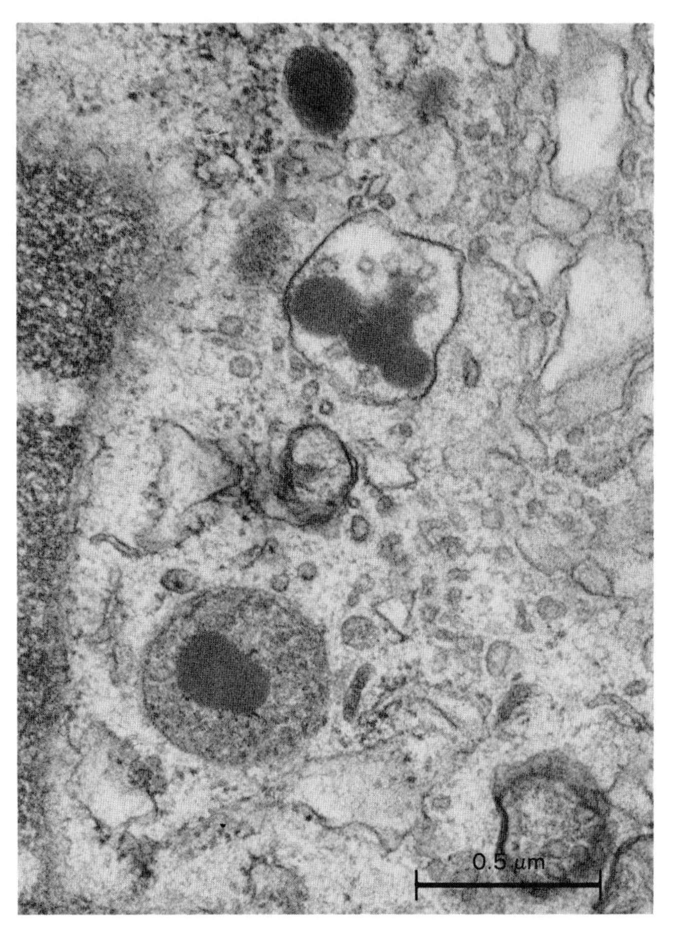

Two views of crinophagy. These electron micrographs illustrate crinophagy in hormone-secreting cells of the pituitary gland. The picture at the left shows a dense secretion granule that has just fused with a multivesicular body believed to be a lysosome. In that at the right, secretory contents are seen inside a multivesicular body and a dense body. Note that the membrane of the secretion granule is not interiorized but is added to the lysosomal membrane (*cis* merger). This fact distinguishes crinophagy from autophagy.

enzymes at first but would acquire such enzymes by subsequent fusion with lysosomes.

Faced with the extent of self-destruction that goes on in living cells, we may well wonder what advantages outweigh such a costly activity. It is likely that, in its primitive form, autophagy, like heterophagy, served an essentially nutritive function: it provided a mechanism for survival during starvation and evolved into a highly efficient process in which cellular constituents are sacrificed in their order of decreasing dispensability. Thanks to autophagy, a cell can do without food for extended periods. It uses up its own substance progressively, but in an orderly fashion, so that it remains organized and functional for a long time. This is true also of organisms. The walking skeletons that came out of Hitler's horror camps survived by consuming the proteins of their nonessential muscles to feed their more important hearts, brains, and blood. Many recovered remarkably well when supplied again with the nutrients needed to rebuild their tissues.

Autophagy remains a primary cell response to food deprivation in most present-day organisms, including the higher animals, where it may also be stimulated by hormones, such as glucagon, that stimulate mobilization of the organism's reserves. But it is obvious that self-support hardly explains the continual and intense turnover of constituents in cells that are perfectly well nourished most of the time. The main advantages that cells gain from their autophagic activity are rejuvenation and adaptability.

Thanks to turnover, cells continually replace their constituents with newly made ones and thereby achieve something very close to perpetual youth. Take the brain cells of an aged individual. They have been there for decades. Yet most of their mitochondria, ribosomes, membranes, and other organelles are less than a month old. Over the years, the cells have destroyed and remade most of their constituent molecules from hundreds to thousands of times, some even more than 100,000 times. These events do, however, leave some trace, in the form of a progressive accumulation of brown, indigestible residues in the lysosomes. This so-called age pigment, or lipofuscin, betrays the age of the cell. It could even be one of the flaws—the result of a slightly less than perfect lysosomal digestive activity—that finally defeat the cell's bid for eternal youth, if, as has been suggested, overloading of lysosomes plays a role in cellular senescence (see p. 72).

The second important advantage of autophagy is adaptability. The cell part that is destroyed can be replaced not only by an identical one, but also, if need be, by another one better suited to the current requirements of the cell. Some of these adaptive changes are evoked transiently, in response to environmental changes of one sort or another. Others are part of the fundamental processes of differentiation and development.

In gland cells, there is a special kind of autophagy called crinophagy (Greek *krinein*, to sift or separate; by extension, to secrete). It occurs through direct fusion between secretory granules and lysosomes (*cis* merger), and results in the destruction of the secretory material. It is an important regulatory mechanism (see Chapter 6).

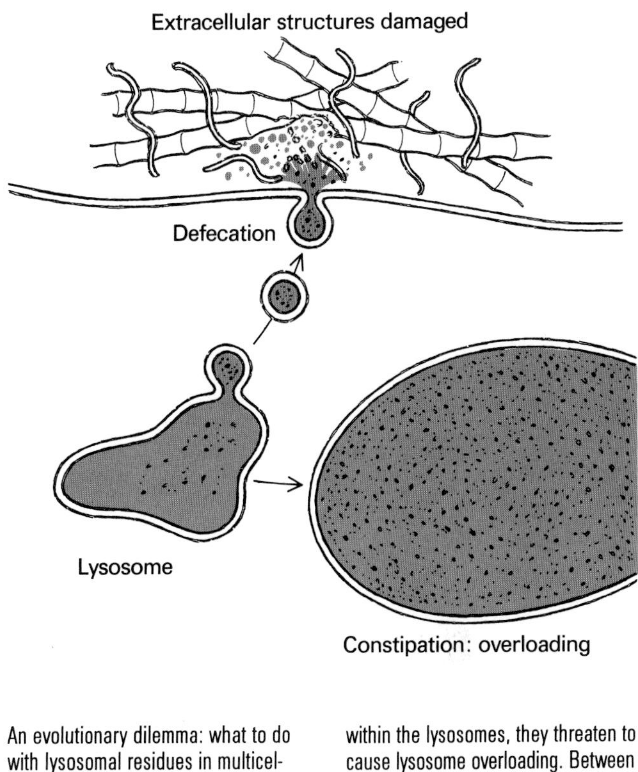

An evolutionary dilemma: what to do with lysosomal residues in multicellular organisms? If the residues are discharged by defecation, as they are in unicellular organisms, the released enzymes may damage extracellular structures. If they are kept stored within the lysosomes, they threaten to cause lysosome overloading. Between these two evils, the second seems to have been selected as less harmful, as indicated by the state of "chronic constipation" of most cells.

Clearing Lysosomes

In the gut, the final products of digestion are cleared by what is called intestinal absorption. They are removed by the mucosal cells, usually with the help of active pumps, and discharged into the bloodstream. Something similar goes on in lysosomes. The various small molecules formed by the digestive process diffuse or are transported across the lysosomal membrane into the cytoplasm, where they are used by the metabolic systems of the cell. It is likely that this clearance requires the help of special carrier systems located in the membrane, but these have not yet been characterized.

But what if digestion does not occur or is incomplete and does not progress to the point where its products can be cleared? What, for instance, if the cells take in some digestion-resistant material or if they suffer some digestive deficiency?

In most protozoa and lower invertebrates, accidents of this sort are of no special consequence, because these cells have the ability to unload the contents of their old lysosomes into the surrounding medium. That process, graphically named cellular defecation, depends on exocytosis. In higher animals, most cells seem to be unable to empty their lysosomes in this manner. They are, so to speak, chronically and persistently constipated. This represents a serious deficiency, which is responsible, as we shall see, for numerous pathological conditions associated with lysosome overloading. The alternative, unfortunately, is just as bad, or probably worse, as also illustrated by pathology. When lysosomes are unloaded in the tissues, the enzymes that are discharged with the residues cause widespread damage to extracellular structures. We may speculate that cellular defecation had to be repressed by natural selection to allow evolution to progress in the direction of greater multicellular organization.

Supplying Enzymes to Lysosomes

Like all enzymes, the lysosomal hydrolases are proteins. Therefore, they should themselves go the way of all the other proteins that enter the lysosomes. They should be digested by their proteolytic congeners. The fact that they do their job properly indicates that such digestion takes place slowly. This is hardly surprising, because without digestion-resistant enzymes there could be no lysosomes. As to what makes for their resistance, presumably it is their ability to keep themselves tightly bundled up in the acid environment prevailing within their natural habitat. As a rule, a protein chain needs to be at least partially unfolded, or denatured (Chapter 2), to become exposed to the clipping action of proteolytic enzymes. Many proteins are denatured by acidity. Obviously, the lysosomal enzymes are not, and this explains their survival in the lysosomal milieu.

They are not eternal, of course, and have to be replaced as they succumb to a chance hit that starts them on the road to breakdown. To find out how this replacement occurs, we have to move to another division of the cell, its export department, of which lysosomal enzyme production is a subbranch. This we will do in Chapter 6. At present, we can only watch the final delivery step. Not unexpectedly, the newly made enzymes arrive membrane-wrapped and are discharged into the lysosomal space by membrane fusion (*cis* merger). Such fresh packages of enzymes are called primary or virgin lysosomes, as opposed to the digestively active secondary lysosomes.

The Digestive Tribulations of a Cell

Dyspepsia, hyperacidity, constipation, and other digestive upsets are the common lot of mankind, the source of a great deal of personal discomfort, as well as of a very profitable alleviating industry. Yet these troubles are nothing compared with the digestive ills that afflict cells. Now that medical investigators are turning into subcellular and molecular detectives they are beginning to find out that a great many diseases are nothing but the manifestations of some digestive disturbance affecting certain cells.

As we have already seen, most of our cells suffer from constipation. It cannot be considered a disease, since it is a natural condition. But it is a serious disability, which leaves our lysosomes very vulnerable to the risk of overloading. The most dramatic examples of lysosome overloading are seen in young children suffering from a genetic deficiency of some lysosomal enzyme. More than twenty-five such deficiencies are known. They make up the group of genetic storage diseases, of which Tay-Sachs disease is a well-known example. In each of them, a lysosomal enzyme is severely deficient. As a result, the lysosomes become filled with materials that require the missing enzyme for their digestion; they swell progressively to enormous sizes and end up choking the cells to death. These lysosomal storage diseases are rare, fortunately, but they are extremely distressing. Children afflicted with them often suffer severe mental retardation and die at an early age.

Besides genetic deficiencies, many other situations may lead to lysosome overloading. It is seen in arteriosclerosis, occurs as a major complication in certain kidney diseases, and can be induced by drugs. The more we get to know about it, the more it appears to be one of the most frequent cellular illnesses, responsible for numerous diseases. In a way it is universal, and an inevitable concomitant of aging, for there is no way in which a cell can completely escape having its lysosomal compartment visited occasionally by an indigestible molecule. Once in, the

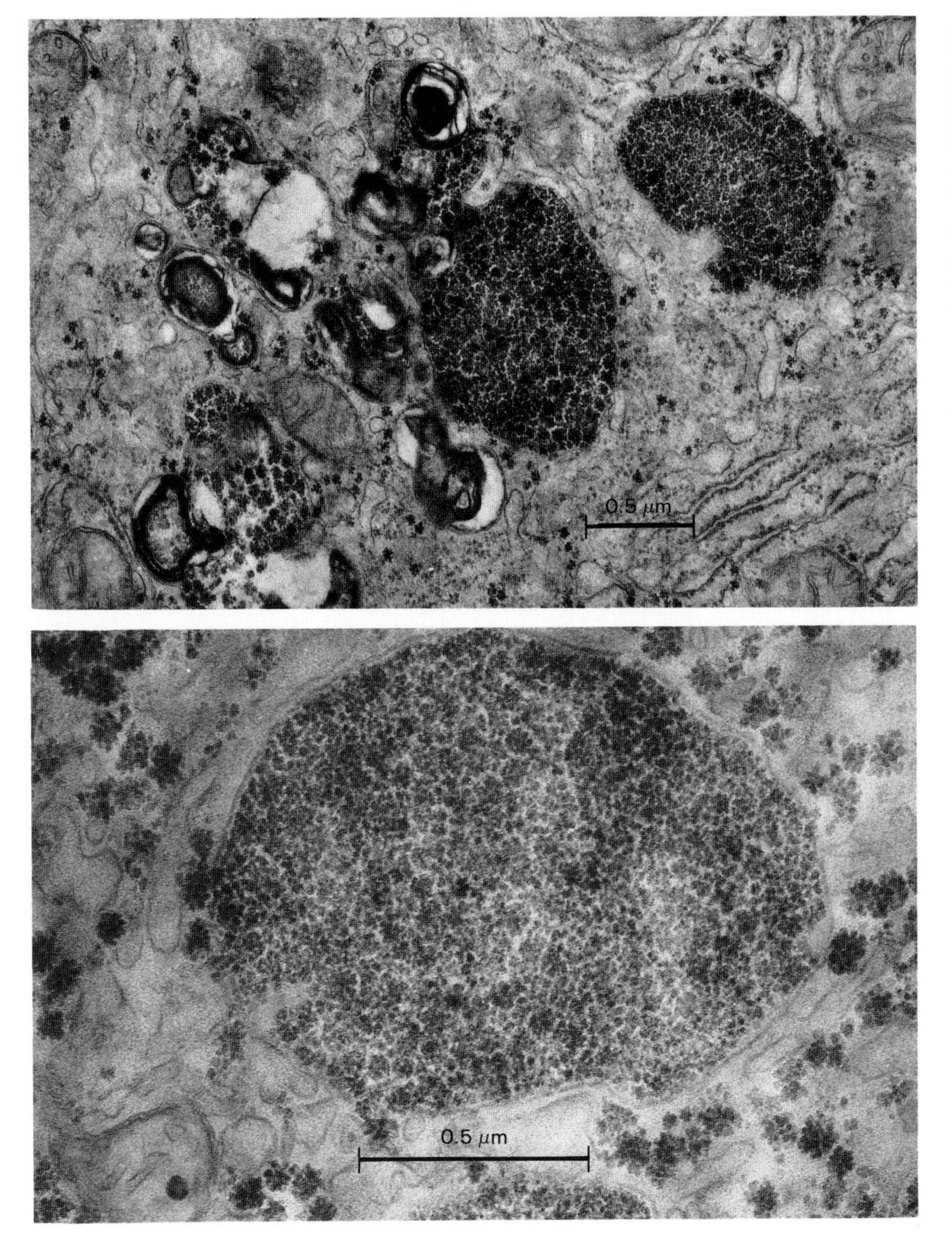

An example of a genetic lysosomal storage disease: glycogen storage disease type II, or Pompe's disease. Due to the genetic deficiency of the lysosomal hydrolase normally responsible for the digestion of the starchlike polysaccharide glycogen, which enters the lysosomal space by autophagy, this material accumulates in lysosomes. In these two electron micrographs of a fragment of liver from a child who died of the disease, glycogen particles can be recognized by their high density and star-shaped structure. The upper picture shows an early stage of incorporation of glycogen into dense lysosomes. In the lower one, a swollen lysosome is completely filled with glycogen.

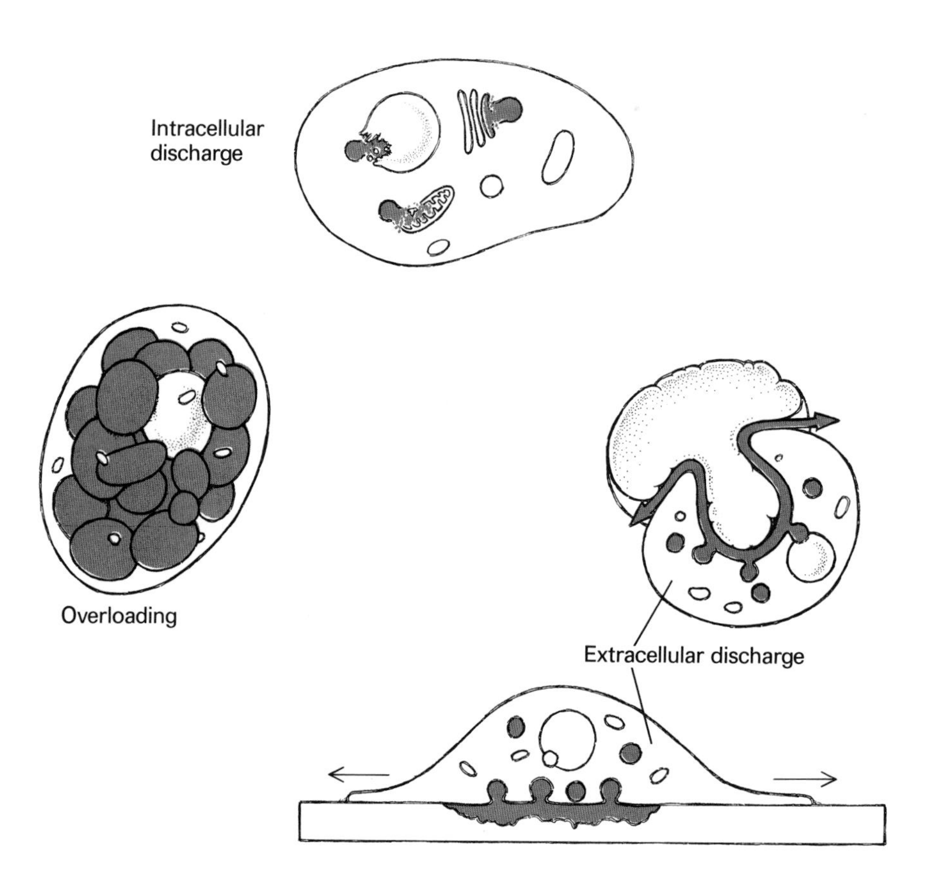

Digestive pathology of cells: the three great lysosomal syndromes.

molecule cannot get out again. So, with time, the lysosomes fill up with residues.

In certain pathological conditions, the constraints that prevent extracellular unloading of lysosomes break down. This may happen, for instance, when cells struggle to take in very bulky objects, such as antigen-antibody aggregates, or if they attempt the impossible task of engulfing a flat surface, such as a basement membrane accidentally opsonized (Chapter 4) by autoantibodies (antibodies against endogenous constituents). Endocytic invaginations then fuse with lysosomes while still open and allow the lysosomal contents to leak out into the extracellular spaces. Rheumatoid arthritis and several other autoimmune diseases illustrate the type of gruesome erosion that extracellular structures may suffer as a result of this kind of discharge, showing that there is a good reason for cellular constipation, in spite of its many drawbacks.

Another frequent lysosomal syndrome is that which results from some injury to the lysosomal membrane. Whatever magic renders this membrane resistant to digestion, it is not foolproof. Predictably, if the lysosomal membrane gives in, the neighboring cytoplasm becomes

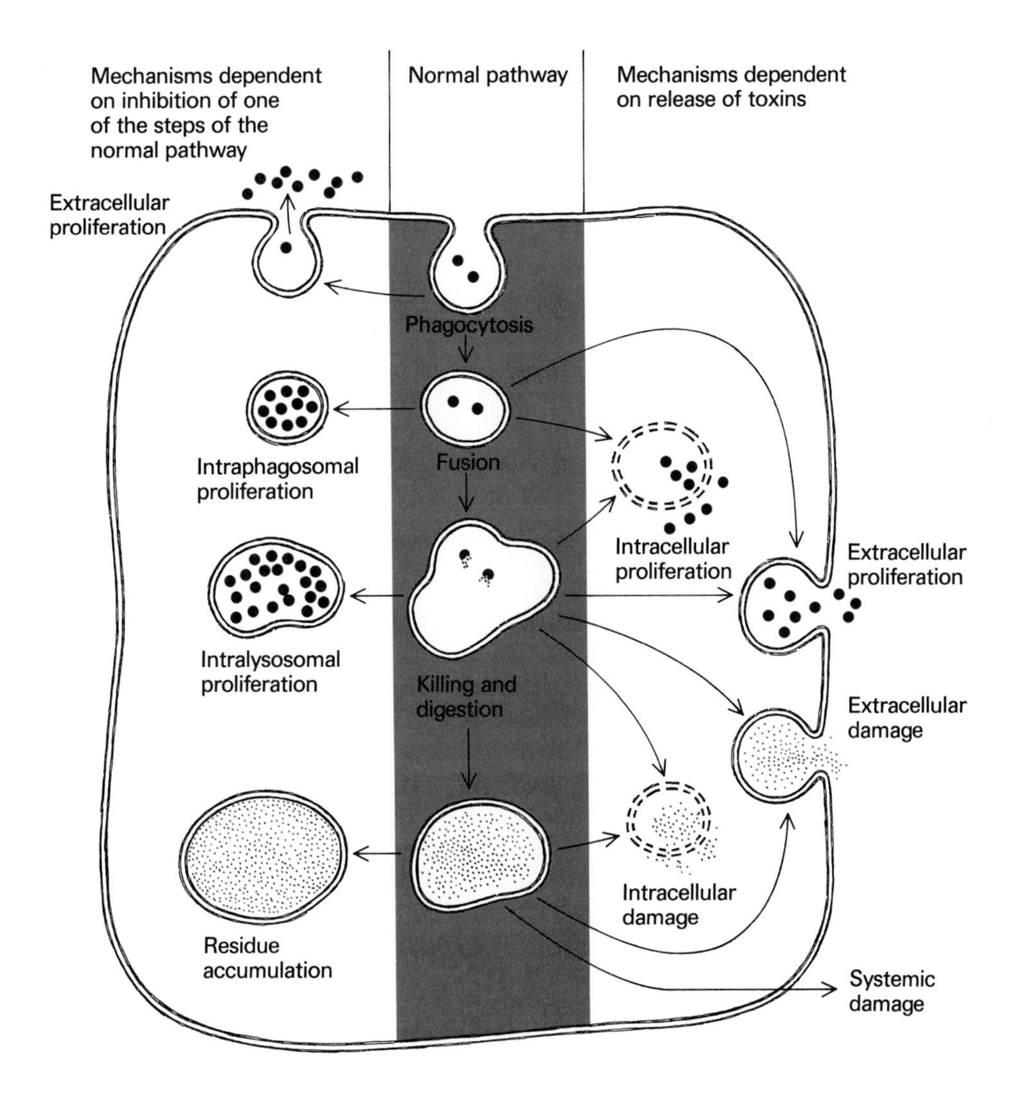

Lysosomes and infection. This diagram illustrates some of the mechanisms whereby pathogenic microorganisms evade, inhibit, or resist destruction by lysosomes, or, alternatively, turn this defense mechanism into a means of causing local or systemic injuries.

exposed to enzymic attack. Extensive cell injuries, and even cell death, may ensue. Gout, asbestosis, and silicosis (miners' black-lung disease) are examples of conditions where such injuries are believed to occur.

Finally, some mention must be made of the role of lysosomes in infection. In a way, most successful infections may be seen as some sort of failure of lysosomal defense. Or to put it differently, a pathogen is a microorganism capable of evading lysosomal destruction. Our bacterial enemies have displayed remarkable ingenuity in this respect. Some slip by our immune system without alerting it and, not being opsonized, avoid being phagocytized. Others inhibit endosome-lysosome fusion—remember the tuberculosis bacillus—or, like the leprosy bacillus, resist lysosomal destruction. Yet others escape by breaking open the membrane of endosomes or lysosomes with an exotoxin (a secreted toxic substance). The most perfidious allow themselves to be killed and then exert a posthumous revenge by poisoning the organism with an endotoxin (i.e., a toxic substance of endogenous origin) produced by lysosomal digestion of their cell wall.

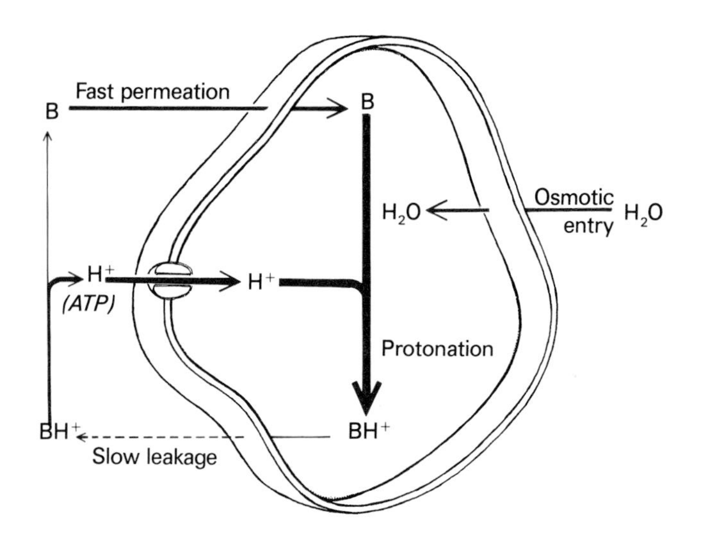

Lysosomotropism of weak bases by proton trapping. Permeant, nonprotonated form B is trapped in the lysosome as nonpermeant, protonated form BH^+. Water is attracted osmotically. The process is powered by ATP through a proton pump, which forces protons inside.

Consequence of lysosomotropism is illustrated by vacuolation. The fibroblast shown here was vacuolated by exposure to chloroquine. For vital staining, see book jacket.

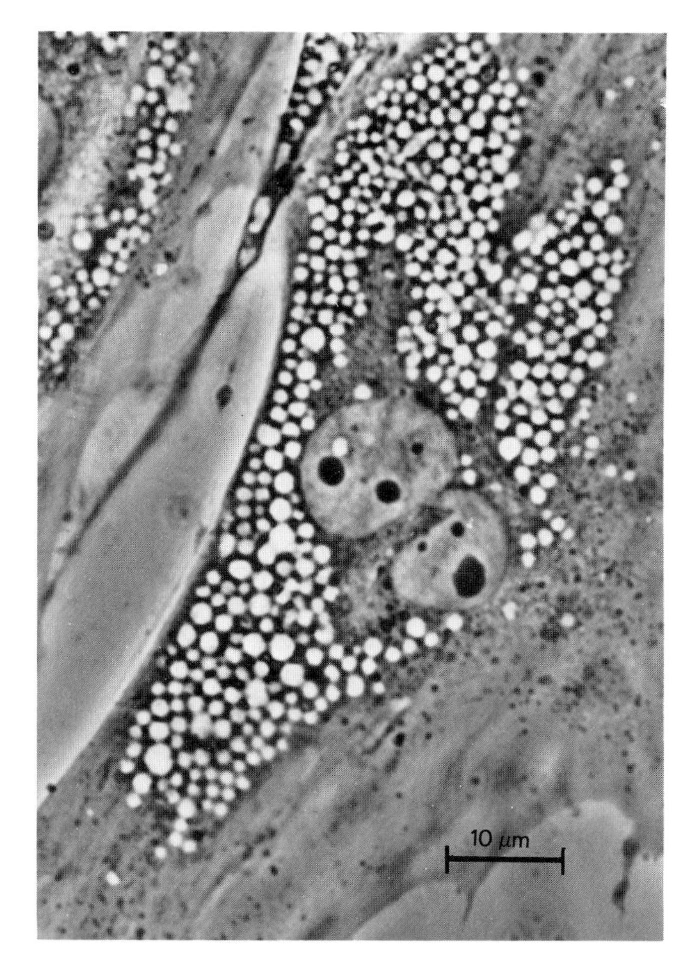

Sending Drugs into Lysosomes

There are many reasons why we may want to send drugs into lysosomes. The most obvious is to correct some local disorder, of which we have just seen there are many different kinds. But, in other instances, lysosomes are not so much a target as a tool of the therapeutic intervention. A large group of substances—among them many drugs, but also a number of dyes, such as neutral red—home spontaneously to lysosomes, where they accumulate very rapidly and reach up to several hundred times the concentration they have elsewhere in the cell. Called lysosomotropic, all these substances are weak bases. They readily cross membranes, including the lysosomal membrane, in their unprotonated form, which is uncharged and adequately lipophilic. When exposed to the lysosomal acidity, they become protonated, acquiring one and sometimes two positive charges that prevent them from crossing the membrane in the other direction.

This phenomenon of proton trapping is not limited to lysosomes. It occurs in endosomes and in any other structure (some parts of the Golgi apparatus?) that is kept acidic by the operation of a proton pump. It leads to all sorts of interesting manifestations. The most spectacular is vital staining, in which the whole endosome-lysosome space lights up like a constellation of brightly colored stars. Known for more than a century, this phenomenon has only recently been explained. Another manifestation of lysosomotropism is vacuolation of the cells, the consequence of osmotic swelling of the invaded structures. In addition, many lysosomotropic substances interfere with lysosomal digestion by neutralizing the local acidity, sometimes also by inhibiting one or more hydrolases. A number of toxic effects and therapeutic actions are explained by such mechanisms. An interesting example is that of the antimalarial drug chloroquine, which is intensely lysosomotropic and is believed to accumulate in the lysosomes of the malarial parasite and there to block digestion of hemoglobin, the parasite's only source of food in the red blood cells where it resides.

Another way of getting into the lysosomes is by endocytosis. This mode of entry holds particularly exciting promises because of its selectivity. Entrance by the endocytic route is governed by surface receptors, and these vary from one cell type to another. Now that we know something of what happens in the lysosomes, we can add a new twist to this tale. Suppose we wish to send a drug selectively to a given type of cell—a cancer cell, for instance, or some pathogenic parasite. Provided our target has receptors that do not occur, or are less abundant, on the cells we wish to protect, we are in business. All we need do is attach the drug to a carrier that is recognized by these receptors, and do so by a chemical bond that will be split inside the lysosomes. There are a few additional conditions. But, basically, this is the kind of concept that is now inspiring the design of what may well be the "magic bullets" of the future.

Lysosomotropism by endocytosis and its application to drug targeting. A drug-carrier complex is taken up selectively by cells possessing a receptor for the carrier on their surface (target cells). In lysosomes, the drug is detached from its carrier by enzyme action and moves out to its target—for instance, the nucleus.

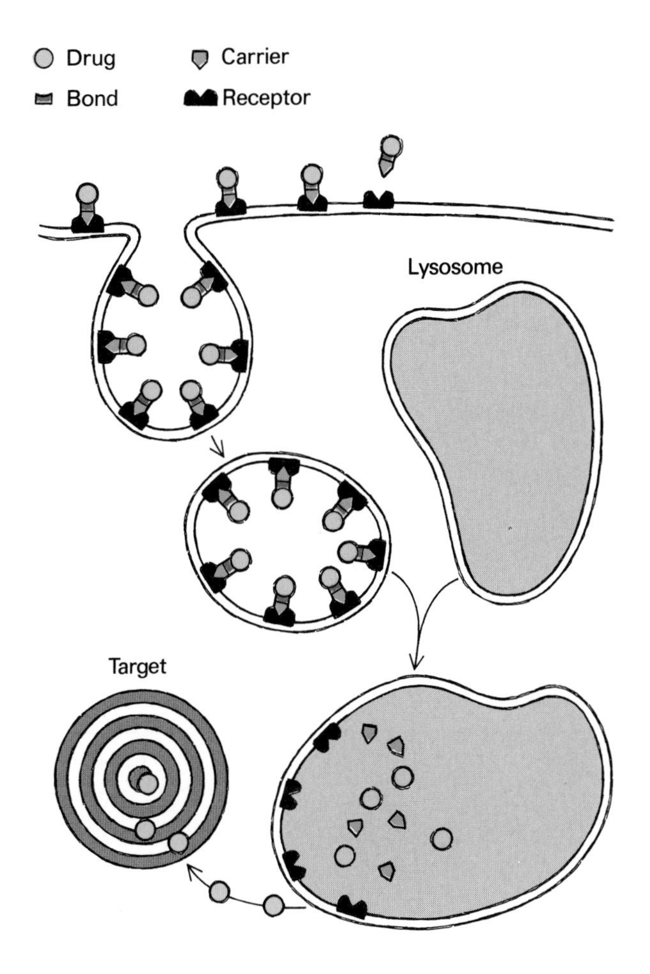

The Importance of Biodegradation

Most students of life have reserved their main admiration for biosynthesis, this wondrous ability of living organisms, shared by even the humblest of bacteria, to manufacture thousands of specific substances, many of which are so complex that even our most sophisticated chemical technology is unable to reproduce them. Degradation, on the other hand, was seen as the natural and inevitable "way of all flesh," a fate constantly menacing the "miraculously fragile" fabric of protoplasm, a threat to life rather than a vital function.

Biosynthesis is indeed a wonderful activity, and a good part of our tour will be devoted to its key energetic and informational aspects. But biodegradation, although much simpler, is equally important, as our plastic age is beginning to find out. Had not the appropriate hydrolases appeared at about the same time as the first proteins, nucleic acids, polysaccharides, and other biopolymers, there would be no biosphere, only a "plastosphere." There would be no adaptation, no differentiation, no evolution, no life. The fabric of life is really not fragile at all. It is made largely of very tough molecules that are no less stable under normal circumstances than are polystyrene or polyvinyl chloride. Digestion, as we have seen, proceeds only in the presence of suitable catalysts.

Biodegradation, therefore, is an essential, life-saving process. But, at the same time, it is by its very nature undissociably life threatening as well. Such tightrope-walking between life and death cannot be entrusted to a random downhill process. Biodegradation is really a strictly controlled activity, fenced in by an elaborate network of protective defenses. That is the lesson we take back with us from our visit to this fascinating lysosomal compartment, which somehow allows the kind of harsh and indiscriminate breakdown treatment that is demanded by biodegradation to be administered with the necessary discrimination and under remarkably safe conditions. But now we must find a way out of the lysosomes.

Escaping out of the Lysosomes

A few years ago, nobody in his senses would have dared to enter a cell by the endocytic route unless there were some strict guarantee that the lysosomal compartment would be bypassed. Once inside a lysosome—so the belief ran—abandon all hope, as in Dante's *Inferno*. Even if you escaped being burned by the acid or cut to pieces by the hydrolases, you would remain forever trapped within a membranous prison, endlessly tossed around from pen to pen by the capricious play of *cis* and *trans* mergers. Indeed, innumerable well-documented examples supported this view, including the many pathologies associated with lysosome overloading.

Yet, as we now know, the view is not entirely correct. There is a way out of lysosomes. There may even be several exits, connected to different intracellular routes that lead either to the plasma membrane or to the Golgi apparatus. All we have to do is to grab hold of some appropriately chosen membrane patch. Soon it will detach and take us out of the lysosomal compartment. The difficulty, apparently, is to hang on to the membrane in the lysosomal milieu. Most substances that bind to plasma-membrane receptors are detached by the lysosomal acidity or by enzyme action. Even some of the receptors are destroyed upon exposure to the corrosive lysosomal contents. Thus, by the time the membrane patch is severed from the lysosomal membrane to be recycled back to the cell periphery, it moves away naked and with its surface scoured. This, at least, is how we interpret the fact that so little is removed from the lysosomes in spite of the continuing recycling of their surrounding membranes. We must assume further that the membrane patches move off in a form—perhaps of very small vesicles—that offers little space for fluid transport. This may make our escape pretty uncomfortable. But we will have to chance it.

Should we accompany the membrane all the way to the cell surface, we would be back where we started. However, some recycling membranes follow a more circuitous route and stop at the Golgi apparatus, perhaps to provide an opportunity for repair of the damage the membrane has suffered while passing through the lysosomal compartment. In addition, there is a direct lysosome-Golgi shuttle associated with the supply of fresh enzymes to the lysosomes. We will take advantage of this service, which is clearly advertised by the occurrence of a specific receptor for lysosomal enzymes on the membrane patches involved. It will bring us conveniently, if not comfortably, to our next objective: the cell's export department.

6 The Cell's Export Industry: Endoplasmic Reticulum, Golgi Apparatus, and Secretion

Living cells manufacture all sorts of export materials, which they assemble, process, package, and transport in a chain of interconnected, membrane-limited enclosures and finally ship out by exocytic discharge. These materials are mostly made of protein or carbohydrate, often combined as glycoproteins or proteoglycans. They fill a considerable catalogue.

A Catalogue of Products

First, we find a variety of substances produced for short-range export and used by cells to organize or clean their immediate surroundings. The walls, the fibers, the matrices, and all the other pieces of the framework that holds the cells together and gives tissues and organs their characteristic architecture are assembled from soluble precursors, such as procollagen (Chapter 2), that are made and discharged on site. The enzymes that trim and consolidate these building blocks are similarly produced and secreted locally by the export machinery of the cells. Specialists of such construction work are the osteoblasts, which erect the scaffoldings of bone; the chondroblasts, which lay down the matrix of cartilage; the fibroblasts, which make connective-tissue fibers.

As is the rule in the living world, construction also calls for destruction. This is accomplished extracellularly by lytic enzymes that are likewise products of cellular export. From the point of view of individual cells, these enzymes serve as a means of invasion; they fragment extracellular structures and clear passages for the cells to

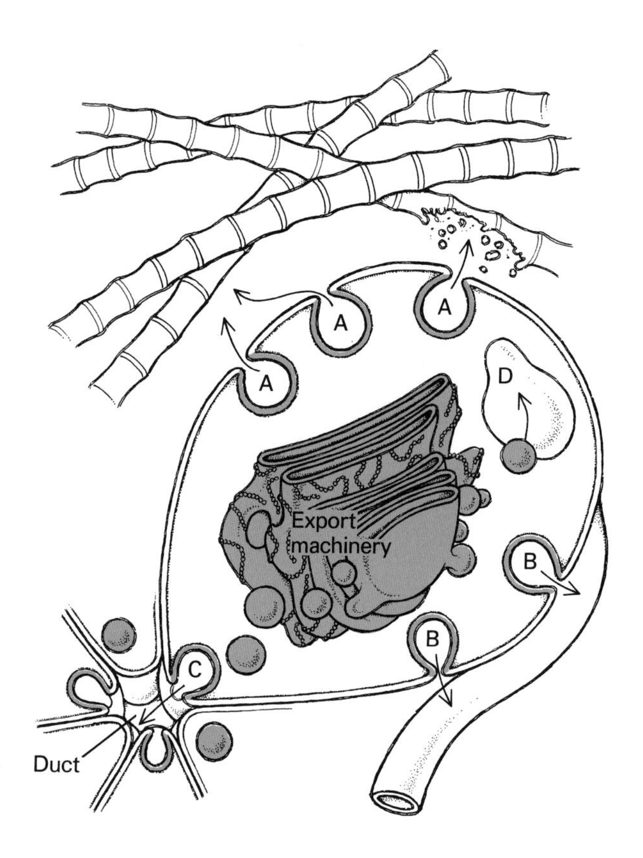

Composite picture showing how different products manufactured by the cell's export machinery are discharged exocytically: (A) into pericellular spaces (building blocks and processing enzymes of extracellular structures); (B) into the bloodstream (plasma proteins, immunoglobulins, endocrine secretions); (C) into a secretory duct (exocrine secretions); and (D) into lysosomes (acid hydrolases). Not all cells make all kinds of products, but most cells make more than one and therefore need a dispatching or sorting system.

move through. Without such a secretion, the leukocyte that led us out of the bloodstream could not have broken through the capillary's basement membrane. Cancer cells may owe their characteristic invasiveness to a particularly aggressive policy of this sort. But such demolition tools are useful also to the resident cellular societies, by providing them with the ability to clean, repair, and remodel their habitat.

While all cells tend to clean their own doorsteps, so to speak, certain specialized cells make the destruction of extracellular structures their lifetime avocation. We have already met the osteoclasts, which burrow their way through the bone matrix with the help of an acid mixture of lysosomal hydrolases (Chapter 5). Fibroblasts, strangely enough, also double as "fibroclasts." When properly stimulated, they will discharge a powerful collagenase, which specifically fragments collagen fibers, and a special protease acting on structural proteoglycans (Chapter 2). Such secretions are physiologically important, but carry grave pathological hazards, as we have seen in the unloading of lysosomes (Chapter 5).

Another major division of the cellular export industry deals with the manufacture of materials for long-distance shipment. Cells involved in this activity are usually arranged into special organs called glands, which are subdivided into two groups. In exocrine glands, the secretory products are collected by ducts and conveyed by them to a specific destination. In contrast, the products of endocrine, or ductless, glands are discharged into the pericellular spaces, from which they are spread throughout the organism by the bloodstream.

The better-known exocrine secretions, such as saliva or pancreatic juice, supply enzymes to the digestive organs. But there are many others. Glands occur in skin, in eyelids, in ear ducts, in the genital tract, and in many other areas. Their products serve for lubrication, protection, waterproofing, disinfection, scenting, and chemical communication, as well as for digestion.

Hormones make up the most important endocrine secretions. But here we have to distinguish between the polypeptide hormones, such as insulin or the pituitary hormones, which are manufactured and shipped out by the machinery that we are about to visit, and a number of other chemical messengers of small molecular size, including thyroxin, epinephrine, acetylcholine, and the sexual steroids, which either diffuse out of the cell by permeation or are released by special devices that we will not have the opportunity to see on this tour.

A special group of cellular export products is represented by the proteins of blood plasma. Many of them are made in the liver, except for the antibodies, which are manufactured by the B-lymphocyte-derived plasma cells (Chapter 3).

Finally, there are the lysosomal enzymes. Technically, these do not deserve the name of export products, because they are mostly used intracellularly. In practice, however, the space into which they are discharged is only one step away from the extracellular space, from which it is derived by the mechanism of endocytosis; its enzymes may even end up physically outside the cell, as mentioned in Chapter 5. In these respects, lysosomal enzymes have the character of secretory products. The cell obviously considers them as such, since it uses its export machinery to supply enzymes to its lysosomes.

Secretion is a specialized activity in that each product is usually made by one or a few cell types. But cells often make more than one product, sometimes for different destinations. This means that the export machinery must include some sort of dispatching system, whereby, for instance, acid hydrolases are sent to the lysosomal space, plasma proteins and hormones to the bloodstream, exocrine products to the appropriate duct, and structural elements and the agents of their remodeling into the immediate cellular surroundings.

Some cellular exports are regulated only by production. The cells dispose of their merchandise as they make it and keep no stocks. The plasma cells, which secrete immunoglobulins, behave in this way; so do the fibroblasts, the main manufacturers of structural collagen. However, other cells, such as those that secrete hormones or the components of digestive secretions, store their products in special granules and discharge them only on command. We will speak of continuous and intermittent secreters to distinguish these two types.

In many cases, the cells export incompletely finished products, leaving the finishing touch, which is usually some proteolytic clipping, to be administered at the site of destination. They do this particularly when the finished product is potentially harmful and can cause local damage. Then they make an innocuous precursor, which is converted into the final product only where and when it is needed. Many digestive enzymes, including some lysosomal hydrolases, are made in this way, in the form of inactive proenzymes, or zymogens. So are certain tissue-degrading enzymes, such as collagenase, and a number of

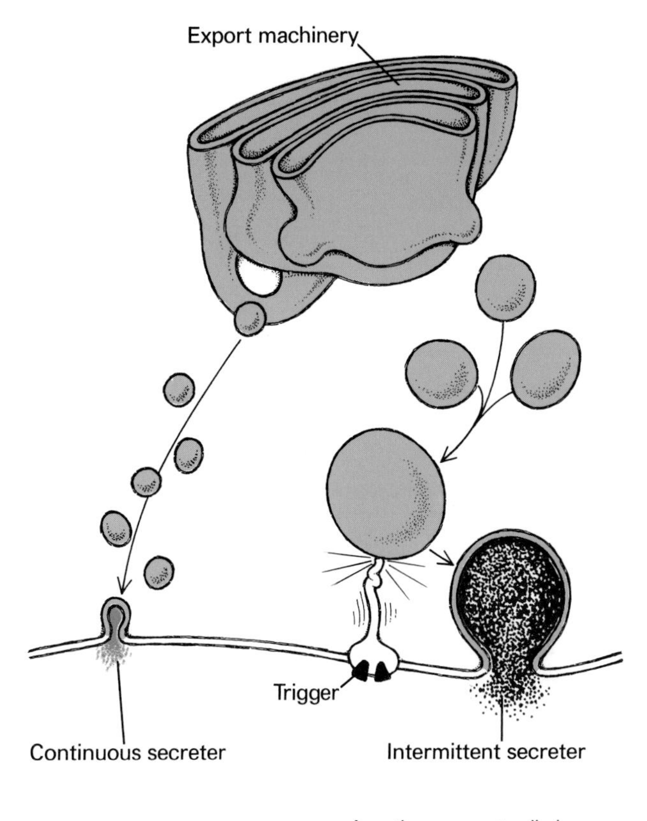

A continuous secreter discharges a product as it is manufactured. An intermittent secreter stores a product and discharges it only when triggered to do so.

blood-borne systems involved in the formation and dissolution of clots, in cell killing, and in other dangerous activities. Structural components destined for assembly into insoluble aggregates are similarly made as larger precursors incapable of forming such combinations spontaneously. An example is tropocollagen, which is secreted as soluble procollagen (Chapter 2). The finishing of these various precursors takes place outside the cells and is mediated by enzymes that are themselves export products. Thus the cells involved are linked by complex networks of mutual interactions.

Sometimes, when particularly strict safeguards are required, a whole cascade of precursor-activation reactions is inserted between the initial trigger and the final process-

After fighting their way upstream against the secretory current, cytonauts emerge into a deep gorge of impressive dimension. It is a cisterna of the endoplasmic reticulum. Tufts of growing polypeptide chains cover its walls.

ing step. Blood clotting is a typical example of such a cascade. It is the result of the proteolytic conversion of a soluble blood protein called fibrinogen into a product called fibrin, which assembles into an insoluble network. The converting enzyme, thrombin, is present as an inactive precursor designated prothrombin. When a tissue is wounded, certain substances released from the injured cells initiate prothrombin activation, but they do so by a complex chain of successive proteolytic events. Activation of fibrinolysin, the enzyme that dissolves clots, likewise depends on a cascade of such events. A particularly striking example of such a mechanism is the activation of complement, the powerful killing system that is attracted to antibody-branded cells (Chapter 3). It consists of no fewer than nine distinct circulating protein molecules, some of which are made of more than one subunit. Binding of one of these components to the Fc tails of bound immunoglobulin molecules triggers a set of consecutive interactions that end up in the construction of an elaborate, derricklike structure by which a life-draining hole is bored through the plasma membrane of the antibody-coated cell. The impression one gains from these and other similar examples is that the number of steps between a triggering event and the final outcome is a direct function of the danger associated with setting off the mechanism accidentally. Proteolytic cascading is Nature's way of constructing multiple-control, "fail-safe" devices.

A Magic Spinnery

The factories in which cells manufacture their export products are not normally accessible to visitors from the outside. The Golgi apparatus, where we landed coming from the lysosomes, lies halfway down the assembly line. If we wish to follow the industrial process from its beginning, we must first force our way upstream. Our unlawful journey is not easy and takes us through a series of narrow caves and convoluted tunnels. But this bit of cellular speleology soon comes to an end, as we emerge into a deep gorge of impressive dimension, part of the endoplasmic reticulum, or ER.

The term reticulum means network in Latin. Its choice reflects the two-dimensional view of the morphologists who observed this system in cross section as a filigree of thin lines. These were recognized as the edges of membranes cut perpendicularly to their plane. Three-dimensional reconstruction showed that the membranes form large, flattened sacs, or cisternae, completely sealed off except for the connections—permanent or intermittent—that link them with each other and with the Golgi apparatus.

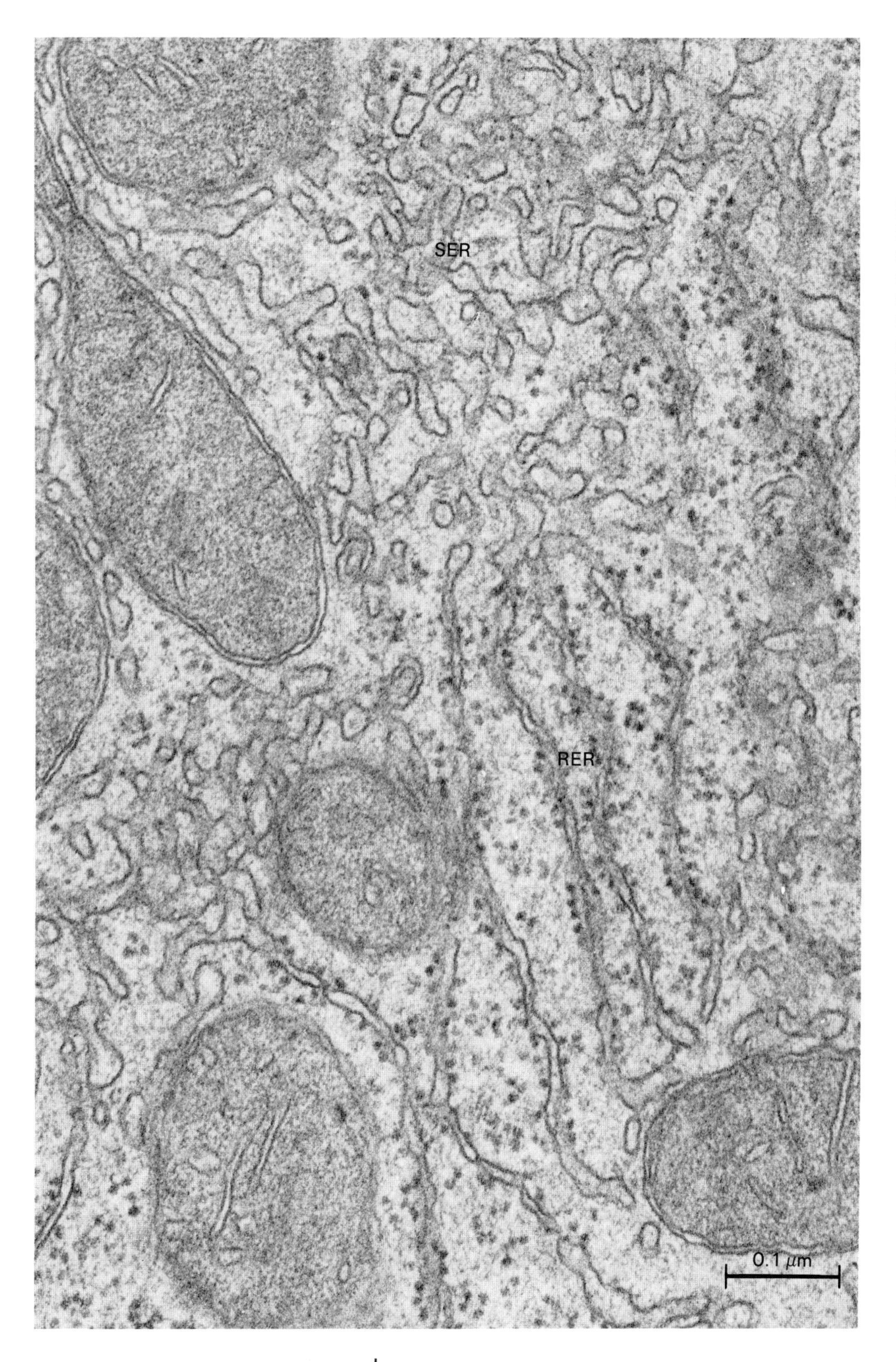

Electron micrograph of a thin section through a rat liver cell shows the endoplasmic reticulum. Narrow, elongated, membranous profiles studded with dense particles (ribosomes) on the cytoplasmic side are cuts through the cisternae of rough-surfaced endoplasmic reticulum (RER). Small, irregular, smooth profiles belong to smooth-surfaced endoplasmic reticulum (SER). Note the continuities between the RER and SER. The large bodies are mitochondria.

Seen from the inside, the flat, membranous walls of ER cisternae appear to be constructed according to the same bilayer-protein blueprint as are the plasma membrane and the boundaries of endosomes and lysosomes. But they are thinner, smoother, more flexible; the phospholipids in the bilayer have a different composition, and there is virtually no cholesterol associated with them. The proteins also are different and are largely devoid of carbohydrate side chains. These endoplasmic membranes lack the characteristic bristly appearance of the plasma membrane. They are covered instead with tufts of silky, tenuous threads, which make them look like shimmering gossamer. The shimmer is not just an optical effect; it is a reflection of movement: the silky threads grow. They grow at a visible rate: roughly 1 nm per second, which at our millionfold magnification amounts to more than 2 inches per minute. They grow steadily until they reach from 50 to 200 nm in length; then they fall off and get carried away by the current. We are enmeshed in a sea of moving silk.

What we are witnessing here is the production of export proteins. The silky threads that grow out of the surrounding membrane are polypeptide chains. They form sinuous rows comprising from ten to twenty threads, characteristically arrayed in an order of regularly increasing length. All the threads of a given set grow at the same rate but with a phase difference proportional to the distance between them. About every 15 seconds, the rearmost thread of a row reaches its full size and drops off, and the tip of a new one emerges through the membrane somewhere in front. Thus, as its individual threads grow out, the row itself moves on, crawling on the surface of the membrane like some strange, hairy centipede. Hundreds of such ghostly creatures trace ever-changing arabesques on the walls that surround us. It is an unforgettable spectacle.

If we now examine the growing threads at the point where they emerge into our cave, we will notice that they are being delivered from behind the membrane through a narrow, proteinaceous tunnel built across the lipid bilayer. At the far end of each tunnel, the outline of a bulbous root can be vaguely distinguished. About 25 nm wide (1 inch at our magnification), it seems to be the actual site where the thread is being spun. Thus, behind each array of growing threads there is a row of bulbs, separated from us by the membrane. By peering through the membrane, we can just discern one more detail: a faint, slender tape strung between the bulbs of each row, and actually slithering through them at the remarkable speed of some 250 nm (10 inches at our magnification) per minute. The movement of this tape is synchronized with the growth of the threads and with the lateral displacement of their points of emergence on the membrane. In fact, the tape, with its string of bulbs, is the real centipede behind the ghostlike creature of which we see only the growing hairs. The name of this hidden demon is polysome, short for polyribosome.

Ribosomes, which will be examined in great detail in Chapter 15, are small particles that serve as the sites of protein synthesis throughout the living world. Their business is to attach amino acids together into polypeptide chains. They receive appropriate instructions as to the order in which to do this from the messenger RNAs, or mRNAs, which are segments of ribonucleic acid (hence the abbreviation RNA) that carry pieces of genetic information transcribed from deoxyribonucleic acid (DNA). These mRNAs are thin, threadlike molecules that run through the ribosomes like a tape through a cassette player. But, instead of informing one ribosome at a time, they run through ten or more simultaneously and thus link them in a string, the polysome. Imagine a tape running through a series of cassette players, each of which is hooked to a separate speaker. The same music will be rendered by each speaker, but in the form of a canon or a fugue, with a delay proportionate to the distance between the cassette players. While the last one in the row starts on the overture, the first one is already into the coda. This is exactly what happens with polysomes, except that they render the tape's message into a specific sequence of amino acids, not of sounds. All the ribosomes of a given polysome read the same text—that is, they make the same polypeptide. They do so at the same rate, but out of phase with each other: the greater the length of tape already translated, the longer the part of the polypeptide already

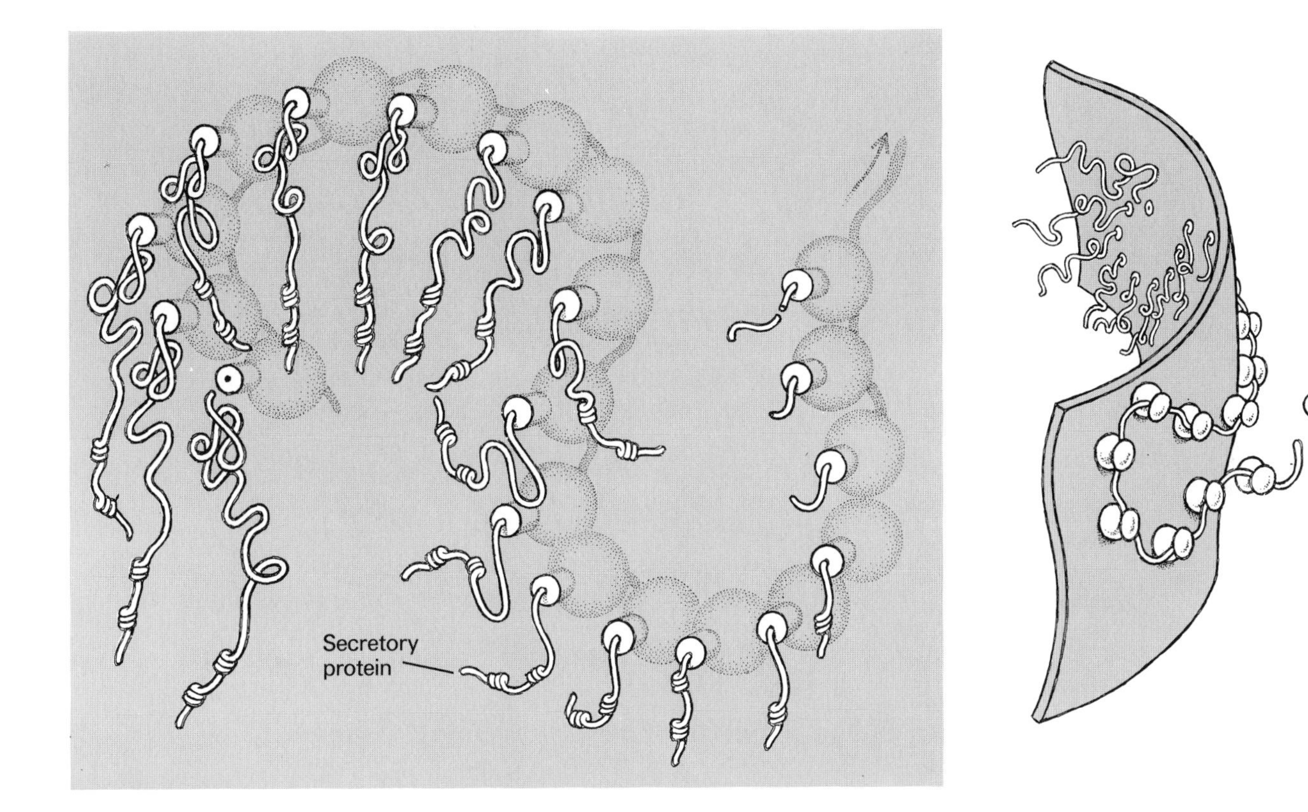

Secretory proteins grow out of the wall of the endoplasmic reticulum. The arrow shows direction of movement of the "centipede," which consists of ribosomes and mRNA. Shadows of these ribosomes and of mRNA can be vaguely distinguished through the membrane.

made. Note that polypeptides grow at their roots (i.e., on the ribosomes), not at their tips. Thus, if you compare two growing chains arising from the same polysome, you will find that the shorter one is identical with the terminal part of the longer one.

We will have an opportunity to watch all this at much closer range when we visit the cytosol (in Chapter 15). In the meantime, we can take advantage of our present location to answer a question that has evoked a considerable degree of speculation among experts. Do all the individual tufts that grow on the walls of a given cisterna consist of the same polypeptides? Or are they different, but all intended for the same final destination? Or do they make up a completely haphazard mixture? Faced with the problem of dispatching different products to different destinations, cells could have found a simple solution to this problem by segregating production, as is generally done in our chemical factories. It seems, however, that their export industry does not operate in this manner. It throws all its products into the same cauldron and sorts them out later.

Processing and Transport

In most instances, polypeptide chains are but the raw secretory products; they require a considerable amount of further processing, trimming, and garnishing before they are ready for shipment. Some are fitted with bulky carbohydrate side chains that contain as many as ten or more sugar molecules (glycosylation). Others are linked with lipids, sometimes many times their weight, as are the plasma lipoproteins made in the liver. All the chains must be folded into their proper configuration, a process that, for some molecules, is sealed by chemical bridges, usually disulfide bonds: S—S. Finally, many molecules are pared down by additional proteolytic pruning.

Some of these changes, such as glycosylation, begin to take place even before the polypeptide chains are actually completed. Others occur as the molecules drift down with the secretory current, sliding along the membranes where the processing enzymes are situated. Complex biochemical reactions are involved, but of these we often see little more than the final steps from our present vantage point inside ER cisternae. Most of the action takes place on the cytoplasmic side of the membranes, where building blocks and energy are in plentiful supply; the ER space itself serves mostly as a collecting channel and assembly line. This organization resembles that of an automobile factory, where engine, body, chassis, wheels, and other parts are made in separate shops and then simply put together on the assembly line with minimum expenditure of manpower.

This we have found to be true of the polypeptide chains, which are synthesized by ribosomes on the cytosol side of the membrane and delivered ready-made into the endoplasmic cisterns. Lipid components and carbohydrate side-chains are likewise assembled on the cytosol side and conveyed through the membrane. How the carbohydrates get through is puzzling, since sugars are notoriously hydrophilic and are not likely to pass readily through a lipid bilayer. A hydrophobic carrier molecule called dolichol is known to be involved in the process, but how it does its job is not clear. We will have more to say about this when we consider biosynthetic mechanisms (in Chapters 8 and 13).

Even though they are restricted, the reactions that take place inside the ER cavities are highly specific and selective. Each polypeptide is processed in a characteristic and reproducible fashion. That this should be so may seem self-evident—indiscriminate, random processing could only lead to chaos—but actually it carries a profound meaning. If polypeptide A, but not polypeptide B, undergoes a given change, it can only be because the enzyme involved "recognizes"—that is, binds in a catalytically effective fashion—something that exists in A and not in B. That something, in turn, can only be an amino-acid sequence, at least at the start. A subsequent step may be commanded by the preceding one, such as the attachment of a given sugar molecule. But, to begin with, the instructions must lie in the naked polypeptide, whose structure is the expression of a genetic message. It follows that this message determines not only the structural and functional properties of the polypeptide at birth, but also the whole concatenation of events whereby its molecule is further processed and modified, routed within and out of cells, combined with other molecules, incorporated in a given cell part, or otherwise handled. In other words, genes govern the whole four-dimensional history of protein molecules. We will revert several times to this all-important point (see Chapter 15).

For now, however, we must continue to accompany the secretory polypeptides on their journey. As we move on, we notice the gradual thinning of the silky growth on the membrane surface. Polypeptide tufts become sparser and eventually disappear, which means that there are no more polysomes crawling on the cytoplasmic face of the membranes. This difference can readily be seen in cross sections. Where polysomes occur, they stand out as dense dots lined up against the membrane edges and give them a rough-surfaced appearance; membranes that do not bear polysomes look smooth. Hence the terms rough endoplasmic reticulum (RER) and smooth endoplasmic reticulum (SER) given to these two parts of the ER. As pointed out, the transition from one to the other is gradual. By the time we enter the smooth part, however, the scene begins to change considerably. We are reaching the end of our roomy endoplasmic cave and approaching the tortuous passages that we traversed on our journey upstream. We are now about to reach the Golgi apparatus, this time lawfully, along with the secretory current.

The shape of the transitional elements that connect the ER with the Golgi apparatus is not easily discerned. At some stages of our journey, we have the impression of squeezing through narrow, twisting ducts; at others, of being carried in a small, sealed capsule. In other words, it is not clear whether the connection through which we are traveling is permanent and tubular in shape or transient and mediated by vesicles that use the soap-bubble trick of fusion-fission to transport secretory products from one compartment to the other. This question is still being debated, and the answer could be different for different cells. Here, again, we encounter the difficulty of having to reconstruct a three-dimensional picture from two-dimensional cross sections. Cut through a convoluted tube or through a string of closed vesicles, and you will see the same picture: a row of ringlike profiles.

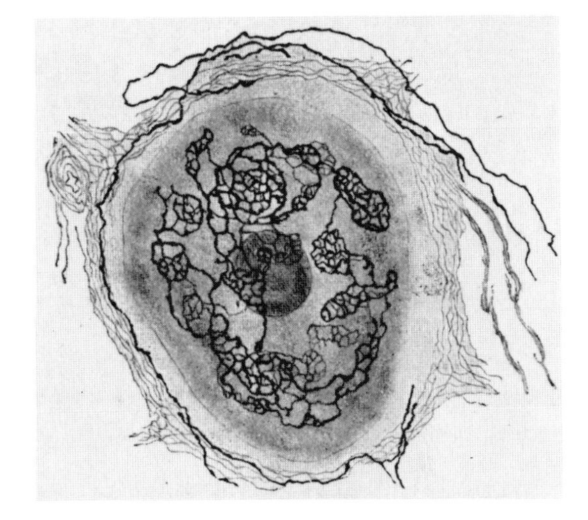

The "internal reticular apparatus," as drawn by Camillo Golgi at the beginning of the century.

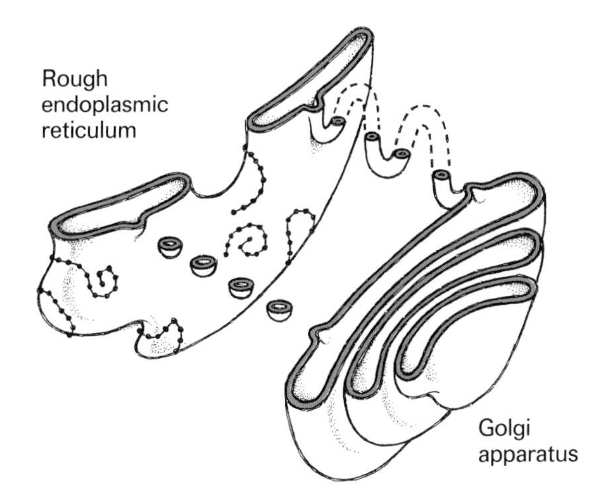

The connection between the endoplasmic reticulum and the Golgi apparatus may be either mediated dynamically by shuttling vesicles or established statically by convoluted tubules. The two possibilities are not distinguishable in thin sections.

Packaging and Delivery

Our present location is named after the Italian histologist Camillo Golgi, who, at the turn of the century, discovered this system in nerve cells that had been impregnated with certain metal salts. He saw it as a fine network and called it internal reticular apparatus.

The exact three-dimensional shape of the Golgi apparatus (often simply called the Golgi) is still being mapped. It has certain characteristic components, but structural details vary from cell to cell, and the exact boundaries of the system are imprecise. It is a sprawling complex of membrane-bounded cavities big and small, with the untidy appearance—which actually hides a considerable degree of order—often seen in warehouses and packaging centers.

The largest and most typical Golgi component is the dictyosome (Greek *diktyon*, network), so named because of the way Golgi envisioned the system. Of course, the

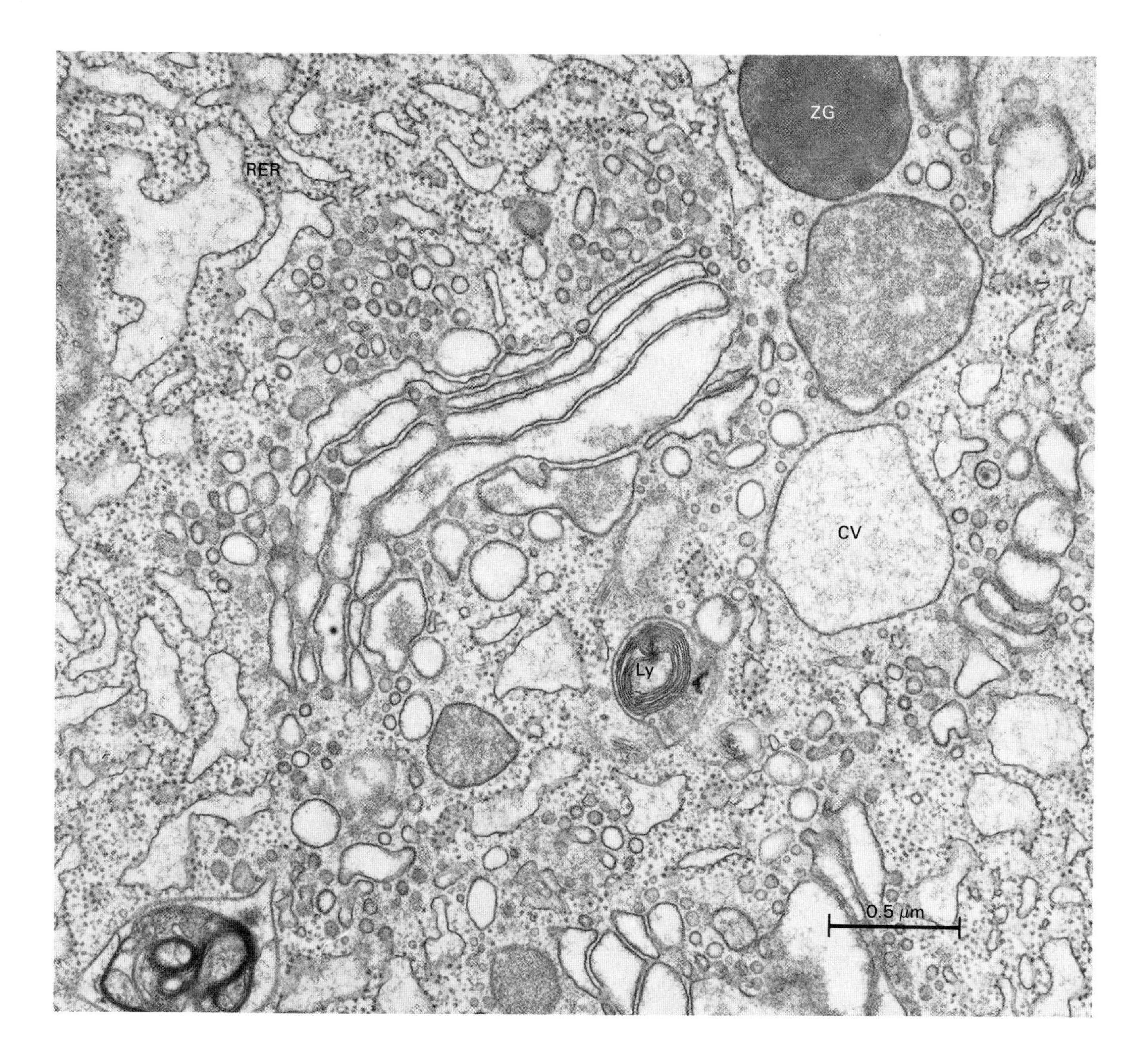

This electron micrograph illustrates the disposition of a Golgi region in a cell of rat exocrine pancreas. A typical stack occupies the center of the picture. Behind its *cis,* or endoplasmic, face (to left) are profiles of RER ending up in SER transitional elements and connecting vesicles. On the *trans,* or exoplasmic, side of the stack, the cytoplasm is filled with vesicles of varying sizes, including three large vacuoles showing successive stages in the maturation of a zymogen granule (ZG) from a condensing vacuole (CV). The small vesicle containing membranous whorls is a lysosome (Ly).

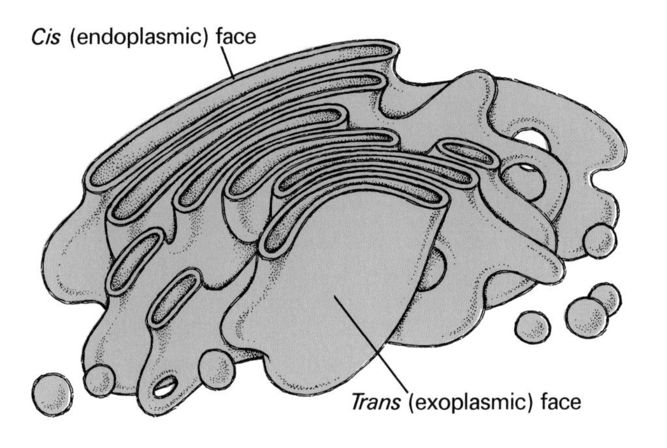

Three-dimensional representation of the Golgi apparatus.

Golgi apparatus is no more a network than is the ER; it only looks like one in cross section. The dictyosome is a stack of about half a dozen large, flat, membranous cisterns, pressed close together and looking somewhat like a pile of large, double-walled dishes. It resembles such a pile also in having curved faces—one convex, the other concave. At present, we are in the first cistern on the convex face of the stack, close to the ER. The walls of this structure do not differ greatly from those of the smooth ER. However, as we wander from cavity to cavity in the direction of the concave face of the stack, we notice that the membranes become thicker and coarser. There are increasing amounts of cholesterol in the lipid bilayer and of carbohydrate side chains on the membrane proteins. Manifestly, our surroundings become more and more plasma-membranelike. This polarity reflects the packaging function of the Golgi apparatus. As secretory products traverse the system, they move over from ER-like containers to containers that are constructed like the plasma membrane. This change is probably required before exocytic discharge by fusion with the plasma membrane can take place.

Exactly how secretory products move through the Golgi apparatus is a hotly debated question. Some believe that the Golgi cisterns move with their contents from one position to the other in the stack, undergo a progressive change of their membranes in the process, and finally fragment into vesicles that carry the products to their final destination. Holders of this dynamic view describe the Golgi stack as having a forming (from the ER) and a maturing face. Others see the Golgi stack as a more static system, through which products move either by way of permanent connections or by vesicular transport, following itineraries that may vary, depending on the product, and that may not necessarily pass through all the cisterns of the stack. In accordance with this conception, they use the purely topological *cis* and *trans* (with respect to the ER) to distinguish the two faces of a Golgi stack. We will use the terms endoplasmic and exoplasmic to avoid confusion with *cis* and *trans* membrane mergers (see p. 95).

On the whole, the static view is probably closer to the truth. The transition reflected in the polarity of the system is not the gradual change postulated by the dynamic view. It involves profound and relatively abrupt differences in which, for instance, whole groups of enzymes are replaced by others. It is clear from all the available evidence that Golgi membranes do not undergo this kind of extensive reorganization, at least at a rate in any way comparable to the rate at which products move through the system. Hence, contents and containers cannot move together, contrary to the view postulated by the dynamic theory.

Secretory proteins undergo a considerable degree of further processing as they move through the Golgi, including partial trimming of some of the oligosaccharide side chains that were assembled in the ER, followed by capping with new sugar molecules; addition of phosphate groups (phosphorylation) or of fatty acids (acylation); and further proteolytic clipping.

A second important function of the Golgi is sorting. How this takes place is known in some detail only for the acid hydrolases destined to become part of the digestive

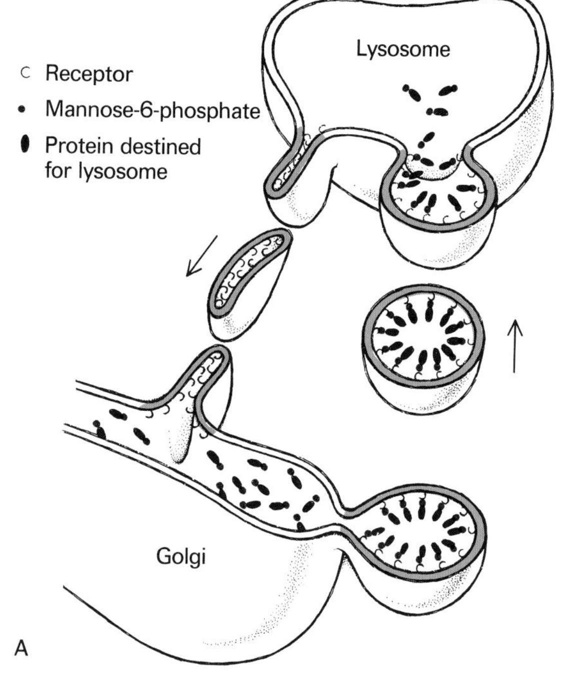

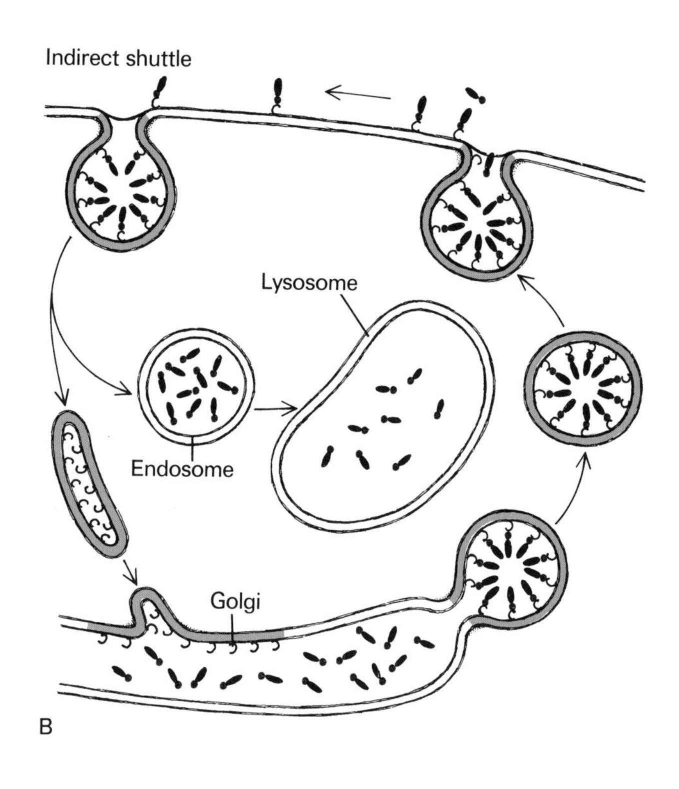

Selective transport of lysosomal enzymes from Golgi to lysosomes.

A. The structure of lysosomal enzyme proteins is so arranged as to allow them to be substrates for enzyme systems that catalyze the attachment of mannose-6-phosphate groupings. Golgi membranes have mannose-6-phosphate receptors clustered in invaginations (probably coated) that give rise to lysosome-directed vesicles. In acidic lysosomal medium, the enzymes detach from the receptors, which are shuttled back to the Golgi.

B. An alternative route takes hydrolase-loaded receptors to the cell surface, from which they return to the cell's interior by endocytosis. Enzymes are unloaded in acidic endosomes, from which the empty receptors are recycled back to the Golgi. The enzymes accompany the endosome contents into the lysosomes.

system of lysosomes. Reconstruction of the events, still partly hypothetic, is as follows. In the Golgi, lysosomal enzymes receive a characteristic addressing label, in the form of terminal mannose-6-phosphate groupings attached to some of their oligosaccharide side chains. The enzyme responsible for initiating this change does not similarly label other glycoproteins; it recognizes a specific structure—presumably an amino-acid sequence or type of sequence—typically shared by all lysosomal hydrolases. Subsequent recognition of the mannose-6-phosphate label is done by specific binding sites clustered on the inner face of certain Golgi membrane patches. These thereby fish out the lysosomal enzymes from the mixture of secretory proteins. The loaded patches then detach in the form of vesicles that selectively move toward the lysosomes (or to the endosomes) and fuse with them, unloading their catch upon contact with the local acidity. Presumably, the empty vesicles are recycled back to the Golgi afterward.

This remarkable sorting mechanism is not perfect; some lysosomal enzyme molecules accompany the main secretory stream and are discharged extracellularly. If they have been properly labeled, these molecules can still be salvaged, because, as briefly mentioned in Chapter 5, the mannose-6-phosphate receptor also occurs on the plasma membrane of many cells. In fact, the possibility that a major part of the loaded receptors may normally travel from Golgi to lysosomes via the cell surface is still being debated, although the existence of a direct intracellular route seems very probable. In any case, because of the occurrence of an extracellular pathway, true secretion of lysosomal enzymes takes place to some extent even under normal circumstances. It may be enhanced considerably in pathological situations in which transport of the enzymes to lysosomes is defective. The consequences of such leakages are, however, less dramatic than those of lysosomal unloading (defecation), because several lysosomal hydrolases, including the particularly dangerous cathepsins (proteases), are synthesized in the form of inactive proenzymes, which are activated only after they reach the lysosomes.

In addition to assigning their destination, packaging of secretory products must also make provision for their mode of delivery, continuous or intermittent. In continuous secreters, the products flow steadily out of the Golgi region, carried by small membranous vesicles that detach from the system by fission and then discharge their contents extracellularly by exocytosis. In intermittent secreters, the products are concentrated in large condensing vacuoles, which give rise to mature secretion granules, the large, densely packed, membrane-bounded structures that are the most typical feature of regulated gland cells. These granules unload their contents by exocytosis, but only upon appropriate stimulation by what is often a complex chain of nervous and hormonal relays that end in the local release of acetylcholine. This messenger unites with a receptor on the cell surface (see Chapter 13), and the resulting conformational change triggers the exocytic discharge of the secretory granules, perhaps by letting in calcium ions. If the stimulus fails to be provided and the cell goes on making more secretion granules, the granules start fusing with lysosomes instead of with the plasma membrane, and the surplus secretory product is destroyed by crinophagy (Chapter 5).

Lysosomal hydrolases are generally conveyed to the lysosomes in a continuous fashion. But there is one striking exception: the polymorphonuclear leukocyte. This white blood cell makes and stores lysosomal hydrolases over several generations in the course of its progressive development and differentiation from stem cells that re-

Exocytic discharge of secretion granules in exocrine pancreas. The zymogen granule (ZG) is approaching the plasma membrane (PM). Fusion of the two membranes by *cis* merger will produce an exocytic invagination (EX).

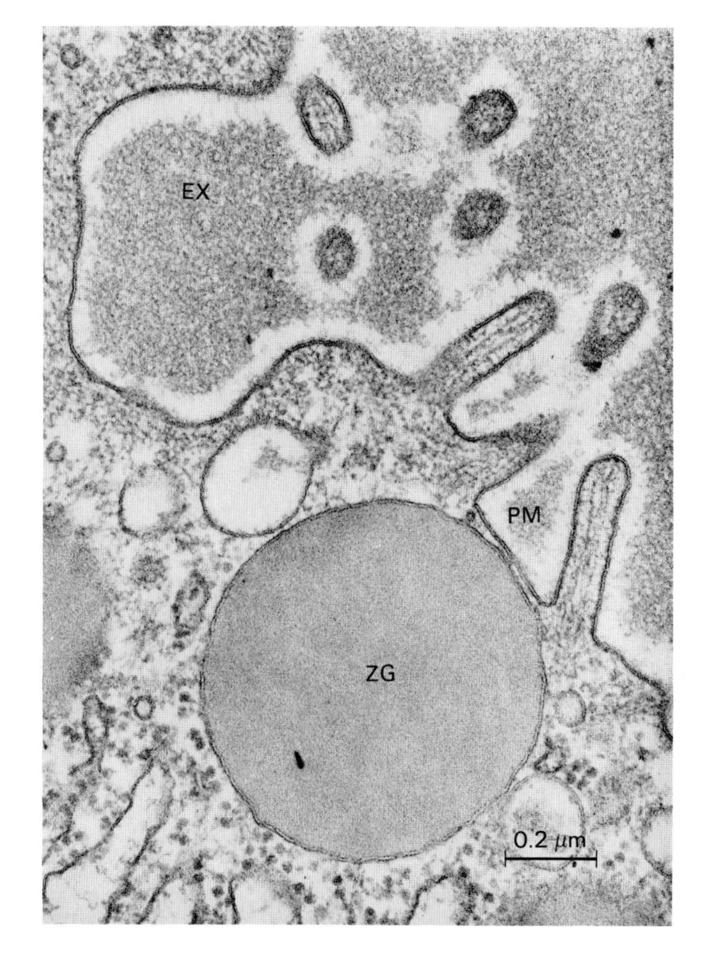

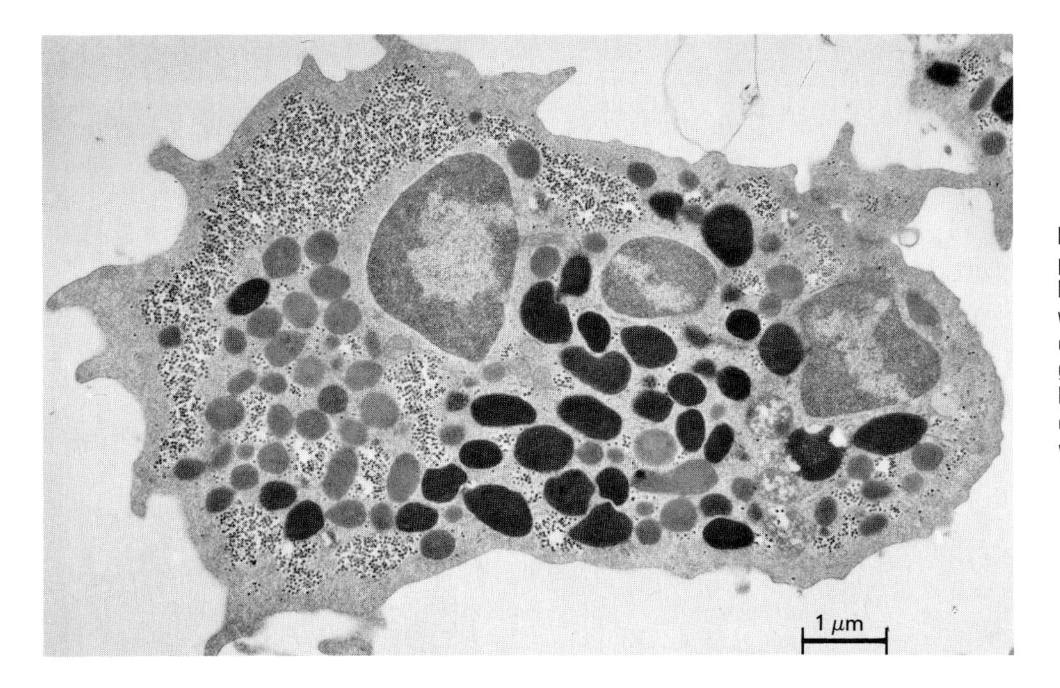

Electron micrograph of a rabbit neutrophil polymorphonuclear leukocyte. The larger, dense granules are azurophils, which are related to the lysosomes of other cells. The smaller, less dense granules, known as specific, are not lysosomes. Both types of granules are discharged massively into phagocytic vacuoles.

side in the bone marrow. It does the same for lysozyme and a number of other microbe-killing agents. It accumulates these materials in two different types of large cytoplasmic granules, known as azurophil and specific. These granules remain essentially inert in the cytoplasm until the cell is seduced by appropriate antibody seasoning (opsonization) into gorging itself with bacteria or other foreign particles (Chapter 4). This triggers a massive discharge of the contents of the granules into the phagocytic vacuoles. Killing and digestion of the prey ensue (Chapter 5). As we know, the leukocyte, unlike other phagocytic cells, does not recover after engaging in such activity. The high degree of specialization that has turned it into a controlled secreter has made it into a one-time secreter as well. At the end of its phagocytic bout, it dies.

The Birth of a Membrane

As far as is known, membranes are never born *de novo*. They always arise from pre-existing membranes by the insertion of additional constituents. This process may be almost as old as evolution. Each generation bequeaths to the subsequent one, mostly through the female egg cell, a stock of preformed membranes from which all the membranes in the organism grow out by accretion, directly or indirectly.

As we will find out when viewing it from the cytosolic side (see Chapter 13), the ER is a major site of membrane biosynthesis. Most phospholipids, as well as cholesterol, are made there. And so are many integral membrane proteins, which are synthesized by bound ribosomes, as are secretory proteins, but which, instead of falling off into the cavity, remain embroiled in the lipid bilayer by some hydrophobic sequences after they are completed. Not all ER proteins arise in this manner. Some are synthesized in the cytosol by free ribosomes and are inserted into the membranes afterward. On the other hand, many non-ER membrane proteins are made in the ER and then translocated to their final abode. By and large, such proteins travel in parallel with secretory products, but by way of the membranes themselves, rather than through the cavities. They move by lateral diffusion along physically continuous membrane domains, cluster on patches that serve in the vesicular transport of products, and thereby hitchhike from one domain to another until they reach their destination, be it in the Golgi, the plasma membrane, the lysosomes, or elsewhere. Like the secretory products, such newly made integral membrane proteins may undergo extensive processing in transit. Examples are the

The secretory pathway. The right-hand side of the diagram shows the itinerary, and concomitant processing, of secretory proteins, from their assembly by membrane-bound ribosomes to their exocytic discharge. The left-hand side shows how an integral plasma-membrane protein made in the ER uses the same pathway to reach its destination.

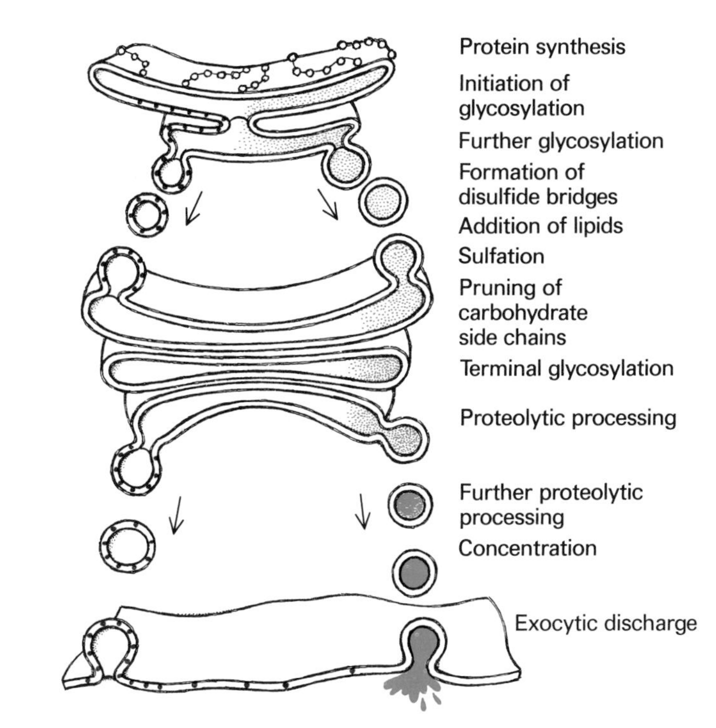

many glycoproteins that characteristically dangle their oligosaccharide side chains on the cell surface. They come from the ER via the Golgi and receive their carbohydrate components as they pass through these structures.

Closing the Circle

Several exits out of the Golgi apparatus are open to us, depending on what kind of recognition mark we wear. Not wishing to return to lysosomes, we will avoid mannose-6-phosphate. Except for that, all other outlets lead to some point outside the cell, and we might as well take the intermittent secreter line, which offers the roomiest accommodation. Even so, the experience is far from pleasant. First, we must enter a condensing vacuole and submit to an almost intolerable degree of compression, as water is being pumped out and the secretion products thicken around us to a nearly solid state. Then comes the waiting, in acute physical discomfort, and with the growing fear that crinophagy may thwart our escape and cast us back into the lysosomal compartment. When release finally comes, it does so with shocking violence. Exocytic unloading of secretion granules is an explosive manifestation. The granules jostle and push against each other, irresistibly drawn toward the periphery of the cell. Often they fuse with other granules lying in their way, creating deep chasms on the cell surface and discharging masses of secretory products. To an outside observer, the phenomenon looks like the sudden eruption of a chain of volcanoes, a veritable "cellquake." To the helpless companions of the secretory products, exocytosis brings a last moment of panic, fortunately of short duration and quickly forgotten in the sweetness of freedom recovered. Back in the balmy extracellular sea, we contemplate again the pitted, moving expanse of the cell membrane, with its twisting excrescences and waving veils. We have completed a circle, entering by endocytosis, exiting by exocytosis, and meandering in between through an endless succession of chambers and corridors.

It has been a fascinating journey, but unreal in some way, like wandering through some surrealistic maze.

The Anxious Journey, by Giorgio de Chirico.

Blank walls everywhere; endless spreads of membranes, always tightly sealed, tantalizingly opalescent, hiding from our view vast sets of machinery which, judging from their manifestations, must be of stupendous complexity. Clearly, we have hardly entered the cell as yet and will have to find a way through this ever-present screen. Before doing so, however, we should take a last, comprehensive look at the whole of the cytomembrane system.

The Vacuome

In the early part of this century, French cytologists proposed the name vacuome to designate what they saw as a complex system of vacuoles and granules occurring in both plant and animal cells. In their view, the vacuome comprised a variety of cytoplasmic structures, including the Golgi apparatus, though not the mitochondria, which they considered a separate system, the chondriome. This distinction was remarkably prescient, and the word vacuome, which never gained general acceptance, deserves to be resurrected. The term vacuolar system, which is sometimes used in the same sense today, is more ambiguous, as it has long served to designate only the import arm of the complex.

Import and export, each with its associated processing activities, represent the main functions of the vacuome. If, however, we consider the anatomical and functional organization of the system, the feature that strikes us most is its transverse division into two domains separated by the Golgi apparatus.

The ER, or endoplasmic domain, lies on one side of the Golgi—the *cis* side, which we call endoplasmic. Geared toward the manufacture of secretory proteins, it has a clear-cut polarity, stretching from the ribosome-binding sites in the RER to the transitional elements between the SER and the Golgi. Except for this asymmetry, ER membranes, which also house a number of metabolic systems unrelated to secretion (see Chapter 13), are largely homogeneous in composition. Traffic through this part of the vacuome is simple and unidirectional. The ER feeds its contents into the Golgi with little apparent reflux. According to all available evidence, secretory products are translocated through an efficient one-way lock even though their transport may depend on a two-way membrane shuttle. This also means that import stops at the Golgi barrier. Indeed, materials taken in by endocytosis do not enter the endoplasmic domain. We succeeded in doing so, but only by forcing our way through the lock.

Things are very different on the other side of the Golgi apparatus, where lies what may be called the exoplasmic domain of the vacuome (not a common appellation, but a very useful one, which we extend to the *trans* face of the Golgi). This domain is composed of a multiplicity of vesicles, vacuoles, and granules of all sizes and shapes, linked with the Golgi, with each other, and with the pericellular environment by a complex network of transport ways. The membranes surrounding these structures share some characteristic features that are also common to the plasma membrane. They are thicker than endoplasmic membranes (about 10 nm, as against 7 nm), largely because of a much greater density of oligosaccharide side chains planted on their luminal face. They also share a preponderance of certain phospholipids, such as sphingomyelin, and a high cholesterol content. Their protein composition, however, is far from homogeneous. Even the plasma membrane is subdivided into different areas.

Traffic in this part of the vacuome is remarkably dense and multidirectional. Out of the Golgi, the main secretory line transports products to their unloading sites at the cell surface, either continuously, by way of a trickle of small vesicles, or discontinuously, with an intermediate stop in secretion granules after concentration in condensing vacuoles. An important side road brings acid hydrolases from the Golgi to the lysosomes. In addition, crowding of the secretory line tends to open the crinophagic detour, which sends excess products from secretion granules to lysosomes.

Phagocytosis, receptor-mediated endocytosis, and perhaps other forms of endocytosis make up the import lines. These converge mainly in the direction of lysosomes, but on their way they may, especially when passing through endosomes, send out side branches that bypass the lysosomes and lead engulfed materials back to where they came from (regurgitation), or across the cell to a separate extracellular region (diacytosis), or to storage granules.

This network of transport lanes appears even more complex when the movement of containers is considered in addition to that of contents. Until a few years ago the magnitude of the container traffic was poorly appreciated. It was generally assumed that membrane material incorporated into the plasma membrane by exocytosis was retrieved by endocytosis, added to the lysosomal membranes by endosome-lysosome fusion, interiorized by autophagic segregation, and finally broken down by lysosomal digestion. In compensation, new membrane was believed to be made in the ER and conveyed to the Golgi to replace the material lost to the plasma membrane by exocytosis. According to this concept, which is known as the "membrane flow" theory and includes the dynamic view of Golgi function (see p. 89), newly made membranes go through a single export-import cycle and then are destroyed. In other words, the cell uses disposable containers.

This view became untenable when the extent of the container traffic was discovered. A gland cell may double its plasma-membrane area with each secretory discharge. A macrophage interiorizes as much as twice its surface-membrane area every hour. Such rates are at least one order of magnitude larger than the turnover rates of membrane constituents. Cellular containers, therefore, cannot be disposable; they must be returnable.

The pathways of membrane recycling are still being mapped out. They probably include a number of direct shuttles, as well as more circuitous routes. The hub of this traffic lies in a region sometimes designated GERL (Golgi-endoplasmic reticulum-lysosomes), of which endosomes are now recognized as a central component. Surprisingly, this rapid and complex recycling of containers does not greatly disturb the essentially unidirectional transport of contents—although a certain amount of "slobbering" does occur, as we have seen (Chapter 4). The cell evidently possesses efficient mechanisms for moving empty containers around, presumably in the form of flattened vesicles. In all likelihood, the acidity main-

Cellular container traffic. The Golgi apparatus serves as a central clearing house and channel between the endoplasmic and the exoplasmic domains: (1) ER-Golgi shuttle; (2) secretory shuttle between Golgi and plasma membrane; (2′) crinophagic diversion; (3) Golgi-lysosome shuttle; (3′) alternative route from Golgi to lyosomes via the plasma membrane and an endosome; (4) endocytic shuttle between the plasma membrane and an endosome; (4′) alternative endocytic pathway bypassing an endosome; (5) plasma membrane retrieval (a) by regurgitation, (b) by diacytosis, (c) via Golgi, (d) via a lysosome; (6) endosome-lysosome pathway; (7) autophagic segregation.

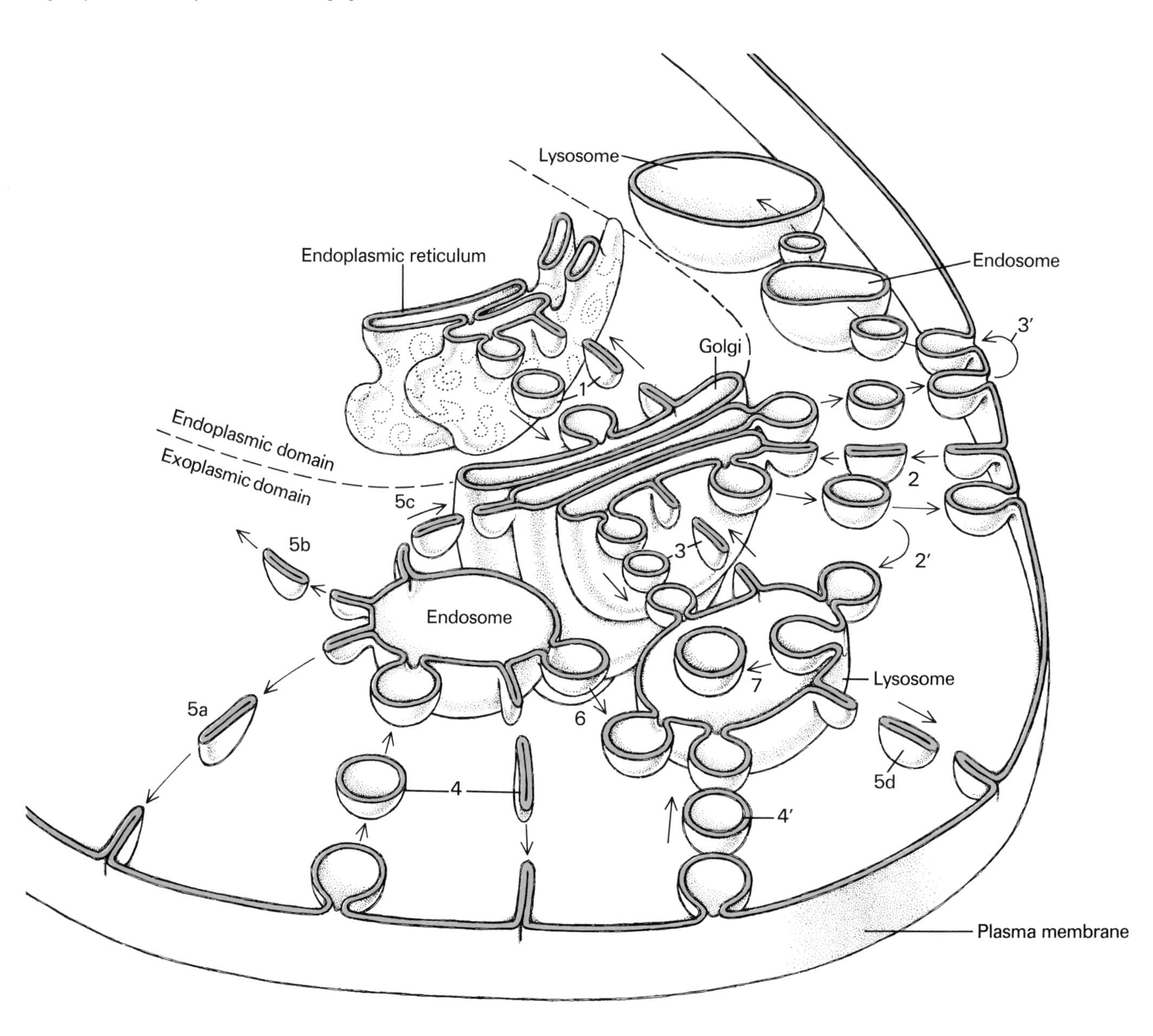

tained in the endosome-lysosome part of the exoplasmic domain plays a key role in stripping membranes of ligands before their recycling.

Even more remarkable is the ability of the cell to preserve the highly differentiated organization of its vacuome. Somehow, in spite of innumerable mergers between membranes of different composition, there is hardly any "running" of components by lateral diffusion across the junction. Sharp discontinuities are maintained, or even created, as in the clustering of occupied receptors, which plays an important role in selective transport. Wherever sorting occurs—on the cell surface, in endosomes, in the Golgi, and perhaps elsewhere—membrane patches bearing specific receptors must be involved. It is likely that the underlying constraints are imposed largely by membrane-anchored cytoskeletal elements, which also provide much of the necessary motile force (see Chapters 12 and 13).

Now that we have taken a more comprehensive look at the vacuome, we understand better our feeling of not being quite inside the cell throughout this part of our tour. Especially in the exoplasmic domain, we were actually looking at an extraordinarily shifting scene, in which a piece of membrane that happened to be on the cell surface at one moment could at another moment be part of the endosome-lysosome system, appear a little later in the Golgi, and finally be back on the surface, all in a few minutes' time. Permeases, pumps, and other transport systems present on the plasma membrane presumably continue to work when interiorized, catalyzing the same exchanges between the cytosol and the vesicle contents as they did on the cell surface between the cytosol and the extracellular medium. (However, the effects may not be the same; remember the proton pump.) We are indeed, functionally speaking, only halfway into the cell—or halfway out of it—when we are inside the exoplasmic domain of the vacuome. So are most of its contents, which either have been freshly brought in by endocytosis or will soon be discharged by exocytosis. We are somewhat deeper inside the cell when in the endoplasmic domain, to the extent that the Golgi lock protects us from invasion by extracellular objects. On the other hand, the irrevocable commitment of the ER contents to delivery into the exoplasmic domain, most often followed by discharge into the pericellular environment, clearly gives ER cavities what may be called a pre-extracellular character.

Prokaryotic Export and the Origin of Eukaryotic Cells

Possession of a vacuome is characteristic of eukaryotic cells. Most bacteria have no intracellular membranes to speak of. Yet eukaryotes arose out of prokaryotes, about one billion years ago. Consideration of the export machinery of prokaryotes may give us an inkling as to how this momentous transformation, crucial to the appearance on earth of all plants and animals, may have taken place.

Most bacteria secrete proteins into their environment. Prominent among these proteins are hydrolytic exoenzymes, which play a digestive role. It is interesting, and probably very revealing, that these exoenzymes are concomitantly made and extruded by polysomes attached to the inner face of the bacterial plasma membrane, in very much the same way as nascent export proteins are delivered through the membranes of the endoplasmic reticulum. The similarity even extends to the molecular mechanisms involved (see Chapter 15), suggesting strongly that the rough ER of eukaryotes originated from the plasma membrane of their prokaryotic ancestors, presumably as a result of some kind of progressive interiorization process. Surface flexibility would be important for such a process to occur, making it likely that our putative distant forbears were devoid of a rigid outer wall. Such naked bacterial cells can be generated by the action of lysozyme (protoplasts) and may also arise occasionally in nature (L forms). They are very fragile, but one can see how, under certain circumstances, their vulnerability might be compensated by a tremendous evolutionary advantage conferred by their surface flexibility. Should this property result in the formation of infoldings of the plasma membrane and, especially, in the intracellular vesiculation of these infoldings, such as occurs in endocytosis, the exoen-

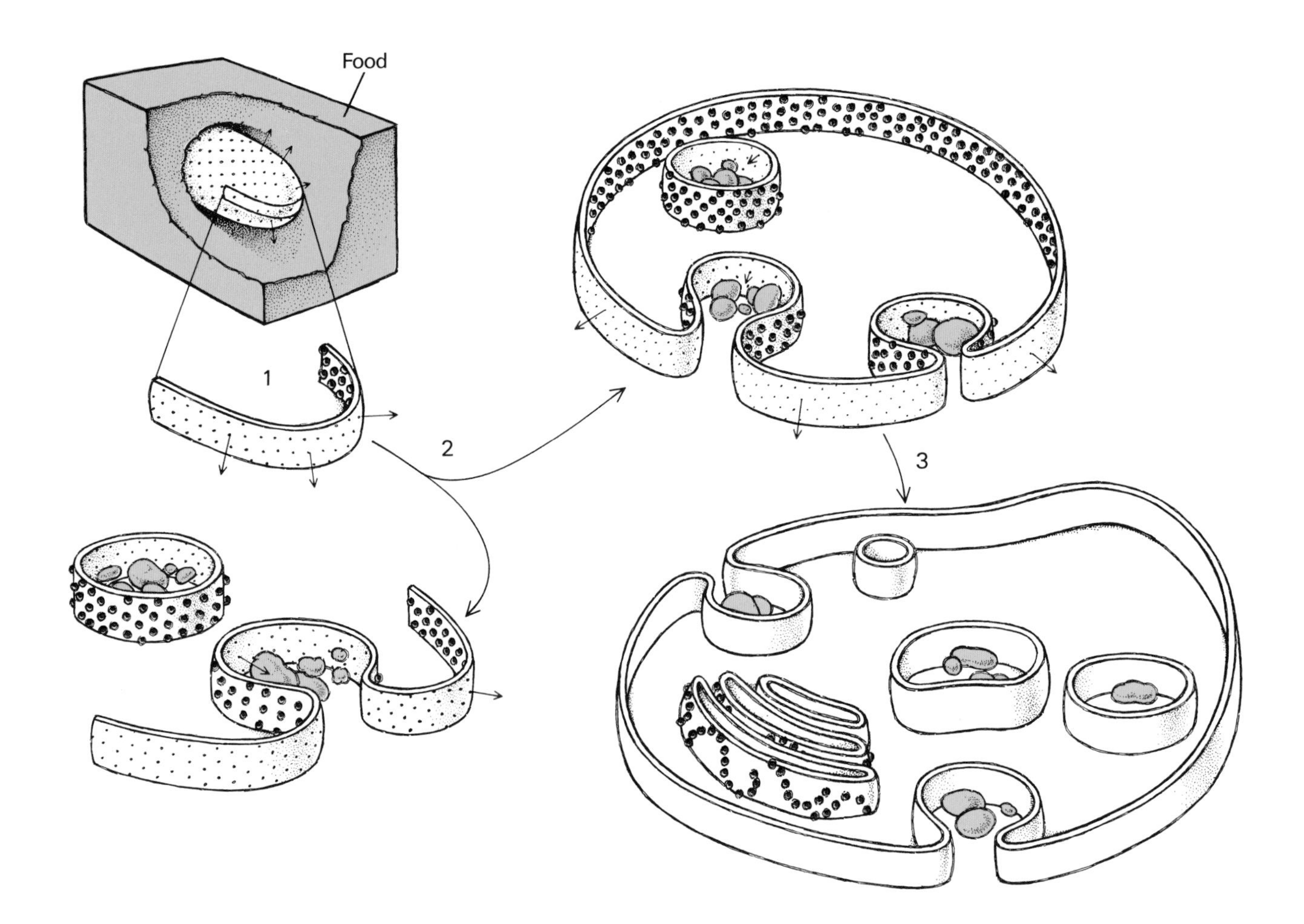

Hypothetical mechanism explaining the evolutionary origin of a primitive phagocyte from a prokaryotic ancestor: (1) like many present-day heterotrophic bacteria, the putative bacterial ancestor relies for nutrition on the extracellular digestion of food by exoenzymes that are made by plasma-membrane-bound ribosomes and discharged outside; (2) invaginations of the plasma membrane in a wall-less form of bacterium create primitive intracellular digestive pockets in which interiorized food particles (primitive endocytosis) are broken down by trapped exoenzymes; (3) differentiation of the intracellular membrane system relegates bound ribosomes to parts of the ER and leads to the formation of different membranous domains, and the cell increases considerably in size.

zymes made by the membrane-bound polysomes would remain trapped inside the interiorized vesicles, where they would be able to act on any extracellular material that got caught in the interiorization process. This would amount to the simplest possible form of intracellular digestion, within vacuoles having the combined properties of endosomes, lysosomes, and rough ER cisternae.

Even in this primitive and haphazard form, such an acquisition would provide a distinctive evolutionary advantage, likely to favor any chance mutation that would result in its further development. Until then, in order to benefit from the digestive activity of their exoenzymes, the cells had to rely on extracellular digestion. Unless they had other means of subsistence, they were practically condemned to reside inside their food supply, like maggots in a chunk of cheese. Henceforth, they would be free to roam the world and to pursue their prey actively, living on phagocytized bacteria or on other engulfed materials. This development could well have heralded the beginning of cellular emancipation.

Another important consequence of the postulated change would be the possibility of growth. Cells depend on exchanges with their environment and are limited in their capacity to grow by the surface area available for such exchanges. We have seen in Chapter 3 the importance of microvilli and other surface projections in this connection. The development of an intracellular vacuolar system, in direct continuity with the outside through endocytic interiorization and lined by a plasma-membrane type of membrane fitted with all the necessary transport systems, would enhance the exchange capacity of the cells manifold and thereby allow them to reach considerable sizes. At the same time, a progressive differentiation of the cytomembrane system might have occurred, with ribosomes becoming progressively relegated to the deeper parts of the system and the more peripheral elements evolving into plasma membrane and the smooth-surfaced parts of the vacuome of today's eukaryotic cells.

According to this hypothesis, the crucial evolutionary step from prokaryote to eukaryote was accomplished by way of a large-size primitive phagocyte. We may never know what really happened. But it is interesting that the most popular theory of the evolutionary acquisition of such important eukaryotic organelles as mitochondria and chloroplasts depends on the occurrence of just such a kind of cell (see Chapters 9 and 10). Also, if this hypothesis is correct, our picture of the vacuome as somehow extending the main boundary between cell and environment to thousands of intracellular recesses would be rooted in evolutionary history.

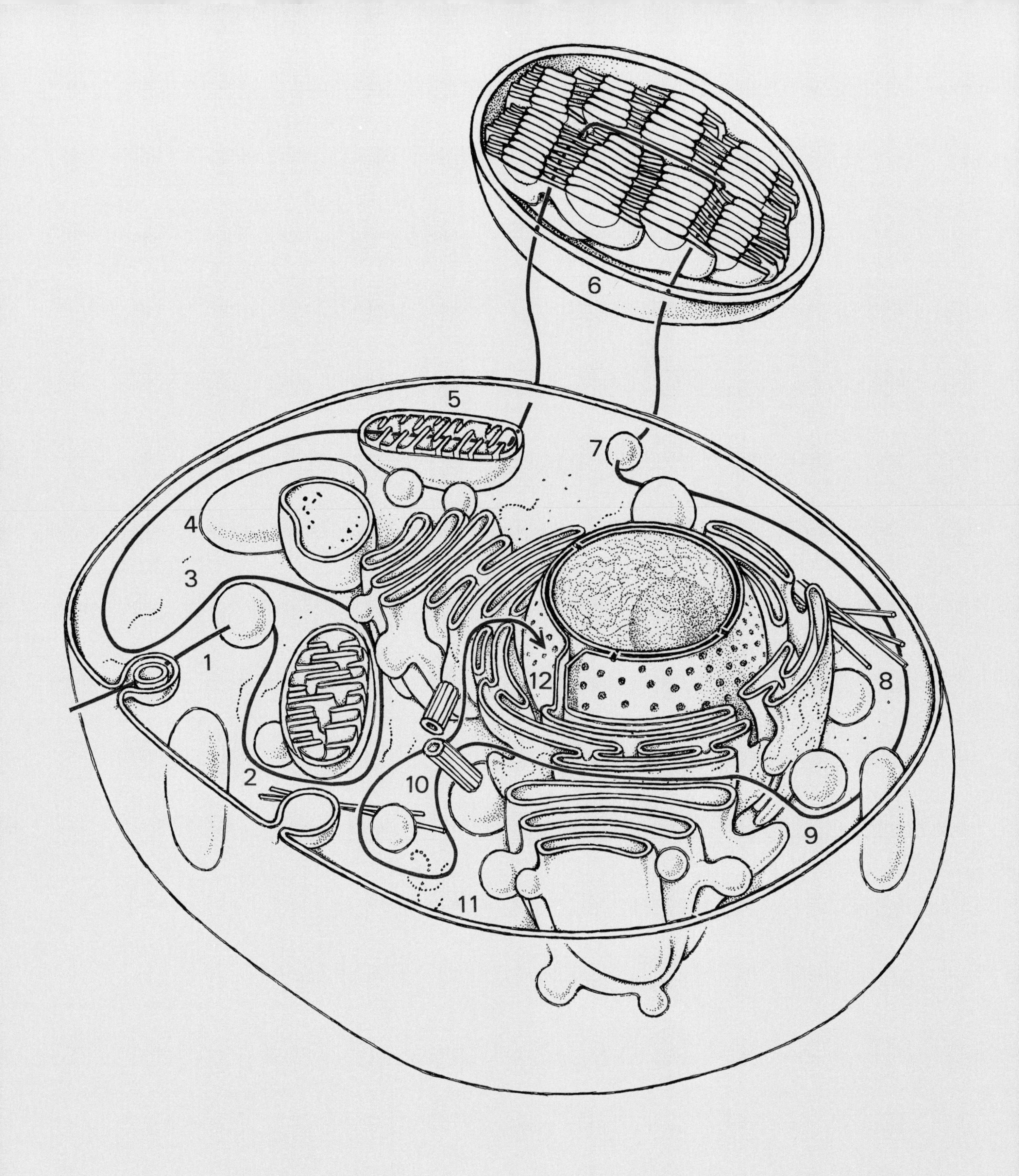

1
2
3
4
5
6
7
8
9
10
11
12

ITINERARY II

The Cytosol and Cytoplasmic Organelles

1. Use virus wrapping to enter cytosol by way of endosome.
2. Survey cytosol and its multifarious contents.
3. Watch anaerobic glycolysis and become acquainted with electron transfer and energy retrieval by phosphorylation.
4. Tour biosynthetic factories and learn about group transfer.
5. Visit mitochondria and examine their protonmotive power generators.
6. Make a side trip to the plant world and see how chloroplasts utilize sunlight.
7. Take a look at peroxisomes and sundry other microbodies.
8. Stop at various skeletal and motile elements.
9. Retrace steps through cytosol, paying special attention to the cytoplasmic faces of membranes.
10. Take last embracing view of cytosol as nexus of metabolic regulation.
11. Contemplate protein assembly on ribosome.
12. Wait near nuclear envelope for mitotic opening.

7 The Cytosol: Glycolysis, Electron Transfer, and Anaerobic Energy Retrieval

At first sight, it would seem that, even reduced a million-fold, bulky creatures of our size could not possibly get through the plasma membrane without causing the cell some irreparable injury. Yet there is ample evidence that it can be done. In a number of infectious diseases, the causal agent, be it a virus, a bacterium, or even a protozoan parasite, can be seen residing and developing freely in the cytoplasm of viable cells of the patient. It came from the outside, and therefore must have succeeded in traversing the plasma membrane without killing its host cell.

Piercing the Veil

The method whereby these intruders do their breaking and entering varies from one to another, but always relies initially on a Trojan-horse type of deception. The attacking microorganism first acts the part of a defenseless prey and allows itself to be engulfed without resistance. Then, when its too-trusting host is just getting ready to dump its catch into the lysosomal acid bath, the wily invader unsheathes its secret weapon, usually a membrane-dissolving enzyme or toxin that opens the endocytic vacuole and frees the imprisoned microorganism before it can be destroyed.

Some viruses—for instance, the causal agent of influenza—use an even craftier technique, prepared at the time the particles get ready to emigrate from their native cell to spread the infection to neighboring cells. As they leave their host cell, these viruses wrap their nucleic acid-protein bodies (nucleocapsids, see Chapter 18) within a spe-

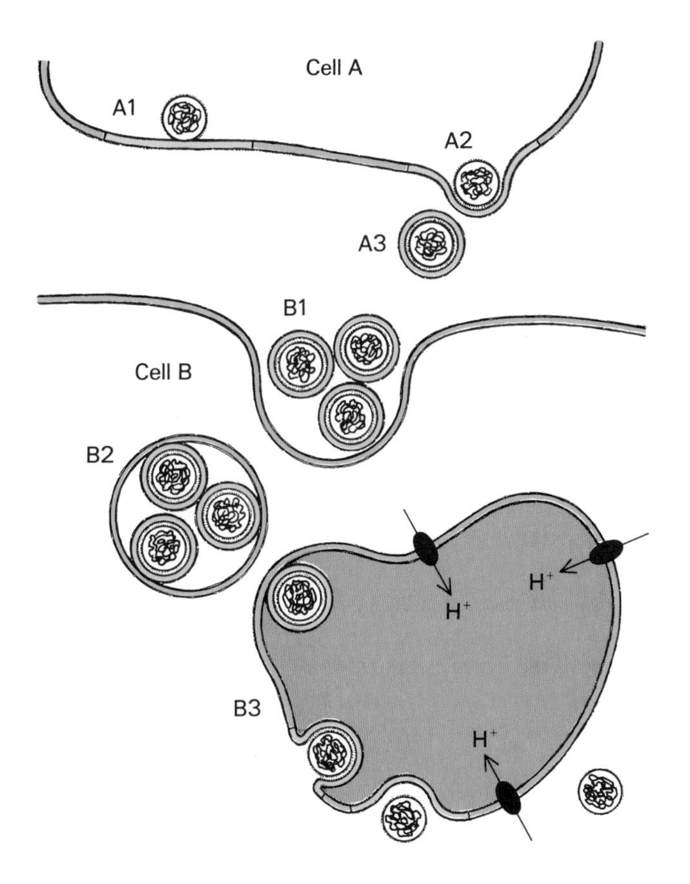

How some membrane-wrapped viruses travel from one cell to another.

A1. A piece of virus-directed membrane is assembled on the plasma membrane of infected cell A. A newly generated viral nucleocapsid approaches the area.

A2. The nucleocapsid pushes the viral membrane patch out to form a bud.

A3. The bud detaches from the cell, releasing a membrane-encapsulated virus.

B1. Viral particles are caught in an endocytic invagination by cell B.

B2. The endocytic invagination closes to form a virus-containing endosome.

B3. The endosome interior acidifies by the action of proton pumps. Exposure to acid triggers the fusion of the viral membrane with the endosomal membrane, projecting the free nucleocapsid into the cytoplasm.

cial piece of plasma membrane assembled under their instructions. Upon meeting another cell, they go through the defenseless-prey routine, with, in this case, the actual dumping into the acid bath included. Now comes the surprise: actuated in some manner by its exposure to the local acidity, the envelope of the virus fuses with the membrane of its vacuolar trap, projecting its infective inmate into the cytoplasm. This mechanism probably operates in endosomes before they fuse with lysosomes.

Once again, therefore, all we need to enter the cell safely is some magic tool borrowed from a microorganism. We have a choice between a chemical membrane-opener and an automatic acid-triggered membrane-ejection device. Let us use the latter. It requires no action on our part that could be botched by faulty timing, and it will remind us of that earlier exciting experience when we donned another type of lysosome-proof jacket and plunged for the first time into the cavernous depths of the cell's vacuolar system.

Behind the Scenes

Our viral vehicle has performed as expected, and we are now in the heart of the cytoplasm. Many strange objects meet our eyes, and it will take quite some time to examine them all in detail. But first, let us take a general look around.

Perhaps our first impression is of being backstage in some theater, seeing the reverse side of the sets with the ropes and pulleys that move them. Endosomes, lysosomes, endoplasmic cisterns, Golgi sacs and vesicles, secretion granules, all those bubble-shaped structures that so far we have seen only from the inside, we now contemplate from the outside. Together they make up thousands of balloons of various sizes and forms, ranging all the way from bulging spheres to flaccid sacs. Milk-white or coral-tinged, but for an occasional ocher-brown lysosome, they glow like opals in the cytoplasmic penumbra. Their surfaces are mostly satin smooth, except in those

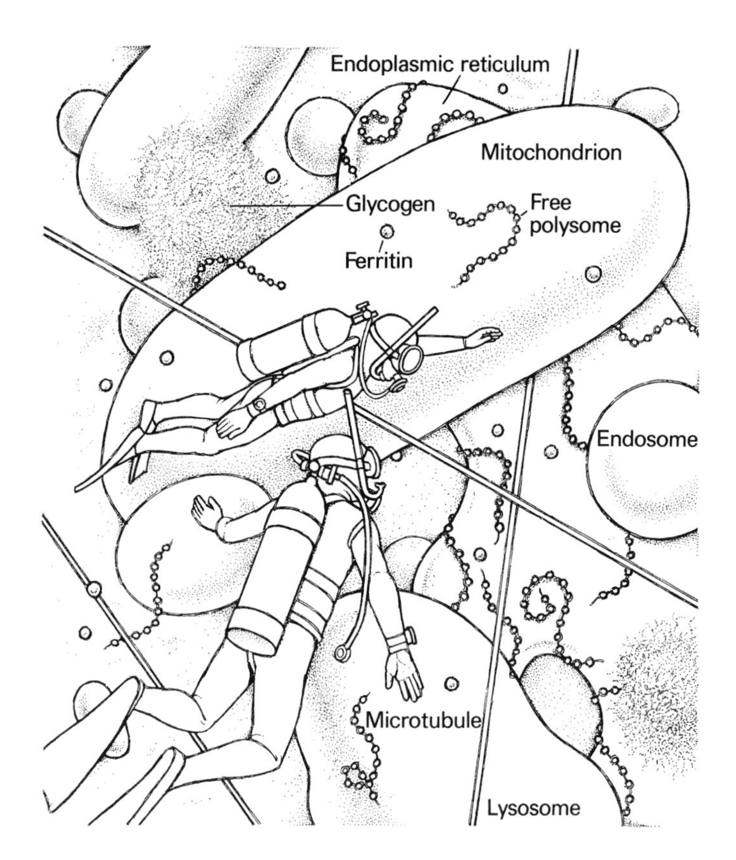

areas where they are roughened by crawling polysomes or raised into small conical or spherical mounds by a trellis of clathrin fibers. We will come back to these intriguing appendages later.

Some of the larger membranous sacs, parts of the endoplasmic reticulum and Golgi apparatus, are joined by coils of tubular connections. They make up massive systems, stolidly immobile, with only surface changes to betray their inner churnings. Others drift sluggishly by in the wake of some cytoplasmic current or jump joltingly ("saltatory" movement), as though pulled by an invisible spring. Flitting between them are shoals of small vesicles, many of them coated. Collisions are frequent in this crowded traffic. As a rule, they are of no consequence; the two partners disentangle themselves, none the worse for wear. But from time to time there results a fusion or fission event, which causes a minor local explosion, sending shock waves reverberating through the cytoplasm.

Interspersed between these now-familiar parts of the import-export machinery, a number of fat, oblong bodies of a vivid pink hue provide an arresting contrast. About the size and shape of a large bacterium, they are in a state of perpetual agitation—twisting, jerking, jostling neighboring structures with the indefatigable automatism of teenagers in a discotheque. Sometimes they split into several parts or join to form weird, hydralike structures. They are the mitochondria, the main centers of energy production in the cell (see Chapter 9).

In some cells—of the liver and kidneys, for instance—we may see yet another type of granule surrounded by a membrane: the peroxisomes. Somewhat smaller than mitochondria, they are of a dull green color, and they are found in clusters, perhaps because they are connected to each other (see Chapter 11).

Should we tour a plant cell instead of an animal cell, the spectacle offered would be similar, but even more variegated and colorful. For, in addition to the components that we have already encountered, we would meet the bright-green centers of photosynthesis, the chloroplasts (see Chapter 10). We might also run into one or more of the numerous membrane-bounded vacuoles and granules within which plants deposit their stores of starch and oil, as well as all those wonderful pigments that adorn their flowers. Such splendor, we might remember, is not an exclusive prerogative of the plant world. If man is a dull animal, relying largely for his color effects on a single black dye, melanin, packed in bodies called melanosomes, many other species—birds, butterflies, fishes—rival the most beautiful flowers.

Together, these objects occupy more than half the cell volume. The spaces between them are filled by a viscous, gelatinous matrix that forms the cytosol, or cell sap, which is the ground substance of the cell. This compartment is limited only by the plasma membrane on the outside and the membranous envelope that surrounds the nucleus on the inside. Thus we can move freely through it, having only to skirt the cytoplasmic bodies or sometimes to squeeze between them when they are pressed close together, as we might do in some aeronautical museum cluttered with multicolored exhibits.

Actually, we are not quite as free as this image might suggest. All sorts of other obstacles impede our mobility.

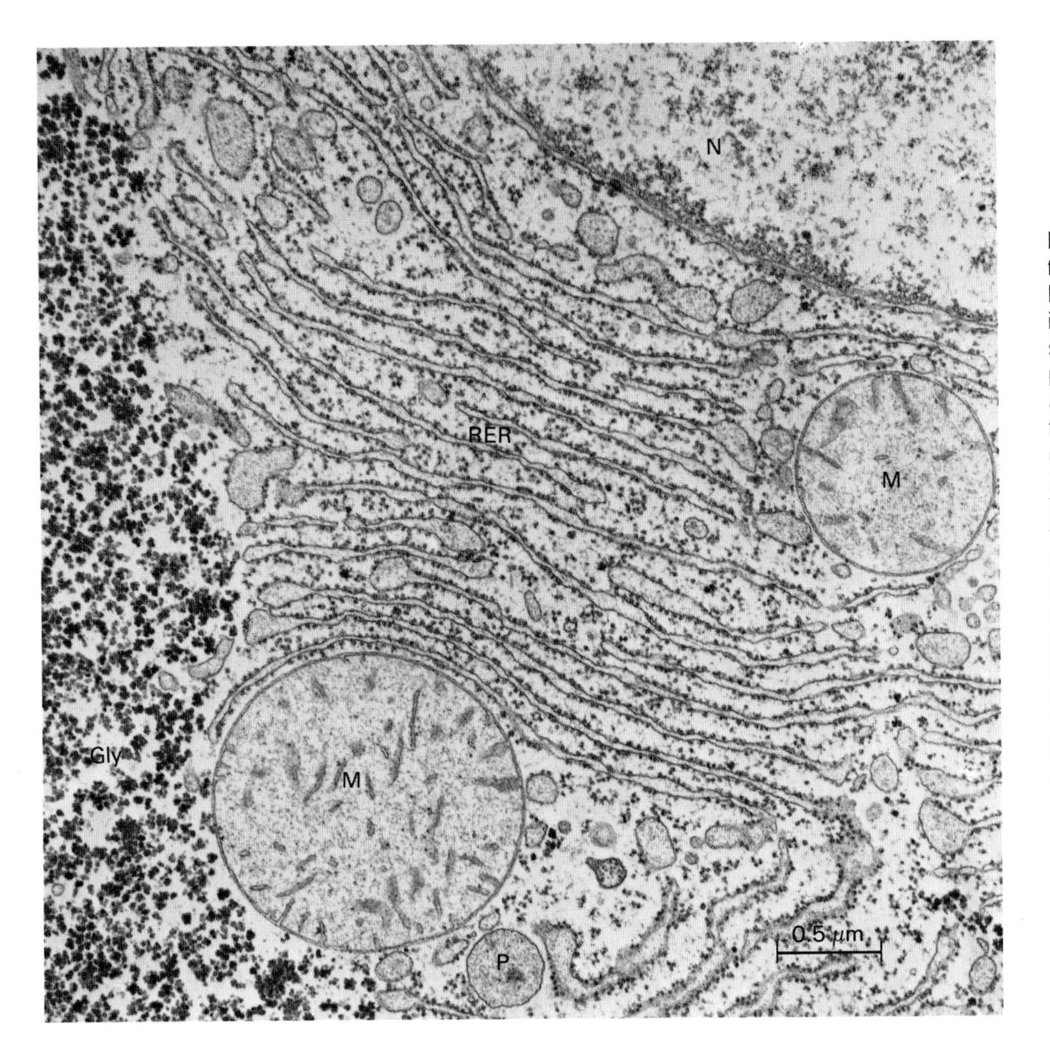

Electron micrograph of a section through a liver cell of a newborn rat. Part of the nucleus (N), surrounded by its envelope (see Chapter 16), can be seen at the upper right. Notice the pores through the nuclear envelope, and the ribosomes attached to it on the cytoplasmic side. Much of the cytoplasm near the nucleus is occupied by parallel arrays of RER cisternae. The bulging ends of the cisternae are transitional elements filled with secretory products, presumably in the process of detaching. Note also the small profiles through similar elements. Nestling between the RER cisternae are two mitochondria (M), as well as a peroxisome (P). The dense area at the left consists of glycogen particles forming a glycogen "lake."

One type we have already met and will explore again in great detail in a subsequent chapter. It is represented by the ribosomes, which generally occur as strings of ten to twenty particles held together by a strand of messenger RNA. So far, we have seen such polysomes only on the surface of ER membranes, but many also float freely in the cytosol and do their job there. With their tenuous threads of growing polypeptide chains, they pose treacherous traps, like invisible seaweeds in a swimmer's way.

In some areas—for instance, near the smooth connections linking the ER to the Golgi apparatus—we may run into an almost impenetrable obstruction: dense clumps of snow-white particles about the size of ribosomes. Looked at through our molecular magnifying glass, however, these particles are very different from ribosomes. Every one is actually a single, treelike structure, branching out into thousands of ramifications, each of which is a chain of glucose molecules linked end to end. These particles are giant macromolecules of a starchy polysaccharide called glycogen, sometimes designated "animal starch." Glycogen trees tend to congregate into small copses, which themselves assemble into forests of fairly extensive size. They are readily seen in the electron microscope as aggregates of small, dense particles, sometimes referred to as "glycogen lakes."

Another impediment in our way is the iron-storage protein, ferritin. It is a compact little particle, much smaller than glycogen, dark brown in color. Although barely discernible in the electron microscope, it is not a harmless object to stumble against: about one-quarter of its weight consists of solid iron hydroxide. Here and there, large droplets of stored lipid may oppose us with their water-repellent surfaces.

Finally, the cytosol is also crisscrossed in many places by an impressive network of girders and cables made up of various microfilamentous and microtubular elements. At any given time, these structures are assembled into a fairly rigid framework that connects the different cell parts with each other and is largely responsible for the shape of the cell and for its attachment to neighboring cells and other extracellular anchoring points. But the framework is not static. Some of its parts are continually dissassembled, to be reassembled in a different manner. Others are made to slide along each other by small molecular motor units. Movements within the cell, and of the cell itself, are largely determined by such changes. Chapter 12 will deal with these various structures, which make up the "bones and muscles" of our cells.

Discounting all these particulate and filamentous elements that clutter the cytosol, we are still left with a considerable amount of material—easily one-third of the total weight of the cell. It consists essentially of water-soluble components and forms the cytosol proper. The architecture of this compartment in the living cell is much debated. Some believe it to behave simply as a concentrated solution of randomly dispersed material. Other investigators see it as a highly organized system in which all constituents are linked by specific interactions or are immobilized by a network of "microtrabeculae." Still others accept the possibility of reversible changes between two such states, a concept that, under the name of "sol-gel transformation," goes back to the early days of colloidal chemistry. Most likely there is a grain of truth in each view, and reality is a compromise. On the whole, however, randomness probably prevails over organization. By all appearances, the cytosol permeates, in an essentially homogeneous fashion, every nook and cranny of the cell that is not occupied by something else. It is the basic filler, the true ground substance of the cell. It is no inert filler, however, but rather a feverishly busy place, where some of life's most important deals are being transacted, in particular those related to energy. It is not the only such site and, for many cells, not the main one. But it is probably the oldest in terms of geological time and, for this reason, has much to tell us about life at its most basic.

Energy, the Perennial Problem

Life is an active process, which depends on the continuous performance of various kinds of work. Consider its most typical character: growth and multiplication. To make a new cell, thousands of proteins, nucleic acids, carbohydrates, fats, and other complex substances have to be constructed, either from scratch or from relatively simple building blocks. To do this, cells need energy, which means that they must be able both to extract energy from their environment and to use it for the accomplishment of chemical work. In addition, living organisms move, manufacture electricity, remodel their surroundings, sometimes even emit light. All this requires energy. As we shall see when we visit the mitochondria and the chloroplasts, evolution has come up with some elaborate solutions to the problem of cellular energy. But those are late inventions, which took more than 1 billion years to develop. Life thrived during all that time. What, then, did it do about energy?

Our guess is that it depended on the kinds of systems that are found today in the cytosol. We have no proof of this, for we cannot go back in time to find out. Nor is there any trace in the fossil record of how early forms of life went about their daily business. What the fossil record does tell us is that as far back as 3.2 billion years ago there existed a microbe, named *Eobacterium isolatum*, that, according to the imprints it has left in some South African rocks, may have been very similar to some of the bacteria we know today. Traces of an even earlier microorganism, *Isuasphaera*, dating back 3.8 billion years, have been uncovered in Greenland. It is almost certain that in those ancient times the earth's atmosphere contained very little oxygen, which is generally believed to be mainly a prod-

Anaerobic fermentations are caused by living microorganisms. This drawing, made by Louis Pasteur, depicts what he saw through the microscope in a droplet of fermenting beet juice.

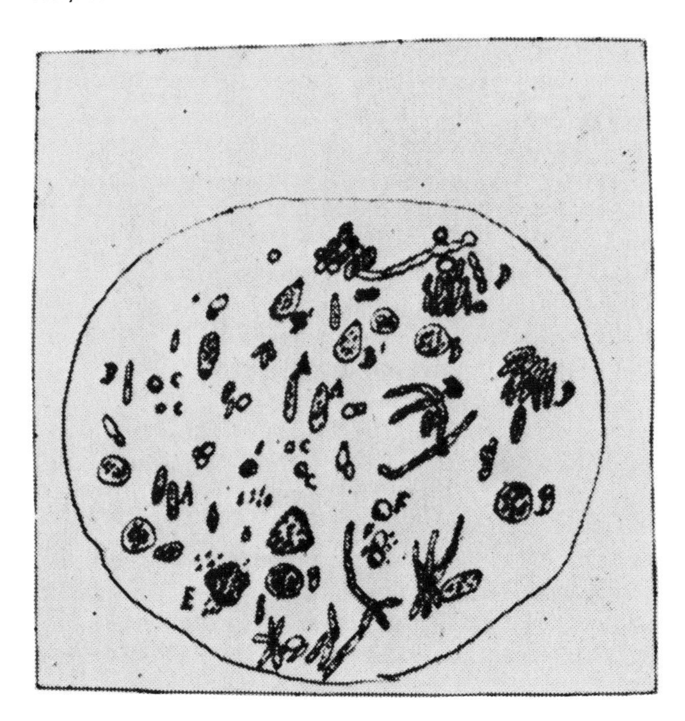

uct of photosynthesis. Thus, *Isuasphaera, Eobacterium*, and many of their descendants must have derived their supply of energy from anaerobic mechanisms—mechanisms capable of supporting life (Greek, *bios*) without (Greek negative: *a*-) air (Greek, *aer*). It so happens that what we find in the cytosol of higher cells and in the cell sap of most present-day bacteria is exactly that: an anaerobic energy-producing mechanism, which is known as glycolysis. Hence our assumption that this system comes to us, perhaps with little change, from those early forms of life that started peopling the earth some 4 billion years ago.

Indeed, glycolysis has the hallmarks of primitiveness, including the quality of relative simplicity. In this respect, the cytosol offers a good introduction to biological energy supply. Even so, for our visit to be profitable, we will need better eyesight than we have used so far. This is not so much to discern complex molecular structures—these will be kept at an absolutely strict minimum—but to apprehend certain key concepts without which we cannot possibly understand life's solutions of its energy problem.

There is another historical aspect to this part of our visit. It will also retrace in some ways the voyage of those early explorers who first discovered the principles of bioenergetics, for anaerobic glycolysis is the cradle of dynamic biochemistry. One might even say that it has been a key element in the development of human civilization. It was discovered many millennia ago in the form of fermentation and was employed by our distant ancestors for the manufacture of leaven, cheeses, and alcoholic beverages. These age-old industries remained purely empirical until 1856, when a Mr. Bigo, a distiller established in the French town of Lille, found himself suddenly threatened with ruin. For some unexplained reasons, his sugar-beet fermentation vats were all going sour, producing lactic acid instead of alcohol. He went to seek the help of a young chemist from Paris who had recently joined the local university staff and who, it was reported, had done some brilliant work. The young man obliged and eventually succeeded in rescuing Mr. Bigo's business. At the same time, he made a discovery that would change the world: anaerobic fermentations are caused by living microorganisms. His name: Louis Pasteur. Later, in 1897, a German chemist, Eduard Buchner, found that a mere "juice" expressed from yeast—none other, in fact, than the yeast cells' cytosol—could bring about the conversion of sugar into alcohol. Buchner thereby demonstrated that the function of the microorganism in alcoholic fermentation is purely chemical and that it does not rely, as was believed by Pasteur, on a special vital force peculiar to living organisms. Buchner also opened the way to a chemical dissection of the glycolytic system present in the cytosol of cells and, in so doing, launched a vast movement that has given us the detailed knowledge of metabolism that we have today.

The term metabolism comes from the Greek word for change (literally "the act of throwing about," from *ballein*, to throw). It covers the sum total of the chemical changes that take place in living organisms. It is subdivided into anabolism (from *ana*, up) and catabolism (from *kata*, down). Anabolism includes all the processes that require energy and are described thermodynamically as endergonic (Greek *endon*, inside; *ergon*, work). Its main function is biosynthesis. Catabolism is made up of the reactions that produce energy and that are called exergonic (*ex*, out) for this reason (see Appendix 2). By necessity, catabolism supports anabolism, as well as all the other forms of work carried out by living organisms (except for such reactions as are powered directly by an outside source of energy, mostly light).

When Buchner made his discovery, only the overall balance of some metabolic changes was known; mechanisms were completely unknown. Glycolysis is the first metabolic process ever to be elucidated. It took all of 40 years and the participation of many of the world's greatest scientists to accomplish this feat, which remains one of the most remarkable and far-reaching pieces of detective work of all time. Yet, all that it brought to light was a dozen chemical reactions, less than one-hundredth of the number of metabolic reactions that were to be recognized in the next 40 years. What makes the elucidation of glycolysis so important is that it was a first. Knowing only the starting point—the simple sugar glucose—and the end-products—lactic acid in one variant, ethyl alcohol plus CO_2 in another—even the most perspicacious of organic chemists could not possibly have predicted the astonishingly circuitous route taken by the natural process. Every step came as a surprise, and their identification required an enormous amount of patience and perseverance—especially with the pathetically primitive tools of the day: a few test tubes, a Bunsen burner, a balance, a light microscope. But once the pathway was clarified, it served as a beacon that illuminated the whole course of subsequent discoveries. It will do the same for us.

Glycolysis, a Power-Giving Snake

Yeast cells convert sugar into ethyl alcohol (ethanol) in twelve consecutive chemical steps that form a reaction chain, a metabolic snake:

$$\text{Glucose} \underset{1}{\longrightarrow} \text{A} \underset{2}{\longrightarrow} \text{B} \dashrightarrow \text{J} \underset{11}{\longrightarrow} \text{K} \underset{12}{\longrightarrow} \text{Ethanol} + CO_2$$

The same pathway is followed, up to the tenth step, by lactic bacilli (those that contaminated Mr. Bigo's vats), as well as by our muscles when they make a sudden effort. Only at step 11 do these diverge from yeast, converting intermediate J (pyruvic acid) into lactic acid instead of CO_2 and alcohol.

Thus, lactic and alcoholic fermentations differ only at the chain's end. Before that, they follow the same universal route, known as the glycolytic chain. This route is not laid out as a visible trail in the cytosol; the snake has no real body. If we put on our high-powered chemical glasses, all that we will see is a chaotic jumble of molecules, A, B, . . . , J, K, mixed with many others, intermediates in other pathways. What joins them, and gives the snake its substance, are the arrows, each of which indicates the occurrence of a specific enzyme (Chapter 2) that catalyzes the chemical transformation shown. The dynamic ordering of the twelve enzymes involved in the glycolytic chain follows automatically from the nature of their substrates and products. The reaction giving rise to D from C must necessarily follow immediately after the conversion of B into C and precede that of D into E. No physical channel is needed to guide the molecules toward their destination. The apparent chaos that we see hides a high degree of order, a dynamic organization generated by the properties of the enzymes present.

This lesson of glycolysis may be generalized. Behind each of the thousands of chemical reactions that take place in living cells there is an enzyme. This is commonplace knowledge today but came to be appreciated only after the glycolytic chain responsible for alcoholic fermentation in yeast began to be unraveled. Enzymology, the study of enzymes, has greatly enriched our understanding both of life and of chemistry and is now beginning to yield important practical results based on the industrial use of enzymes extracted from natural sources. The exigencies of our tour will allow only occasional references to this important branch of biochemistry. But we must at least keep in mind that every activity we observe, what-

In a chaotic jumble of molecules A, B,..., J, K, each enzyme picks out its own specific substrate and converts it into the next intermediate in the metabolic chain. Order is created out of the chaos by the specificity of the enzymes.

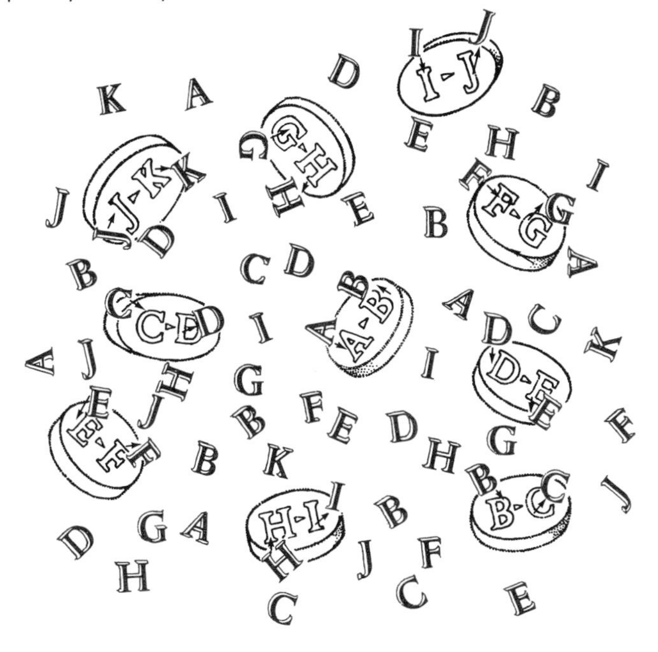

ever its nature, depends on the catalytic participation of enzymes.

Enzymes are often assisted by accessory substances called cofactors, or coenzymes. In glycolysis, two such cofactors require our attention. One is called NAD, which stands for nicotinamide adenine dinucleotide. Biochemists, you will notice, are greatly addicted to abbreviations. But they have an excuse. Most of the substances they deal with are too complex to be represented explicitly every time they are referred to. NAD is one, and we will not even bother to look at its chemical structure here (those interested will find it in Appendix 1). It is, however, worth noting that the nicotinamide part of the molecule is a vitamin known as vitamin PP, which stands for *pellagra preventiva*. Deficiency of this vitamin in the diet causes pellagra, a severe nutritional disease formerly widespread on the American continents. This is not an isolated example. Most vitamins either act as coenzymes or are part of one, which explains why an organism cannot do without them. We will see the function of NAD a little later.

The other cofactor that we must look at is designated ATP, for adenosine triphosphate. Eventually, we will have to consider the chemical structure of ATP. But right now, all we need know is that it can be hydrolyzed (split with the help of water) into adenosine diphosphate (ADP) and inorganic phosphate (P_i) and, conversely, that it can arise (provided energy is supplied) from the condensation of ADP and P_i with removal of water:

$$\text{ATP} + \text{H}_2\text{O} \underset{\text{(endergonic)}}{\overset{\text{(exergonic)}}{\rightleftharpoons}} \text{ADP} + \text{P}_i$$

The critical function of ATP in glycolysis was revealed when it was found that the breakdown of glucose is coupled to the assembly of ATP: for every molecule of glucose converted into lactic acid or ethanol, two molecules of ADP are phosphorylated to ATP. The link is an obligatory one. If ATP synthesis cannot occur, as when there is a lack of ADP, glycolysis stops.

What this remarkable phenomenon actually means became clear when the energetics of the process were considered. Glucose fermentation releases free energy: about 47 kilocalories (kcal) per gram-molecule of glucose broken down. On the other hand, the assembly of ATP from ADP + P_i requires free energy: about 14 kcal per gram-molecule of ATP formed. Thus, of the 47 kcal released by the breakdown of glucose, $2 \times 14 = 28$ kcal, or 60 per cent, are utilized to form ATP, instead of being dissipated as heat. Glycolysis powers ATP synthesis; coupling of the two processes is an energy-retrieval mechanism.

Here again, glycolysis provided a first. When other catabolic processes were subsequently discovered, they also were found to be coupled with the assembly of ATP. Not only glycolysis, but the whole of catabolism powers ATP formation: coupling is a universal energy-retrieval mechanism.

But what of ATP itself? What use its synthesis? The answer to this question, or rather an inkling of it, was first given in the early 1930s, when it was found that a muscle rendered incapable of glycolysis by a poison (mono-iodo-acetic acid) could still perform a small amount of work at the expense of stored "phosphate-bound energy." Eventually the chemical reaction directly connected with the contractile machinery was identified as the hydrolysis of ATP to ADP and P_i. In this way, ATP was recognized as the missing link between glycolysis and muscular work. Glycolysis powers ATP formation; ATP breakdown powers muscular work:

$$\begin{array}{ccc} \tfrac{1}{2}\text{ Glucose} & & \text{ADP} + \text{P}_i \\ & \times & & \rightleftarrows \text{Work} \\ \text{Lactic acid} & & \text{ATP} + \text{H}_2\text{O} \end{array}$$

This was a tremendous discovery, which opened one of the main doors to the understanding of life. For not only muscular work, but virtually every kind of work performed by living organisms is powered by ATP. Scratch the surface of any kind of bioengine, be it an ion pump in a membrane, a contractile fiber in a flagellum, a light generator in a glowworm, or any of the multiple synthetic reactions whereby living organisms manufacture their own constituents: almost invariably you find ATP as a source of the required energy. It is the main fuel of life, and the function of catabolism is explained by its ability to support the restoration of ATP:

$$\begin{array}{ccc} \text{Foodstuffs} & & \text{ADP} + \text{P}_i \\ & \times & & \rightleftarrows \text{Work} \\ \text{Breakdown products} & & \text{ATP} + \text{H}_2\text{O} \end{array}$$

Epitomizing this relationship is the historical example of anaerobic yeast. This organism grows and multiplies, accomplishing tremendous feats of chemical engineering and information transfer in order to make new little yeast cells similar to their progenitors. It does all this with the conversion of sugar into alcohol as sole source of energy. Between this crude form of energy and the thousands of different processes it powers lies ATP, just as electricity lies between the burning of coal or oil and the wonders of modern technology. We will encounter ATP again and again on our tour.

The last lesson we learn from glycolysis concerns the mechanism of the coupling. Here we must look more closely at the anatomy of our glycolytic snake. This part of the visit may be a little rough for some of our fellow tourists. Unfortunately, there is no easier way. The cell is a chemical machine. Its workings cannot be understood without at least an elementary acquaintance with certain physicochemical concepts. (To help those in need of a refresher course, some basic notions have been summarized in the appendices.)

Obligatory coupling compels cell to use 60 per cent of the energy released by anaerobic glycolysis to manufacture ATP from ADP + P_i. Breakdown of ATP, in turn, powers various forms of biological work.

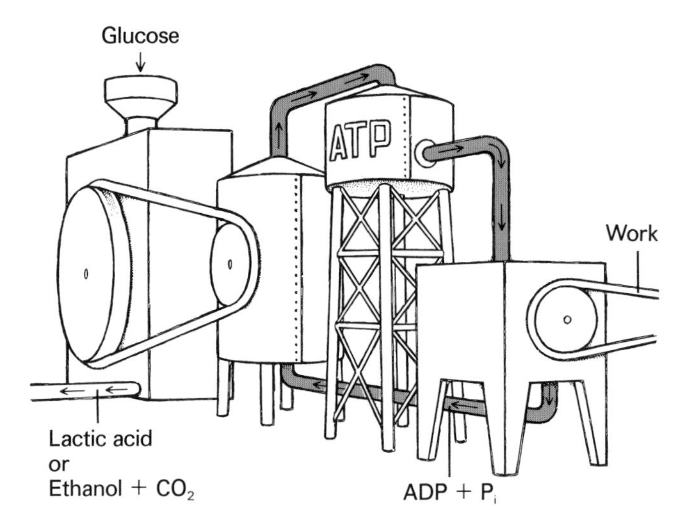

Burning without Air

From the chemical point of view, glycolysis consists simply in a halving of the glucose molecule and rearrangement of its constituent atoms, with neither gain nor loss of matter:

$$\underset{\text{(Glucose)}}{C_6H_{12}O_6} \longrightarrow 2\ \underset{\text{(Lactic acid)}}{CH_3\text{—}CHOH\text{—}COOH}$$

or:

$$\underset{\text{(Glucose)}}{C_6H_{12}O_6} \longrightarrow 2\ \underset{\text{(Ethanol)}}{CH_3\text{—}CH_2OH} + 2\ CO_2$$

Count the atoms on either side of the arrows, and you find the same number of carbons, hydrogens, and oxygens. From head to tail, the snake neither gains nor loses weight. In between, however, it goes through some fairly elaborate contortions that are all geared to one central function: the making of ATP. This is accomplished at steps 6 and 7 by a complex process of oxidoreductive phosphorylation.

The role of the five steps that precede this central reaction is to ready the glucose molecule for participation in it. This is quite a job, and an expensive one, since it means converting the 6-carbon glucose molecule into two molecules of a phosphorylated 3-carbon compound called phosphoglyceraldehyde. The two phosphate groups required for this purpose are supplied by ATP, so that we have an apparently paradoxical situation: a reaction destined to make ATP starts by spending it. This is not unusual. Many foodstuffs need to be activated, with expenditure of energy, before they become susceptible to degradation. It is one more of the chores the cell's maid-of-all-work, ATP, is saddled with. Of course, this initial energy investment is subsequently reimbursed with interest. Otherwise it would be useless.

With the making of phosphoglyceraldehyde, we come to the key reaction in glycolysis, which is the oxidation of this compound to phosphoglyceric acid, and the coupled condensation of ADP and P_i to ATP. We will not consider all the details of this reaction, since that would get us involved in some fairly complicated chemistry. But one aspect deserves our attention, namely the actual significance of the word oxidation. We are all familiar with the saying that we get our energy from "burning" our food. But the analogy with some sort of combustion engine that this image evokes is misleading and should be corrected. In strictly descriptive terms, the conversion of phosphoglycer*aldehyde* into phosphoglyceric *acid* consists in the addition of an oxygen atom to the aldehyde group (CHO) to make a carboxylic acid group (COOH). In this sense, it resembles the type of oxidation that accompanies combustion. Unlike what happens in combustion, however, the extra oxygen atom does not come from atmospheric oxygen; there is no such oxygen. Glycolysis is an anaerobic process, which can take place in the complete absence of oxygen. We are dealing here with a case of "burning without air."

If not atmospheric oxygen, what then is the source of the extra oxygen atom acquired by phosphoglyceraldehyde? The answer is *water*, though not regular water picked up from the medium. It is a water molecule that arises from the coupled condensation of ADP and P_i and is transferred directly to the oxidative reaction in chemically bound form. We indicate this fact by putting H_2O in parentheses; then, if we represent the rest of the phosphoglyceryl radical by R, we may formulate the reaction as follows:

$$\mathrm{ADP} + \mathrm{P_i} \longrightarrow \mathrm{ATP} + (\mathrm{H_2O})$$
$$\mathrm{R{-}CHO} + (\mathrm{H_2O}) \longrightarrow \mathrm{R{-}COOH} + 2\ \mathrm{H}$$

What this equation tells us is that the oxidative step actually consists in the removal of hydrogen. It is a dehydrogenation. But here again we must watch out. The hydrogen removed in this reaction is not hydrogen gas, H_2, which is a stable diatomic molecule. It is the reactive hydrogen atom, H, which never occurs in free form but is always transported or exchanged through the mediation of carriers, either as such or in the form of a stripped electron (e^-), with which it freely equilibrates according to the relationship

$$\mathrm{H} \rightleftharpoons e^- + \mathrm{H}^+$$

In aqueous media, the protons, or hydrogen ions, H^+, that participate in this equilibrium are readily available, thanks to the dissociation of water:

$$\mathrm{H_2O} \rightleftharpoons \mathrm{H}^+ + \mathrm{OH}^-$$

Thus the oxidative reaction of glycolysis can be written also as a removal of electrons:

$$\mathrm{R{-}CHO} + (\mathrm{H_2O}) \longrightarrow \mathrm{R{-}COOH} + 2\ e^- + 2\ \mathrm{H}^+$$

What is true of glycolysis is true also of most other oxidative reactions that occur in living organisms. As a rule, biological oxidations take place without direct participation of oxygen through removal of hydrogen atoms or electrons, a type of reaction no doubt inherited from those early times when life first appeared and went on to evolve for hundreds of millions of years in the absence of atmospheric oxygen. When the oxidized molecule acquires oxygen, as in the example considered, the source of the extra oxygen is water or some water-generating reaction.

As to atmospheric oxygen, which is obviously essential to all aerobic organisms, including ourselves, its role is to pick up the electrons released by the oxidative reactions:

$$\tfrac{1}{2}\,O_2 + 2\,e^- + 2\,H^+ \longrightarrow H_2O$$

The end result is the same as in ordinary combustion. But the mechanism is different. This can be verified readily with the help of the heavy isotope of oxygen, ^{18}O, which can be distinguished from the prevalent isotope ^{16}O by means of a mass spectrograph. If glucose is burned in a furnace in the presence of $^{18}O_2$, the isotope will be found in the CO_2 formed, indicating that the glucose carbon has joined with the atmospheric oxygen. But, if the glucose is burned by a living organism breathing $^{18}O_2$, the isotope is recovered not in the exhaled CO_2, but in H_2O. This seemingly trivial difference actually makes all the difference between life and death. No organism has developed a way to retrieve energy in biologically usable form from a combustion type of oxidation. In contrast, there are countless devices for extracting energy from electron transfer. We will come back repeatedly to this important topic. Note that addition of oxygen does take place in some biological reactions. These are termed oxygenations, to distinguish them from oxidations. They are not directly involved in oxidative energy retrieval.

The reverse of an oxidation—the gaining of hydrogen atoms or of electrons—is called a reduction. Since electrons and hydrogen atoms cannot circulate in free form in an aqueous medium, neither reaction can ever happen without the other. Electrons cannot be abandoned unless they can be picked up; whenever a substance is oxidized, another is reduced. The reactions, therefore, are always oxidation-reduction reactions, electron-transfer reactions. Enzymes catalyzing such transfers are called electron transferases, or oxidoreductases.

In the central oxidative step of glycolysis, the electron acceptor is NAD^+, oxidized form of the cofactor NAD, which owes its role in this and countless other metabolic processes to its ability to act as an electron carrier:

$$NAD^+ + 2\,e^- + H^+ \longrightarrow NADH$$

The complete electron transfer reaction may therefore be written schematically as follows:

$$R\text{—}CHO + (H_2O) + NAD^+ \longrightarrow R\text{—}COOH + NADH + H^+$$

This is the reaction that is coupled with the assembly of ATP (from which you may remember it derives its hidden water molecule). Whatever the mechanism of this coupling, it means that the transfer of one pair of electrons from phosphoglyceraldehyde to NAD^+ (1) releases enough free energy to power the assembly of one molecule of ATP and (2) is subjected to a constraint such that it can proceed only if ATP is made at the same time.

Formally, the system may be seen as an electrochemical transducer that converts a flow of electrons into chemical work. We may therefore apply the general theory of electricity, which says that the maximum amount of work—the real work is always less owing to inevitable losses as heat—that can be performed by an electric machine is given (in joules) by the quantity of electricity (in coulombs) passing through the machine, multiplied by the potential difference (in volts) of the electric generator.

In the present case, we know the work. It takes 14 kcal (58,600 joules) to make one gram-molecule of ATP. We also know the quantity of electricity passing through the system: 2 electrons per molecule of ATP formed, or 2 electron-equivalents or Faradays ($2 \times 96{,}500 = 193{,}000$ coulombs) per gram-molecule of ATP. If we divide the joules by the coulombs, we find the minimum voltage of our electricity source: 0.3 volt, or 300 millivolts (mV). In other words, the transfer of electrons between phosphoglyceraldehyde and NAD^+ must occur across a potential difference of at least 300 mV. Otherwise it could not power ATP synthesis. (For additional theoretical background, see Appendix 2.)

Of course, the glycolytic snake is hardly constructed like a conventional electric generator. It has no outlets into which we can plug a voltmeter to verify our conclu-

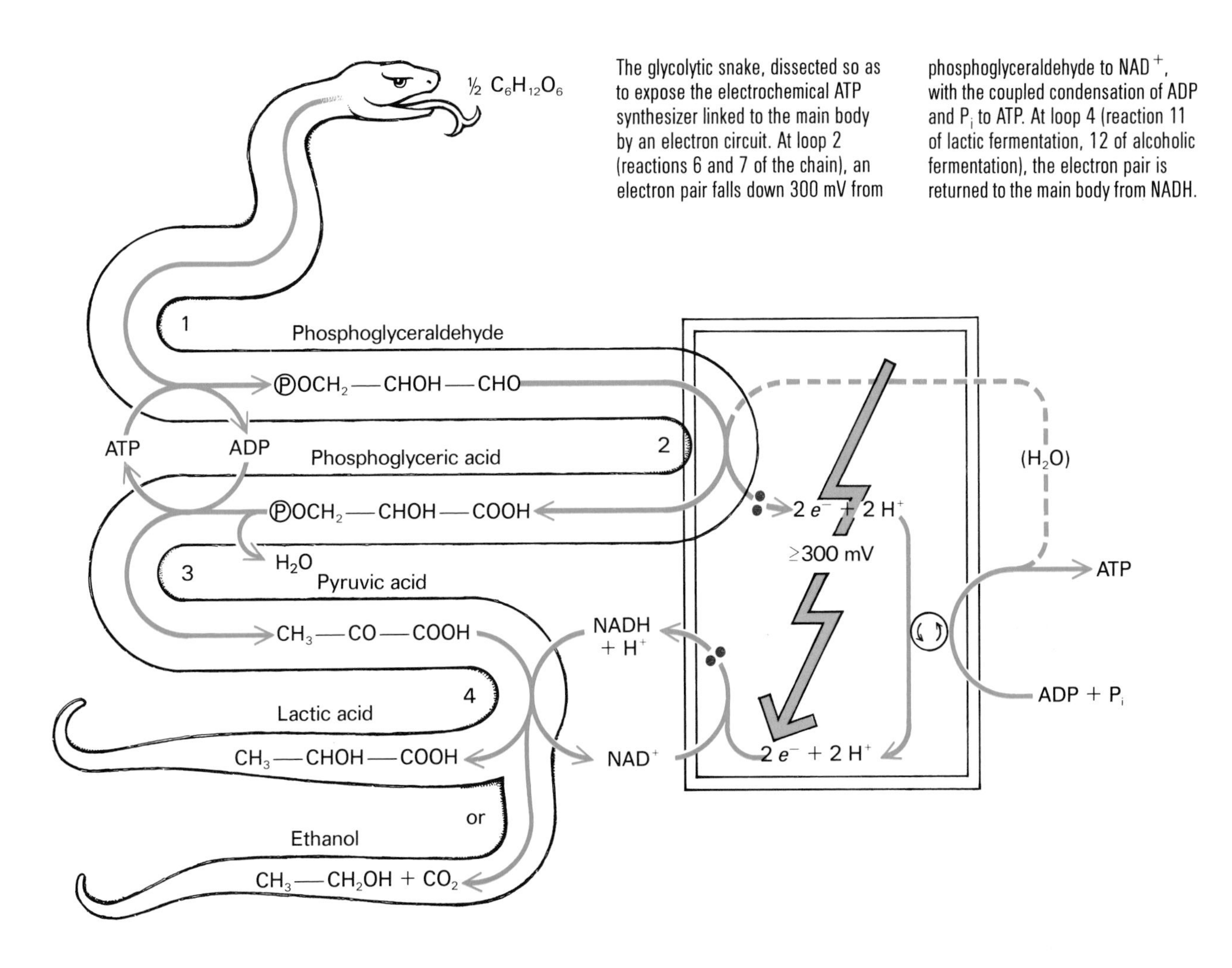

The glycolytic snake, dissected so as to expose the electrochemical ATP synthesizer linked to the main body by an electron circuit. At loop 2 (reactions 6 and 7 of the chain), an electron pair falls down 300 mV from phosphoglyceraldehyde to NAD^+, with the coupled condensation of ADP and P_i to ATP. At loop 4 (reaction 11 of lactic fermentation, 12 of alcoholic fermentation), the electron pair is returned to the main body from NADH.

sion. But it has an electron donor (phosphoglyceraldehyde) and an electron acceptor (NAD^+). For the transaction to take place, the affinity for electrons of the acceptor must be stronger than that of the donor. This affinity can be measured. It is called the oxidation-reduction potential, or redox potential, and is expressed in volts. Each redox couple (Red/Ox, e.g., phosphoglyceraldehyde/phosphoglyceric acid; NADH/NAD^+) has its characteristic redox potential. What we have established by the preceding calculation is that, if the transfer of an electron pair is to support the formation of one molecule of ATP from ADP and P_i, the difference between the redox potentials of the donor and acceptor couples must be at least 300 mV, which it indeed is for the two couples in glycolysis.

Once its main business of making ATP has been successfully completed, the glycolytic chain still has to wind up its affairs and balance its books. This is the main function of the latter part of the chain. There are three accounts to be settled: (1) phosphate, which was initially donated to the chain from ATP; (2) water, which entered the chain at the oxidative step to supply the extra oxygen of phosphoglyceric acid; and (3) electrons, which left the chain at this same step. These imbalances are now compensated. First, a water molecule is given off. Then the phosphate group is returned to ADP to regenerate the ATP that was invested at the beginning. And, finally, NADH gives back its electrons. In lactic fermentation, the electron acceptor is pyruvic acid, which is reduced to lactic acid:

$$CH_3\text{—CO—COOH} + \text{NADH} + H^+ \longrightarrow CH_3\text{—CHOH—COOH} + NAD^+$$

In alcoholic fermentation, the electron acceptor is the product of decarboxylation of pyruvic acid, acetaldehyde, which is reduced to ethanol:

$$CH_3\text{—}CO\text{—}COOH \longrightarrow CH_3\text{—}CHO + CO_2$$
$$CH_3\text{—}CHO + NADH + H^+ \longrightarrow CH_3\text{—}CH_2OH + NAD^+$$

These final electron transfers take place across very small potential differences, with no energetic benefit. They are necessary only to make the system self-contained. Cut off the snake's tail, while substituting another acceptor to collect the electrons from NADH, and the system can still perfectly do its job of generating ATP, except that it is now an oxidative system and its final product is pyruvic acid:

$$\underset{\text{(Glucose)}}{C_6H_{12}O_6} \longrightarrow 2\ \underset{\text{(Pyruvic acid)}}{CH_3\text{—}CO\text{—}COOH} + 4\ e^- + 4\ H^+$$

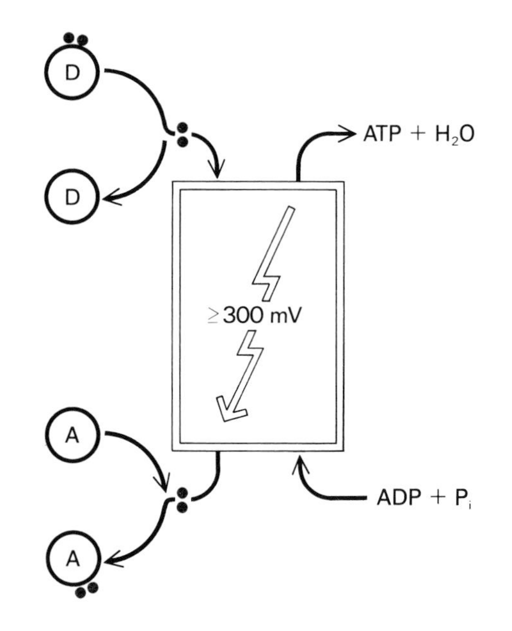

The oxphos unit is the universal energy generator. It receives a pair of electrons from a reduced donor (D) and returns it to an oxidized acceptor (A) at a potential level at least 300 mV lower. The electron pathway through the unit is so designed as to force ADP and P_i to join together into ATP with removal of water. This device is often reversible, allowing an electron pair to be lifted to a potential level at most 300 mV higher, at the expense of the hydrolysis of ATP.

Oxphos: Life's Golden Energy Gadget

Several features of the glycolytic snake are of general significance and apply to the whole of energy metabolism. The most universal of these generalizations concerns the production of ATP. Throughout nature, this central piece of energy currency arises, as in glycolysis, through a coupled electrochemical reaction that links the phosphorylation of ADP to the transfer of electrons across a difference of electric potential. There is virtually no exception to this rule. Animals, plants, fungi, bacteria, all living beings, including man, derive their ATP from the operation of such coupled reactions. As might be expected, a great many different reactions of this type occur, and their chemistry is often complex. So as not to have to go into such details, while retaining the possibility of gaining some insight into the working of the cell's main energy centers when we visit them, we will refer to systems that catalyze an oxidoreductive phosphorylation as *oxphos* units and will represent them schematically by the "boxed lightning" symbol shown at the right. Note that this terminology and symbolism do not belong to the standard language of biochemistry. They are introduced here for the sole purpose of helping us during our tour.

The basic design of an oxphos unit is that of an electrochemical transducer coupling the assembly of one molecule of ATP from ADP and P_i to the fall of two electrons down a 300-mV potential difference. The electrons are fed into the oxphos unit by a donor (D), which passes from the reduced to the oxidized state, and are collected at the other end by an acceptor (A), which changes from the oxidized to the reduced state. Protons may or may not accompany the electrons, depending on the nature of the molecules involved.

In glycolysis, the donor couple is represented by phosphoglyceraldehyde/phosphoglyceric acid, and the acceptor couple by NAD^+/NADH. But this is just one particular case. Other systems use other donors or other acceptors or both, allowing for a large number of different oxphos units that operate by a wide variety of mecha-

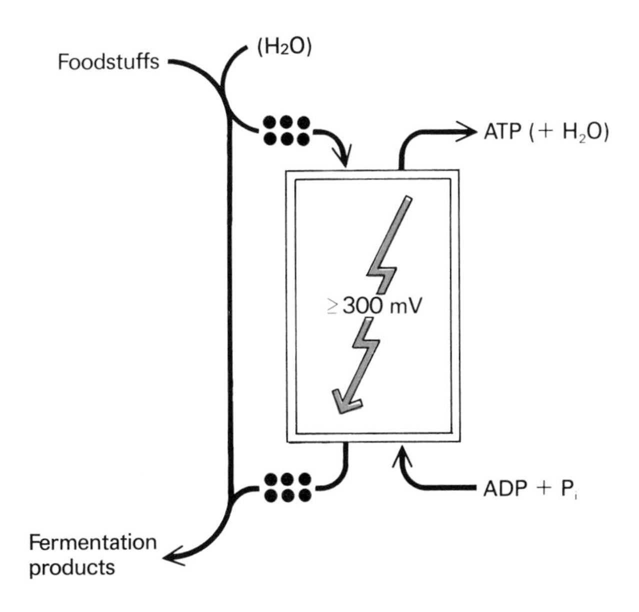

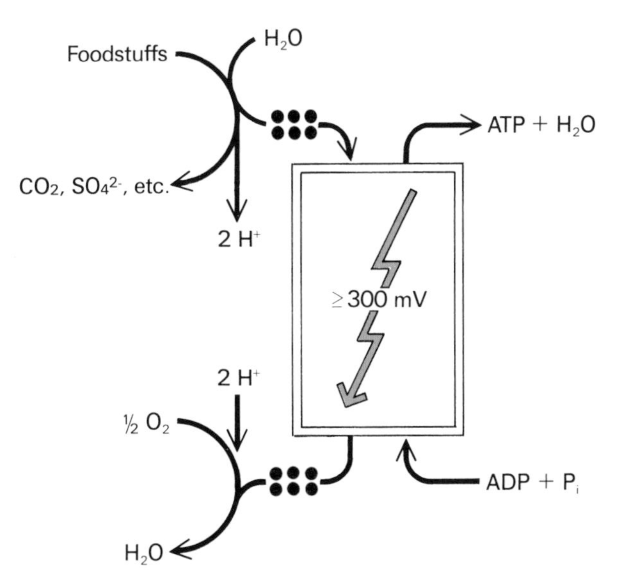

In anaerobic fermenters, a single flow of matter suffices to support electron flow and ATP assembly through oxphos units because the electron acceptor arises metabolically from the electron donor. In glycolysis, for example (see the illustration on p. 113), pyruvic acid (or acetaldehyde) arises from phosphoglyceraldehyde.

In aerobic heterotrophs, a dual flow of matter is needed to support electron flow and ATP assembly through oxphos units. One feeds electrons into the oxphos units at the expense of foodstuffs, which, with the help of water, are oxidized to their final waste products. The other, supported by oxygen, serves to pick up the electrons from terminal oxphos units.

nisms. We will encounter many in our tour. Some, like the one in glycolysis, receive their electrons from metabolic substrates and are said for this reason to catalyze substrate-level phosphorylations. Many others, including the most important ones, such as are in mitochondria and chloroplasts, are fed with electrons from NADH and other carriers. They catalyze carrier-level phosphorylations and operate by mechanisms that are entirely different from those involved in substrate-level phosphorylations. We will examine these mechanisms in Chapter 9.

Oxphos units are powered by a flow of electrons. To function, they need to be connected both to a source of electrons and to an electron collector. Glycolysis and the other anaerobic fermentations have the peculiarity that their electron flow is supported by a single flow of matter, thanks to the fact that the metabolic chain generates its own final electron acceptor. The main advantage of this kind of metabolism is that all it requires from the environment is an appropriate foodstuff—glucose, for instance. But it is tremendously wasteful, because the final products of fermentations, such as lactic acid or ethanol, are energy-rich molecules. As will be seen, they leave the cells with more than 90 per cent of the potential energy of glucose unused. This need not be so; in fact, it is quite exceptional. In the more usual situation, the electrons released by oxidative reactions are collected directly or indirectly by an exogenous electron acceptor, in which case there is no need for the cell to reject valuable materials, and catabolic degradation of the substrate can proceed further. But now a dual flow of matter is necessary to maintain the flow of electrons.

Substances capable of serving as electron acceptor abound in nature—for instance, the sulfate ion, SO_4^{2-} (which can be reduced right up to the level of sulfur, S, or of hydrogen sulfide, H_2S); the ferric ion, Fe^{3+} (which is readily reduced to the ferrous state, Fe^{2+}); the nitrate ion, NO_3^- (which will go to nitrite, NO_2^-, and further, up to ammonia, NH_3); CO_2 (which can be reduced all the way to methane, CH_4); and even the simple proton, H^+ (which will yield hydrogen gas, H_2). The most widespread and efficient electron acceptor is molecular oxygen, O_2, which is reduced to water, H_2O (or occasionally to hydrogen peroxide, H_2O_2).

Every one of these substances, and many others, have been adopted as electron acceptor by some organism through the development of appropriate enzymes. Their reduction accounts for many of life's manifestations, including the stench of sulfurous fumes, the remodeling of ferruginous silts, the recycling of atmospheric nitrogen, and the mysterious emanations that send ghostly will-o'-the-wisps flitting across the surface of marshes. Between the bacteria responsible for these phenomena and the innumerable living organisms, including man, that respire atmospheric oxygen, there is a hidden common bond: they all support their energy-yielding metabolic oxidations with the help of an exogenous electron acceptor.

Here comes another important generalization. Not just glucose, but every possible kind of foodstuff used by a living organism to support its energy needs acts by supplying electrons to ATP-generating oxphos units. There simply is no other source of metabolic energy for heterotrophic organisms—those that feed on the products of the biosynthetic industry of other organisms (Greek *heteros*, other; *trophê*, food). The autotrophs (*autos*, self), also feed electrons into oxphos units, but from other sources.

"Burning" food for energy really means breaking down the foodstuffs and enriching them with oxygen at the expense of water, in such a way as to produce electrons that are fed into ATP-generating oxphos units from which they are collected by oxygen (or by some other acceptor).

As we will see when we visit the mitochondria, life has displayed remarkable ingenuity in the exploitation of this electron flow, intercalating up to four consecutive oxphos units on the pathway of metabolic electrons, many of which cascade down potential differences of 1 volt or more. As much as 80 per cent of the free energy released by the oxidation of foodstuffs may be retrieved in this way and used for the assembly of ATP. We will come back to this topic in Chapter 9.

Electron Flow: A Generalized View

Not all electron exchanges take place through ATP-generating oxphos units. Quite often, electrons are transferred across small potential differences, with little change in free energy. Occasionally, they may hurtle down a major potential difference, but without the kind of constraint that would allow the cell to make use of the energy released. Biological electrons resemble rivers in this respect. Like waterfalls, precipitous electron falls are infrequent, and not every one of them is harnessed to a power station.

This hydrodynamic image is useful, provided we keep in mind that electrons do not actually flow inside living cells in the way they do through an electric conductor; they are exchanged in discrete steps, between a reduced donor and an oxidized acceptor. The donor is oxidized in this transaction and thereby becomes able to act as acceptor to some other donor that occupies a higher potential level. The acceptor, on the other hand, having become reduced, can now serve as donor for an acceptor of lower potential level. Electrons tumble down in this way from one carrier to another until they reach their final acceptor, usually oxygen.

Electrons, therefore, do not flow down grades of variable slope, as rivers do most of the time. They fall down a succession of abrupt steps of variable heights. The electron-flow map of living cells resembles not so much a natural network of waterways as the kind of artificial system of interconnected reservoirs that seventeenth-century engineers constructed with such relish in the gardens of the rich. Instead of exploiting the terrain for esthetic enjoyment, however, natural selection has favored energy, utilizing the contour in such a way as to generate as many electron falls as possible that have the right height to power an oxphos unit.

The hydrodynamic analogy helps us appreciate an important aspect of electron transfer, which so far has been mentioned only in passing, namely the absolute level of potential at which electrons are either donated or ac-

The electron cascade. Each reservoir contains a redox couple. Connections between reservoirs are mediated by electron-transferring enzymes. Altitude of the reservoirs determines the direction of flow and height of the falls. Many electron falls of 300 mV or more (for electron pairs) are harnessed to energy-retrieving machines (oxphos units).

cepted. In our image, it corresponds to the altitude of the reservoirs, their height above sea level. Once you have this information, you can predict accurately, from the difference between their two altitudes, the direction of water flow between any two reservoirs, as well as the maximum work that can be obtained from the fall of a given quantity of water from the higher reservoir to the lower one (or, conversely, the minimum amount of work that must be accomplished to pump a given quantity of water up from the lower reservoir to the higher one).

The equivalent of altitude for electron reservoirs is the oxidation-reduction potential (in volts) of the relevant redox couples. In possession of that information, one can readily estimate the corresponding energy changes, as explained in Appendix 2. A more convenient way of evaluating the potential energy of electrons is to give directly the result of such a calculation—that is, the free energy of the reaction—for the particular case when one pair of electron-equivalents is transferred to oxygen with formation of water. We choose oxygen as acceptor because of its universal function as final electron acceptor for all aerobic organisms. Take, for example, the NADH/NAD^+ couple. We consider the reaction

$$NADH + H^+ + \tfrac{1}{2}\,O_2 \longrightarrow NAD^+ + H_2O$$

The free energy of this reaction, $\Delta G_{ox\ (NADH/NAD^+)}$, expressed in kilocalories per pair of electron-equivalents, is a direct measure of the maximum amount of work that can be obtained when electrons fall from the NAD reservoir all the way down to what, for most organisms, is their zero level of energy: water. It truly expresses the electron potential (not to be confused with the oxidation-reduction potential) of the NADH/NAD^+ couple.

Like all free-energy changes, electron potentials vary with the state of the system (Appendix 2). In the present example, the concentrations of NADH and of NAD^+, the hydrogen ion concentration (pH), the partial pressure of oxygen, and the temperature come into play in fixing the exact value of $\Delta G_{ox\ (NADH/NAD^+)}$. Obviously, we cannot make these fine adjustments and in fact lack the information to do so in most cases. All we can do is try to approximate as best we can the conditions that prevail in living cells. The ΔG_{ox} values estimated in this manner will be called "physiological" electron potentials, the quotation marks serving to remind us that we are dealing with approximate values subject to a certain amount of fluctuation, even under perfectly normal conditions.

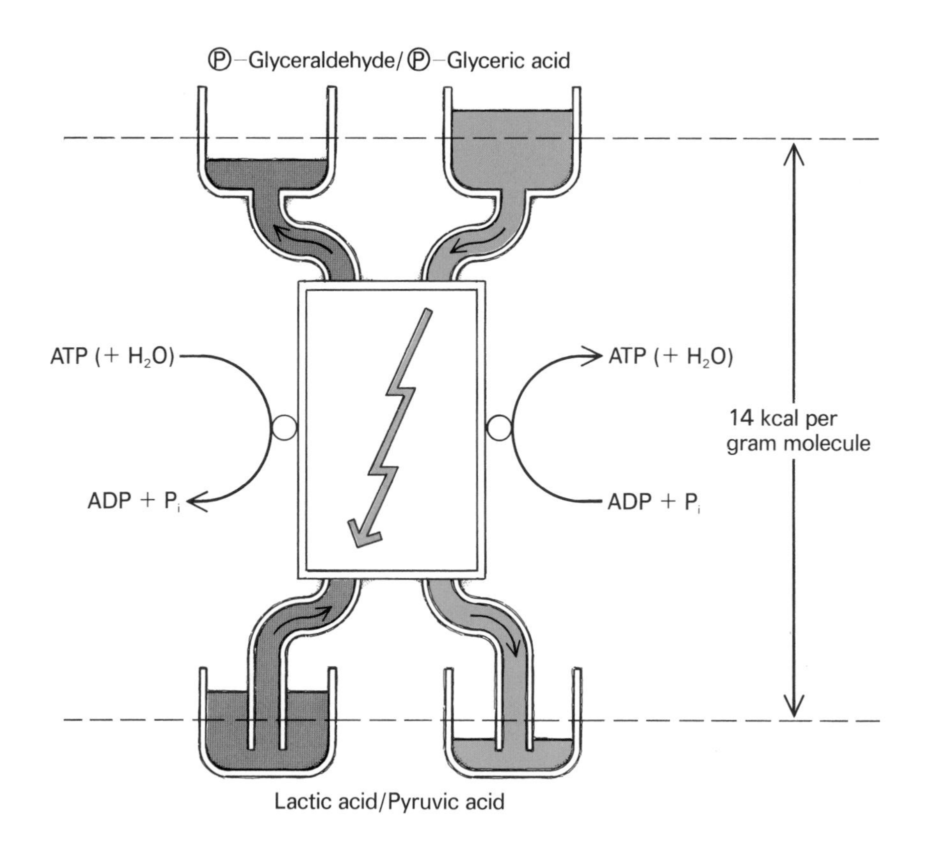

Graphic representation of electron flow through glycolytic chain. Under "physiological" conditions (dashed lines), the system is at equilibrium. The right-hand part (blue) depicts a situation in which downward electron flow, coupled to ATP formation, is elicited by an increase in electron potential of the phosphoglyceraldehyde/phosphoglyceric acid couple and a decrease in potential of the lactic acid/pyruvic acid couple. Such changes might result from corresponding changes in the ratio of the concentration of the reduced form to that of the oxidized form of each couple. The left-hand part (red) indicates a reverse situation in which upward electron flow, supported by ATP hydrolysis, is favored.

In glycolysis, the "physiological" electron potentials are of the order of −63 kcal per pair of electron-equivalents transferred to oxygen for the phosphoglyceraldehyde/phosphoglyceric acid couple and of −49 kcal per pair of electron-equivalents transferred to oxygen for the $NADH/NAD^+$ couple, as well as for the ethanol/acetaldehyde and lactic acid/pyruvic acid couples. These values, which are given negatively to indicate the exergonic nature of the reactions (see Appendix 2), indicate that glycolysis operates close to thermodynamic equilibrium. Between phosphoglyceraldehyde and NAD^+, the difference in potential is 14 kcal per pair of electron-equivalents, just enough to power the assembly of one gram-molecule of ATP. Between NADH and either acetaldehyde or pyruvic acid, the difference in potential is negligible. This means that the system is easily reversible and that the direction of electron flow depends on small perturbations. When glycolysis serves catabolically, as considered in this chapter, the level in the upper reservoir is somewhat higher, or that in the lower reservoir lower, than indicated, and the electrons can flow down and support ATP assembly. But, if the electron levels change in the opposite direction and ATP is supplied from another source, the flow of electrons is reversed and glycolysis has an anabolic role. This happens in liver, for instance, when carbohydrate is made from noncarbohydrate sources (gluconeogenesis), and in plants (with NADP instead of NAD), where the required ATP is provided by a light-powered mechanism (see Chapter 10).

Note that the cost of making ATP is no more constant than any other "physiological" free-energy value. It is itself subject to fluctuation, depending on the intracellular concentrations of ATP, ADP, and inorganic phosphate. If, for example, the concentration of ATP goes down and that of ADP goes up, as might occur in the course of heavy work, ATP formation will require less than 14 kcal per gram-molecule, and the equilibrium conditions of the glycolytic oxphos unit will be correspondingly altered. As we will see in Chapters 9 and 14, a fundamental regulating mechanism depends on this kind of interaction.

8 The Cytosol: Group Transfer and Biosynthesis

As ATP flows out of oxphos units, where does it go? Not an easy question to answer, for ATP rushes along hundreds of invisible trails—that is, diffuses down hundreds of concentration gradients—to wherever work is being performed and ATP consumed. Some of these trails lead to membranes, to which they bring fuel for transport mechanisms—for instance, the sodium-potassium pump. Others go to contractile fibers, to support mechanical work. Most of them, however, stop right here in the cytosol, with ATP becoming entangled with some local molecule. As a result of the scuffle, a piece of the ATP is appropriated by the encountered molecule. This liaison is usually short-lived and soon succumbs to a new collision. One or two additional affairs may follow. But eventually, this molecular group swapping comes to an end with the sealing of a stable bond between two building blocks used by the cell in the construction of its constituents.

Trailing ATP

Whatever their starting point, however circuitous their course, most cytosolic ATP trails lead to the same central biological function: biosynthesis. By charting them, we find the answer to a key question that was briefly evoked in the preceding chapter: How does a cell succeed in making thousands of different compounds with, as sole source of energy, the splitting of ATP to ADP and P_i, the central process that is repaired by the operation of oxphos units? Let us now define the problem in somewhat more precise terms.

Most biosynthetic reactions are dehydrating condensations between two molecular building blocks:

$$X—OH + Y—H \longrightarrow X—Y + H_2O$$

There are many different X's and many different Y's. They include the amino acids, which combine with each other to make proteins; the simple sugars, which associate into polysaccharides and other carbohydrate components; the mononucleotides, which polymerize into nucleic acids; the fatty acids, which join with glycerol and other alcohols to form lipids; as well as a host of other, more specialized, molecules. Putting them together correctly requires two conditions: information and energy.

Biosynthetic assemblies do not occur in haphazard fashion. They rely on the right kind of X becoming linked to the right kind of Y. As a rule, the instructions that allow the proper selection of biosynthetic partners are encoded in the specificity of the enzymes involved. When it comes to making the enzymes themselves, the instructions come from the genes, which close the circle by also providing the information for their own duplication. We will not consider this aspect of the problem further at present, because our tour will end with a detailed visit to the cytoplasmic network of information transfer and its controlling data centers in the nucleus.

The energy requirement of biosynthesis is explained by the fact that a dehydrating condensation cannot occur spontaneously in an aqueous medium. The overwhelming abundance of water drives the equilibrium of such a reaction far in the opposite direction, that of hydrolysis. To reverse the process, work must be performed, and therefore free energy must be supplied to the system from some outside source. In living cells, this source is represented in the last analysis by the hydrolysis of ATP, which, as noted, yields some 14 kcal per gram-molecule:

$$ATP + H_2O \longrightarrow ADP + P_i$$

Depending on the type of reaction, one or more molecules of ATP are consumed for every molecule of X—Y made. The overall free-energy balance is always negative, as required by energetics (see Appendix 2); most often markedly so, thereby making the biosynthetic process essentially irreversible under all conditions. But the question is: How is the energy transferred from one reaction to the other? Splitting the ATP first and then using the energy released by this process to join X with Y is not going to work. All we can get from the hydrolysis of ATP is heat—that is, random molecular motion, which cannot, under the conditions prevailing in living cells, be channeled to power a specific process. ATP splitting and X—Y formation must be *coupled,* so that one can provide the driving force for the other. The secret of this coupling is simple: never break a bond as such; always exchange one for another, by the mechanism of group transfer.

Group Transfer: Life's Second Golden Energy Gadget

Group transfer lies at the heart of biosynthesis. Its manifestations are infinitely varied and often highly involved, but its basic principle is remarkably simple. Essentially, it consists of the transfer of a chemical radical or group from a donor to an acceptor. We represent such a process schematically as follows:

$$A—B + C \rightleftharpoons [A\cdots B\cdots C] \rightleftharpoons A + B—C$$

in which A, B, and C each stand for some kind of molecular grouping, B in particular being the group transferred. As shown by this scheme, group transfer involves the participation of some sort of unstable ternary intermediate (shown between brackets), in which the group is transiently shared between its former and its new partner. It is a typical example of the eternal triangle at the molecular level: A—B forms a happy enough pair until C comes along and, after some fuzzy sharing of partners, takes off with B. Consistent with the villainous role of C, the reaction is also described as an attack by C on B or as a lysis (splitting) of A—B by C. (Example: hydrolysis, when water is the attacking agent.) The deprived victim of the attack, A, is called the leaving group. These roles are reversed when the reaction proceeds from right to left: A is

Group transfer requires two conditions: (1) the presence of an enzyme (transferase) capable of bringing molecule A–B and attacker C together in a position that allows the transfer of B; and (2) B must have more affinity for C than for A (i.e., the group potential of the B–C bond must be lower than that of the A–B bond).

the attacker or lytic agent, and C is the leaving group. The reaction remains a B transfer in both directions, making the term transfer the preferred designation.

For a drama of this sort to unfold in the human sphere, two conditions must be obeyed. First, there must be opportunity for the dramatis personae to interact in sufficiently intimate fashion. Next, B must have a greater penchant or affinity for C than for A, making the B—C bond stronger than the A—B bond. Group-transfer reactions—with due allowance for the distance between the world of humans and that of molecules—are subject to the same two conditions.

Opportunity is reflected in the kinetic condition. It generally requires the participation of a specific enzyme, or transferase, capable of bringing A—B and C (or B—C and A) close enough together to allow destabilization of the existing bond and formation of the ternary intermediate. Cells contain hundreds of such group transferases. Together with the electron transferases they make up more than 90 per cent of the total enzymic equipment of any living organism.

The second condition is thermodynamic. If the B—C bond is stronger than the A—B bond, more work must be done to break B—C than A—B, which is equivalent to saying that more free energy is lost when B binds to C than when it binds to A. On the energy scale, therefore, the B group lies lower in B—C than in A—B and, given the opportunity, will fall to the lower level. Just so in human relations; having fallen for one person does not always prevent one from falling more deeply for another. In the molecular world, however, the strength of the two bonds is not the only factor involved. The relative abundance of the four parties (A—B, C, A, and B—C) is also important. Molecular infidelities are mass events involving large numbers of individuals. A bond capable of resisting ten attackers may well yield to the assault of 10,000 because of the influence of concentration on chemical potential (see Appendix 2).

Most biological group transfers rely on what is known as a nucleophilic attack, by which is meant that the attacking agent has an affinity for positively charged radicals (the atomic nucleus is positively charged). Electrophilic attacks are rarer, except in their most naked form of electron transfer.

Nucleophilic attacks are generally perpetrated by negatively charged ions or by their protonated counterparts. In the former case, the reaction is particularly simple:

$$\mathrm{A{-}B} + \overset{\ominus}{\mathrm{C}} \rightleftharpoons [\overset{\ominus}{\mathrm{A}} \cdots \overset{\oplus}{\mathrm{B}} \cdots \overset{\ominus}{\mathrm{C}}] \rightleftharpoons \overset{\ominus}{\mathrm{A}} + \mathrm{B{-}C}$$

If protonated reactants participate, protons are exchanged with the medium, as in electron transfer:

$$\mathrm{A{-}B} + \mathrm{CH} \underset{\mathrm{H^+}}{\longleftrightarrow} [\overset{\ominus}{\mathrm{A}} \cdots \overset{\oplus}{\mathrm{B}} \cdots \overset{\ominus}{\mathrm{C}}] \underset{\mathrm{H^+}}{\longleftrightarrow} \mathrm{AH} + \mathrm{B{-}C}$$

The unstable ternary intermediate has the same structure in both formulations. It consists of two negatively charged groups vying to share an electron pair with the same positive radical B^+:

$$[\overset{\ominus}{\mathrm{A}}\!:\cdots \overset{\oplus}{\mathrm{B}} \cdots :\!\overset{\ominus}{\mathrm{C}}]$$

The winner, as we have seen, is the one that accepts the group at the lower energy level. It is obviously very useful to know the energy level occupied by a given group in its various combinations, just as it is to know the energy level occupied by electrons, since one can then predict the spontaneous direction of the group's transfer between any two partners and, at the same time, evaluate the maximum amount of work that can be powered by this transfer (which is also the minimum needed to reverse the transfer).

The most convenient way of evaluating the energy level of a transferable group in a given combination is by the free energy of hydrolysis, ΔG_{hy}, of that combination—that is, the free energy released when the group is attacked by water or by a hydroxyl ion, OH^-. For example:

$$\mathrm{A{-}B} + \mathrm{H_2O} \longrightarrow \mathrm{AH} + \mathrm{B{-}OH}$$

or:

$$\mathrm{B{-}C} + \mathrm{OH^-} \longrightarrow \mathrm{CH} + \mathrm{B{-}O^-}$$

The free energies of these hydrolysis reactions are called group potentials. Estimated for "physiological" conditions, approximating those prevailing in living cells, the group potentials effectively measure the energy levels of transferable groups, just as the electron potentials do for transferable electron pairs. In both cases, we measure these levels with respect to the most evident natural base line: H_2O (or OH^-) as acceptor for group potentials; O_2 (to H_2O) as acceptor for electron potentials. (Note the central role of water.)

According to the convention above, the "physiological" values of $\Delta G_{hy(A-B)}$ and $\Delta G_{hy(B-C)}$ represent the group potentials of the B group in its combinations A—B and B—C, respectively. As explained in Appendix 2, the "physiological" free energy of the B transfer between A—B and C is then readily computed from the difference between the two group potentials:

$$\Delta G_{transfer} = \Delta G_{hy(A-B)} - \Delta G_{hy(B-C)}$$

If the (negative) potential of A—B is greater in absolute value than that of B—C, the ΔG of the transfer is negative: B falls from a higher energy level in A—B to a lower one in B—C; its transfer from A—B to C can occur spontaneously. In the opposite case, this transfer is endergonic, and free energy must be supplied if B is to be lifted from its lower level in A—B to its higher level in B—C.

Group potentials also serve conveniently for evaluating the free energies of coupled biosynthetic reactions. If, for example, the dehydrating assembly of one molecule of X—Y (see p. 120) is supported by the hydrolysis of n molecules of ATP to ADP + P_i, the following relationship is valid, irrespective of the mechanism of coupling:

$$\Delta G_{biosynthesis} = n\,\Delta G_{hy(ATP \rightarrow ADP)} - \Delta G_{hy(X-Y)}$$
$$= n(-14) - \Delta G_{hy(X-Y)}$$

As first pointed out by Fritz Lipmann, one of the founders of modern bioenergetics, the bonds found in natural substances fall roughly into two classes: high-energy bonds, represented by a "squiggle" (~); and low-energy bonds, represented by a simple line (–). The terminal phosphate bond of ATP is the archetype of high-energy bonds. Many of the bonds found in natural constituents (ester, amide, peptide, glycoside) are low-energy bonds, with "physiological" group potentials of some –6 to –8 kcal per gram-molecule. It is the difference between the two that allows the hydrolysis of ATP to power biosynthetic assemblies. As to how this is actually accomplished, the answer is: with the help of Janus.

Introducing Janus, the Double-Headed Intermediate

The ancient Romans had a god named Janus—the month of January is dedicated to him—who was believed to have two faces, one looking into the past, the other into the future. Biochemistry has rediscovered Janus as the coupling demon of biosynthesis. It arises from a nucleophilic attack by an oxygen-containing building block (X—OH or X—O$^-$) on ATP or some related energy-rich molecule, which we will designate provisionally as A—B, in order to avoid going into complex chemistry. The attack is spearheaded by the oxygen atom:

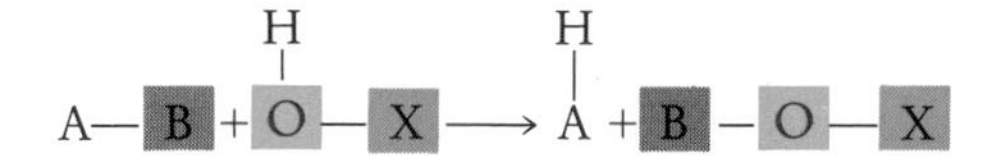

This reaction is a transfer of the B$^+$ radical (B-yl group) from A—B to X—O$^-$. What gives B—O—X its double-headed character is that it can also engage in a transfer of the X$^+$ radical (X-yl group); for instance, when attached by building block Y—H or Y$^-$:

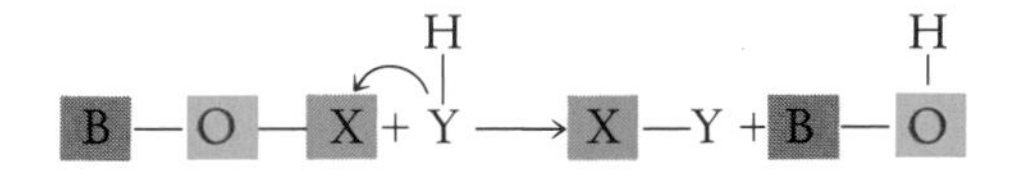

Now watch what happens when the two reactions are allowed to proceed sequentially:

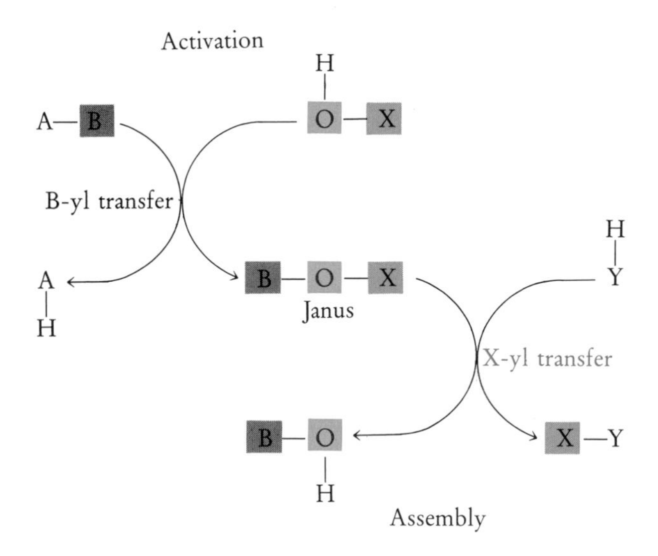

The half-reaction on the left of the above diagram reads (follow both left-pointing arrows):

A—B ⟶ A—H + B—OH

That on the right:

X—OH + Y—H ⟶ X—Y

We observe hydrolysis of A—B and dehydrating condensation of X—Y. But water is nowhere to be seen. It is transferred in hidden form from X—OH to B—OH, through the central oxygen of Janus, while simultaneously a greater or lesser portion of the group potential of A—B is retrieved in the X—Y bond. Janus thus serves both as conveyer of group-linked energy and as bearer of hidden water; it is Mercury and Aquarius all in one, if such freedom may be taken with mythology.

One can readily verify this mechanism experimentally by providing the cell with X—OH molecules labeled with the heavy isotope of oxygen ^{18}O and analyzing the products of the reaction with a mass spectrograph. The ^{18}O is found in B—OH, not in water, as it would be if the reaction were an authentic dehydrating condensation.

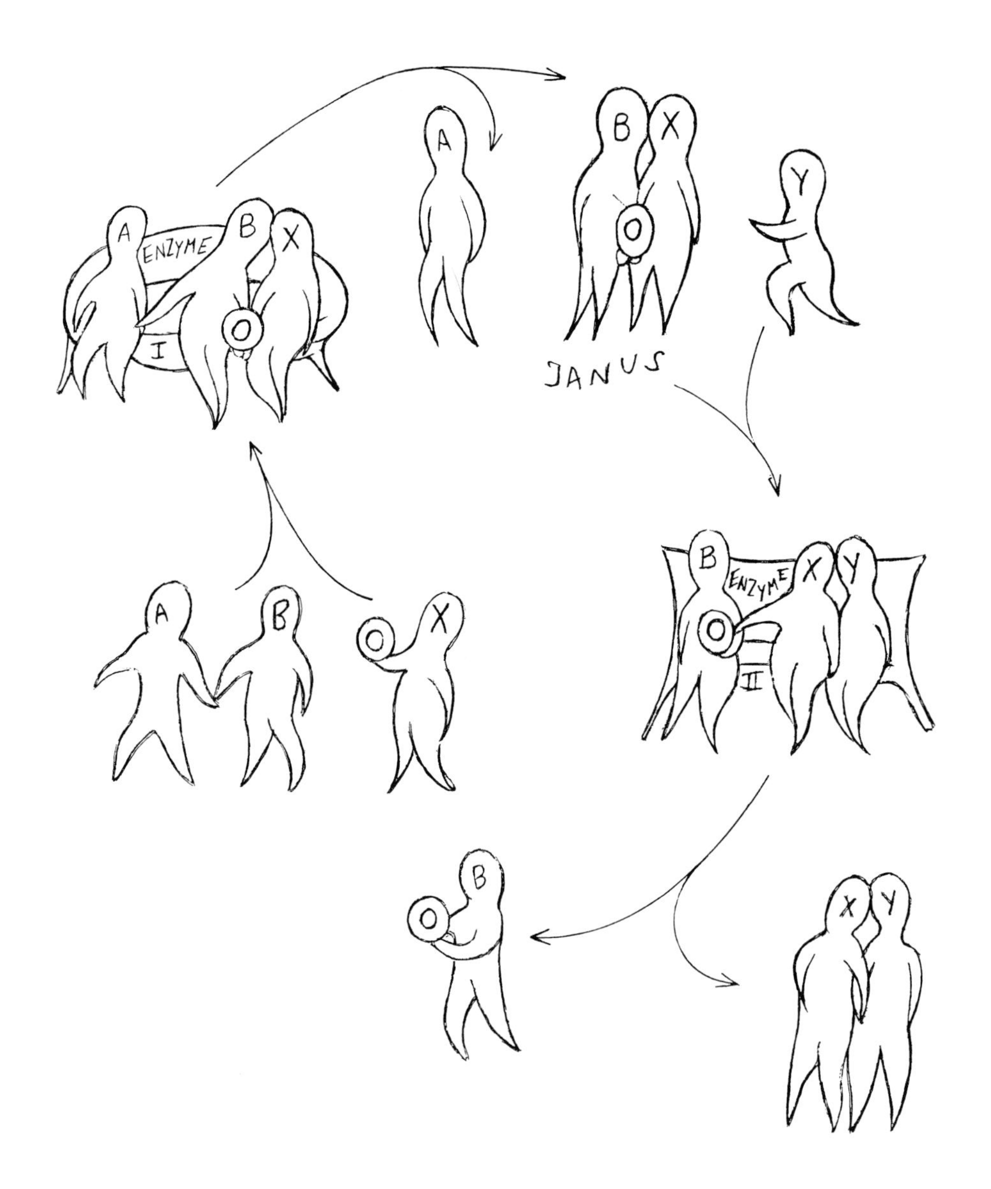

Sequential group transfer. In the first reaction, B is separated from A by an oxygen-proferring attacker X. But new attacker Y comes on the scene, removing X from B, which is left holding the oxygen by which it was tempted. The B–O–X intermediate is double-headed Janus. It consists of the two transferable groups participating in reactions I and II, joined by the oxygen atom that changes hands in the transaction.

Such, basically, is the mechanism of biosynthesis. It takes many guises. But fundamentally it always relies on sequential group transfer, linked by a double-headed intermediate. The essential anatomical features of this intermediate are two transferable groups connected by a central oxygen atom. When approached from the left, it offers an energy-rich B-yl group attached to an X—O$^-$ carrier. When seen from the right, it appears just as convincingly as an X-yl group proferred by a B—O$^-$ carrier.

According to this general scheme, biosynthesis always proceeds in at least two steps, connected by Janus. The first step, which depends on some sort of group transfer from the energy donor (ATP or some related molecule), serves to lift the X-yl group from its zero level of energy (X—OH) to the high-energy level it occupies in Janus. This step is called activation. The final step, or assembly, in which the X-yl group is transferred to its natural acceptor Y, proceeds downhill, from the high-energy activation level to the X—Y level.

In an important variant of this basic two-step mechanism, the Janus intermediate donates the activated group to a carrier, which itself transfers it to the final biosynthetic acceptor, as shown in the sequence of reactions on the facing page.

Janus, the double-headed intermediate, consists of two transferable groups linked by a central oxygen. Attacked on the left, it yields the B^+ group; on the right, the X^+ group. In each case, the remaining group is left with the oxygen.

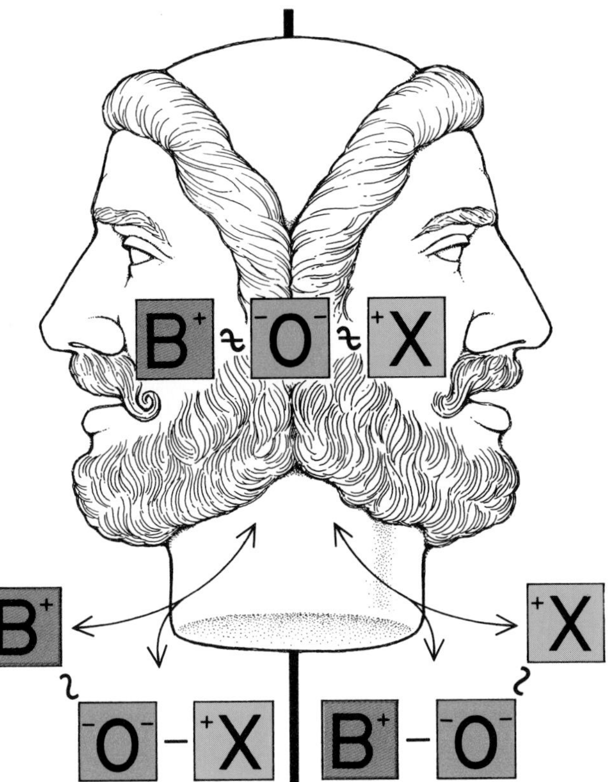

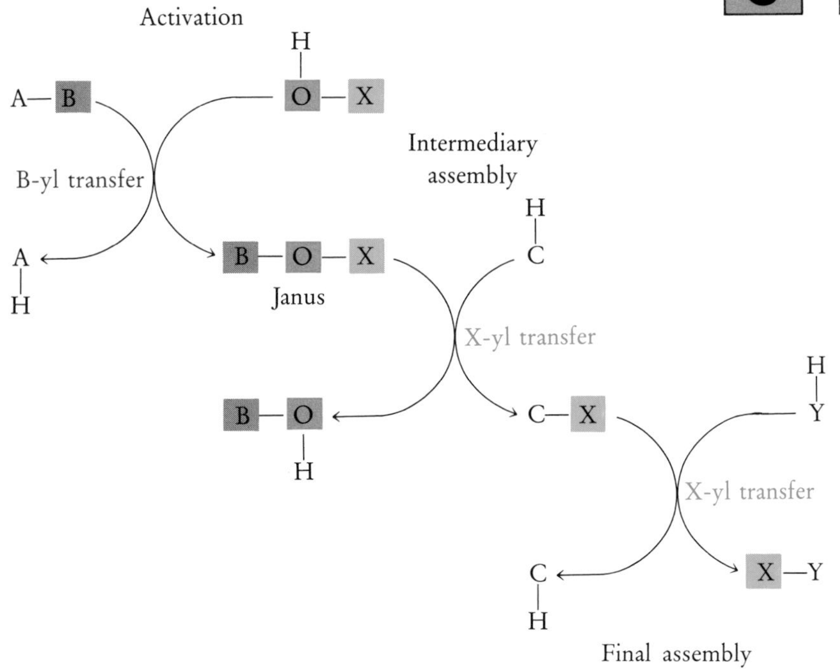

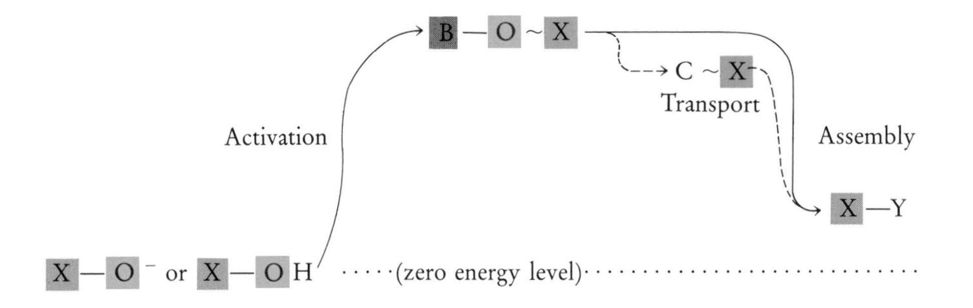

In such three-step mechanisms, the X-yl group remains at a relatively high energy level in its combination with the carrier, leaving the main energy drop to occur at the final assembly step (shown above), as befits a process in which a stable product is to be formed. The cell thus operates very much like a builder who first hoists his materials high up with a crane, moves them around in a horizontal plane, and finally lowers them into place.

In a number of cases, activation and assembly are carried out by the same enzyme, which then catalyzes some sort of concerted process in which the Janus intermediate remains enzyme-bound. Such bifunctional enzymes are called synthetases or ligases (Latin *ligare*, to bind). Without them, many biosynthetic processes would be very inefficient or even could not occur at all. Janus intermediates are often highly unstable molecules that would not survive very long if they were let loose. Frequently also, they are compounds of such high group potential that they cannot, with the coupled splitting of the group donor A—B as sole source of energy, be produced at a concentration high enough to permit efficient diffusion between two physically separated enzymes. Keeping such intermediates enzyme-bound and strategically placed so that they can immediately be trapped by the exergonic assembly process as they are formed helps overcome these difficulties.

On the other hand, it is often very useful to the cell, sometimes even indispensable, to have the two steps of biosynthesis take place at different sites. Activation requires energy and the participation of ATP; it is usually carried out in the cytosol or in some site closely connected with the cytosol and as amply supplied with ATP—for instance, the cytosolic face of a membrane. Assembly, on the other hand, frequently depends on an accurate supply of information, which is more readily secured on a structured substratum, such as is provided by the ribosomes in the synthesis of proteins (see Chapter 15) or by chromosomal scaffoldings in that of nucleic acids (see Chapters 16 and 17). Another advantage of the physical separation of the two steps is that it allows centralization. A single activation reaction suffices for each X-yl group, which can then be transported in ready-for-use form to any number of assembly sites. In actual fact, the economy is even greater: the cell often makes use of the transport phase to modify or otherwise process the X-yl group in various ways, so that a single activation reaction may serve to energize several biosynthetic building blocks that are chemical modifications of each other.

When assembly is separated from activation, a stable transport form of the activated building block is needed, and the activation step must be sufficiently exergonic in itself to produce appreciable concentrations of it. A number of Janus intermediates answer these requirements; their B—O^- part then acts as carrier of the X-yl group. In other instances, the requirements are met thanks to the use of special carriers. Almost invariably in such cases, activation of the building block and its attachment to the carrier are accomplished by a single, ligase-type enzyme. Several important coenzymes function as group carriers.

The role of activation as a prerequisite to metabolic processing is not restricted to biosynthesis. In fact, a considerable part of metabolism requires prior activation of the substrate. This is even true of many catabolic reactions, as was illustrated earlier by the example of glycolysis. The group carriers thus also serve as handles whereby the attached molecules are presented to their modifying enzymes.

But the time has come to put some chemical flesh around the bare bones of schematic abstractions. In the organization of this tour, an effort has been made to bypass chemical details as much as is feasible. But there is a limit to what can be understood of an essentially chemical

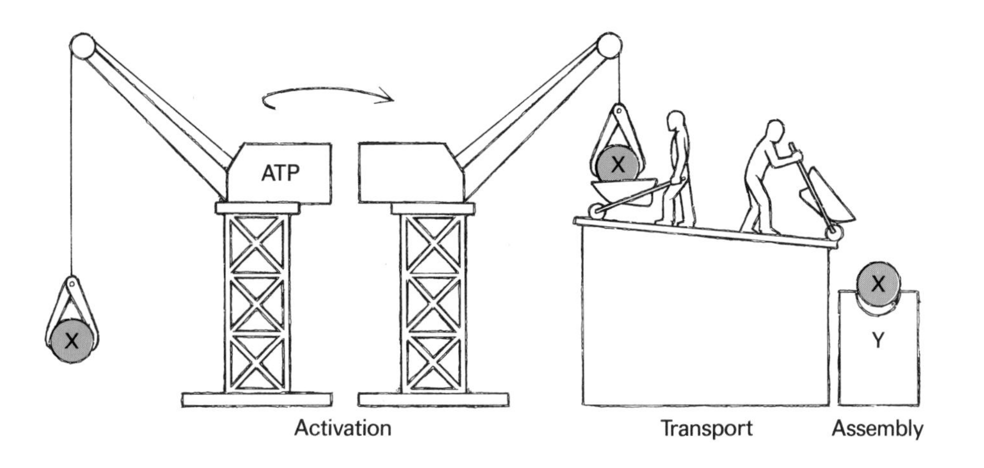

In biosynthesis, the activation step serves to lift building block X to a high-energy level (group potential) with the help of ATP. In the assembly step, the group is transferred to final acceptor Y, with a distinct drop in potential. In a number of cases, the activated group is first transferred, with little drop in potential, to a carrier C, from which it is then transferred to its final acceptor Y.

machine without the language of chemistry. Those who find the next part of the visit too arduous should, however, not lose heart. Even if they skip much of it, they should still be able to catch up with us later without too much difficulty. On the other hand, those with a better grounding in biochemistry may derive some enjoyment, and perhaps some illumination, from the kind of bird's-eye view of biosynthesis that will be provided. Let everyone tag along, therefore, and stay with the group.

The Source of Group-Transfer Energy

In the abbreviation ATP, A stands for adenosine. It is a nucleoside, which is defined as the combination of a base—adenine in the present case—with carbon atom number 1 of ribose, a 5-carbon sugar, or pentose (see Appendix 1). Three other important bases engage in similar nucleosidic combinations with ribose: guanine, which, like adenine, belongs to the group of purines, and cytosine and uracil, members of the pyrimidine family. The corresponding nucleosides are called guanosine (G), cytidine (C), and uridine (U).

We will consider the detailed structure of the bases when we look at the anatomy of nucleic acids and at the genetic language. In the meantime, let us concentrate on the other end of the nucleoside molecule, which is occupied by carbon atom number 5 of ribose (numbered 5′ to distinguish it from carbon 5 of the base). This carbon bears a hydroxyl group OH, which in most natural combinations of nucleosides carries a phosphoryl group. Such nucleoside monophosphates are called nucleotides; they are designated as adenylic, guanylic, cytidylic, or uridylic acid or by the abbreviations AMP, GMP, CMP, UMP, in which MP stands for monophosphate.

To this terminal phosphoryl group of the nucleotides, one or two additional phosphoryl groups may become attached by the kind of linkage found in pyrophosphoric acid (pyrophosphate bond) to produce the nucleoside diphosphates ADP, GDP, CDP, and UDP, and the nucleoside triphosphates ATP, GTP, CTP, and UTP.

Base	Nucleoside	Nucleoside monophosphate	Nucleoside diphosphate	Nucleoside triphosphate
Adenine	Adenosine (A)	Adenylic acid (AMP)	ADP	ATP
Guanine	Guanosine (G)	Guanylic acid (GMP)	GDP	GTP
Cytosine	Cytidine (C)	Cytidylic acid (CMP)	CDP	CTP
Uracil	Uridine (U)	Uridylic acid (UMP)	UDP	UTP

In summary, representing a nucleoside by the symbol N (which stands for A, G, C, or U) and making explicit its 5′-hydroxyl group, we have:

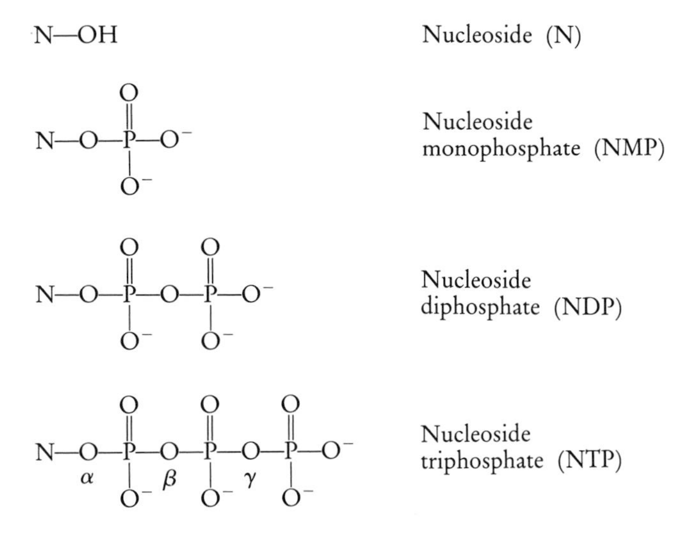

These NTPs are really super-Janus types of molecules, triply double-headed. Their three phosphoryl-bond oxygens (α, β, γ) each separate a distinct pair of transferable groups. This character makes the NTPs susceptible—at least theoretically—to as many as six distinct types of nucleophilic attacks, which will be designated α_p, α_d, β_p, β_d, γ_p, and γ_d, in which α, β, γ stand for the bond under attack, and the subscripts p and d for proximal and distal (with respect to N):

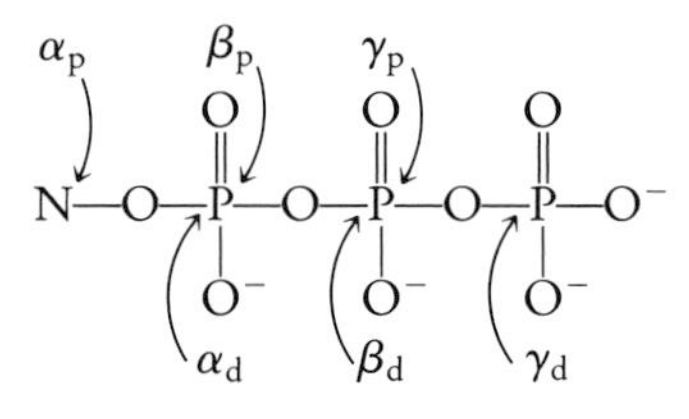

In practice, these possibilities (which are displayed explicitly on the facing page) are exploited very unequally. With the exception of a few rare α_p or β_d approaches, all biosynthetic attacks on NTPs are either β_p or γ_d. As far as is known, α_d or γ_p attacks are never used.

Most biosynthetic processes can be classified as regular two- or three-step mechanisms dependent on one of the above attacks. As might be expected in such a complex field, the main theme occasionally undergoes some variations. But to the cognoscenti—which, by now, should include all the cell tourists who are still with us—it remains easily recognizable. One seemingly atypical variant occurs when an NTP acts in Janus capacity, as donor of a group in a final assembly process. Such single-step mechanisms disobey the two-step rule only in appearance. The group donated by the NTP still required prior activation—by group transfer from some other NTP, or by the operation of oxphos units, or by both—before it could be transferred exergonically. In such cases, the activation step coincides with the repair phase of the ordinary two- or three-step mechanisms (see pp. 130–131).

The group potentials that are brought into play in the different types of attacks on NTPs are not equivalent. As already mentioned, the "physiological" free energy of hydrolysis of the γ bond of ATP is of the order of -14 kcal per gram-molecule. All terminal phosphoryl groups in NTPs and in NDPs have the same group potential. In contrast, the "physiological" free energy of hydrolysis of the β bond of NTPs is considerably higher, partly because of a difference in standard free energy of hydrolysis (about 3 kcal per gram-molecule) and, more importantly, because most cells contain highly active pyrophosphatases that hydrolyze inorganic pyrophosphate as it arises. Therefore, hydrolysis of the β bond is followed by hydrolysis of the pyrophosphate formed:

$$\begin{array}{llll} \text{NTP} + H_2O & \longrightarrow & \text{NMP} + \text{PP}_i & \Delta G_{\text{hy}(\beta\ \text{bond})} \\ \text{PP}_i + H_2O & \longrightarrow & 2\ \text{P}_i & \Delta G_{\text{hy}(\text{PP}_i)} \\ \hline \text{NTP} + 2\ H_2O & \longrightarrow & \text{NMP} + 2\ \text{P}_i & \Delta G_{\text{hy(total)}} \end{array}$$

The same end result can be achieved by hydrolyzing the γ bond first, and then the β bond, in which case the "physiological" free energies of the reactions are known:

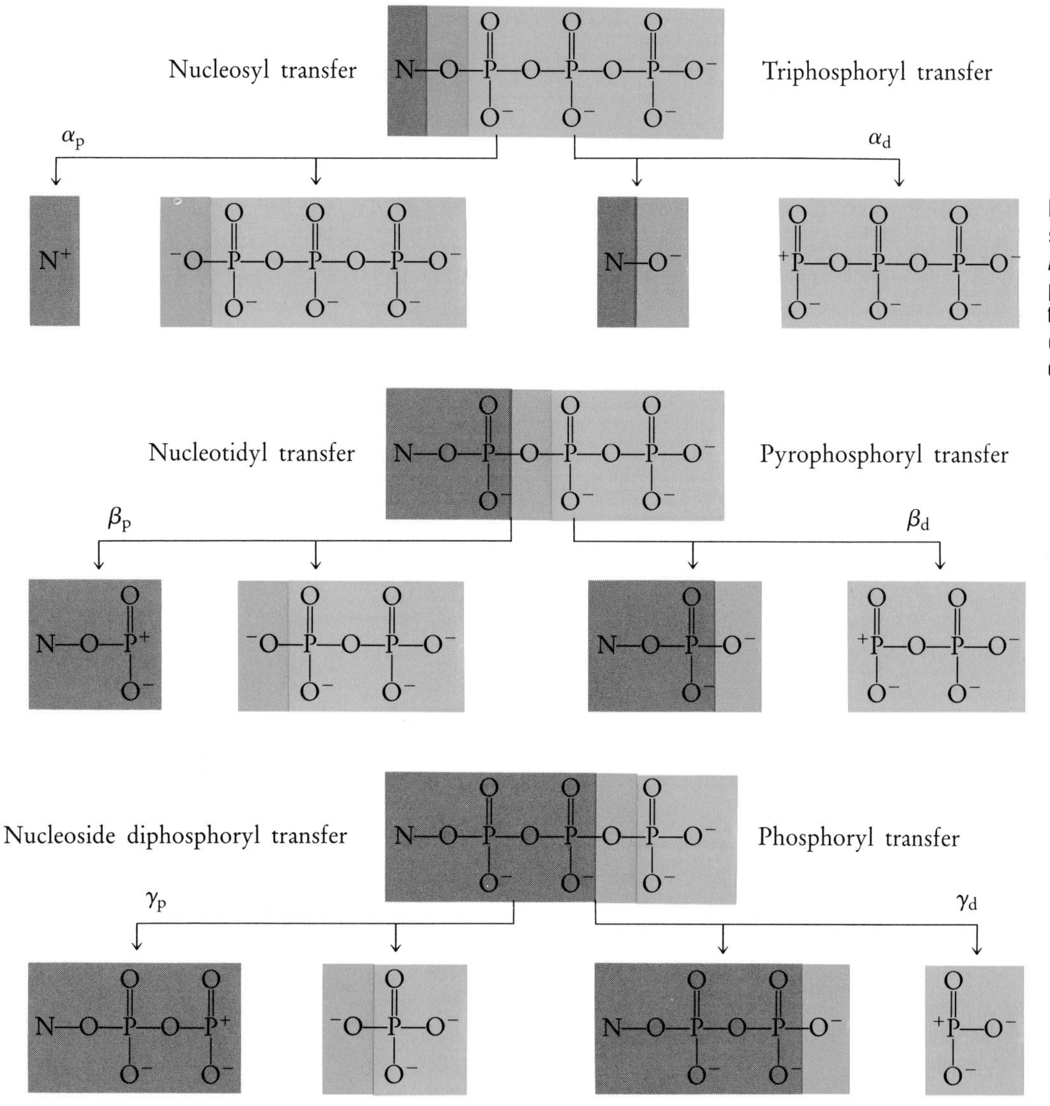

Diagrammatic representation of the six group-transfer reactions in which ATP and other NTPs may theoretically participate, depending on which of the three phosphoryl-connecting oxygens, α, β, or γ, serves as the central Janus oxygen.

$NTP + H_2O \longrightarrow NDP + P_i$	-14 kcal per gram-molecule
$NDP + H_2O \longrightarrow NMP + P_i$	-14 kcal per gram-molecule
$NTP + 2\ H_2O \longrightarrow NMP + 2\ P_i$	-28 kcal per gram-molecule

Whether we start with the β bond or with the γ bond, the total free-energy change associated with hydrolysis of the two bonds must be the same. Therefore, by elementary bookkeeping rules:

$$\Delta G_{\text{hy}(\beta\ \text{bond})} = -28\ - \Delta G_{\text{hy(PP}_i)}$$

The "physiological" free energy of hydrolysis of the β bond of NTPs depends on the value of $\Delta G_{\text{hy(PP}_i)}$—that is, on how close to equilibrium the action of pyrophosphatases maintains the concentrations of PP_i and P_i. This is not known with any accuracy. But it is a fair assumption, in view of the high activity of the enzymes, that a state very near equilibrium is maintained—in other words, that $\Delta G_{hy(PP_i)} \simeq 0$. Accordingly, we will adopt for the "physiological" free energy of hydrolysis of the β bond the (maximal) value of −28 kcal per gram-molecule. It is worth noting in this connection that some microorganisms do not maintain a very low pyrophosphate concentration but operate with a pyrophosphate-based economy instead. Apparently, the extra energy expenditure imposed on β mechanisms by pyrophosphatase action is not a vital necessity.

As to the α bond, its "physiological" free energy of hydrolysis is of the order of −7 kcal per gram-molecule in NMP. For the reason just mentioned (hydrolysis of PP_i), it approaches −21 kcal per gram-molecule in NDP, and −35 kcal per gram-molecule in NTP (extending the reasoning to PPP_i).

The table below summarizes the values of group potentials that will be used in our subsequent analyses. It will be remembered that these values are subject to fairly wide fluctuations, depending on the conditions prevailing in the cells (see Appendix 2). But they suffice to help us understand the main energetic features of biosynthetic mechanisms.

Hydrolysis reaction	"Physiological" ΔG_{hy} (kcal per gram-molecule)
$NTP \longrightarrow NDP + P_i$	−14
$NDP \longrightarrow NMP + P_i$	−14
$NMP \longrightarrow N + P_i$	−7
$NTP \longrightarrow NMP + PP_i\ (2\ P_i)$	−28
$NTP \longrightarrow N + PPP_i\ (3\ P_i)$	−35
$NDP \longrightarrow N + PP_i\ (2\ P_i)$	−21
$PP_i \longrightarrow 2\ P_i$	~0
$PPP_i \longrightarrow PP_i + P_i\ (3\ P_i)$	~0

When a bond in an NTP has been sacrificed for the benefit of biosynthetic work, it must be repaired. If the bond is the γ bond of ATP, some oxphos unit takes care of the repair. If any other, it is repaired at the expense of one or more γ bonds of ATP, thanks to the occurrence of transphosphorylating enzymes that catalyze the following reactions:

$$ATP + N \longrightarrow ADP + NMP$$

$$ATP + NMP \rightleftharpoons ADP + NDP$$

$$ATP + NDP \rightleftharpoons ADP + NTP$$

The first reaction is irreversible because of the large difference in "physiological" free energy of hydrolysis between the γ bond of ATP and the α bond of NMP. The other two reactions exchange bonds of equal energetic value and are freely reversible. The cost of these transfers is itself borne by the operation of oxphos units, which, therefore, end up paying the full energy bill. Note, however, that this bill covers only that part of the biosynthetic work—often the major one, or even the only one, but not always—that depends on group transfer. Other processes, especially electron transfer from high-potential donors, also may come into play. Biosynthetic reductions are particularly important in autotrophic organisms (see Chapter 10).

Putting the various repair reactions together, we end up with the condensed diagram at the top of the facing page, which henceforth will be referred to as the central repair machinery. This machinery also provides for the activation of such building blocks—for example, P_i or an NMP—as are donated in single-step biosynthetic processes. Note that PP_i cannot be incorporated in an NTP as such, but must first be hydrolyzed. (The same is true for PPP_i, not shown on the diagram because its appearance, if it occurs at all, is very rare and fleeting.)

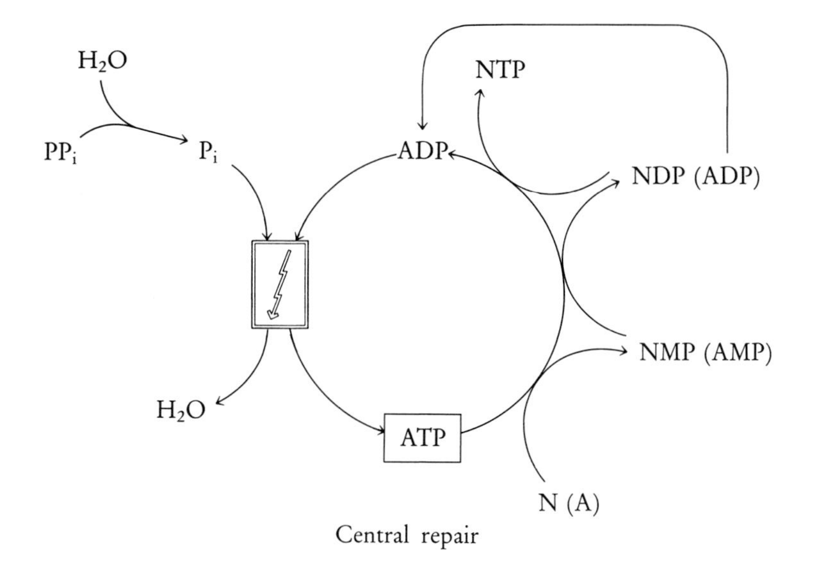

Central repair

An Optional Excursion

We have arrived at a point in our tour where many participants are probably eager for a well-deserved break. Some, however, may wish to put their newly acquired knowledge to a test. A side excursion has been arranged for these more adventurous tourists. It can be skipped by the others. It will take us browsing through a biochemistry textbook and help us recognize a few simple basic patterns behind the enormous complexity and infinite diversity of the chemical mechanisms whereby living cells manage to construct thousands of substances, most of which are still beyond the possibility of synthesis by our most advanced technology. During this excursion, the participating building blocks will be represented systematically in ionized form (X—O^- and Y^-), unless they are known to be protonated. The movement of protons has been indicated when necessary. Detailed chemical structures will not be shown. They can be found in Appendix I.

Single-Step Processes

In these reactions, some part of an NTP is transferred to a final biosynthetic acceptor. The transferred group arises from a precursor, usually P_i or an NMP, which may be seen as the X—O^- building block of our general two-step scheme, previously activated and incorporated into the NTP by the operation of what has been called the central repair machinery. The donating NTP has the character of a double-headed Janus intermediate.

Reactions dependent on γ_d transfer. The NTP involved is almost invariably ATP and acts as an NDP carrier bearing a phosphoryl group:

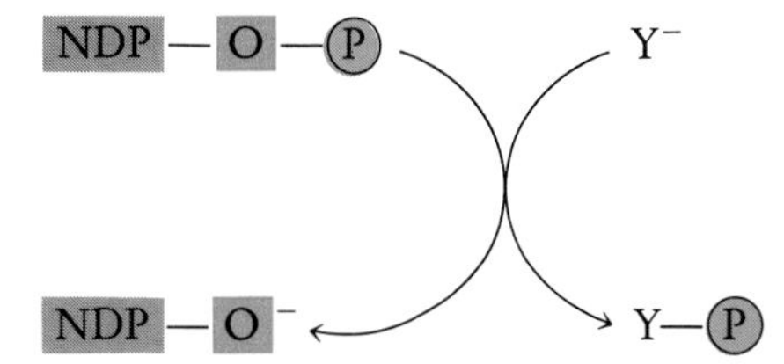

Most of the phosphorus contained in natural substances—a diversified group that includes the nucleoside phosphates, several coenzymes, the nucleic acids, the phospholipids, and numerous metabolites—first enters into its combinations by this kind of reaction. Exceptions are the terminal phosphoryl group of ATP itself, which is incorporated by oxphos action, and a number of cases in which inorganic phosphate attacks a pre-existing bond (phosphorolysis).

The cost of the biosynthetic transaction is readily computed:

$$\Delta G_{\text{biosynthesis}} = -14 - \Delta G_{\text{hy(Y—P)}}$$

Common phosphate esters, including NMPs, have "physiological" group potentials of about −6 to −8 kcal per gram-molecule. The free-energy loss associated with their assembly thus varies between 8 and 6 kcal per gram-molecule, which suffices to make the reaction essentially irreversible. A number of other phosphate compounds, however, have "physiological" group potentials of the order of -14 kcal per gram-molecule, which makes the transphosphorylation with ATP freely reversible ($\Delta G \simeq 0$). Among them are all the NDPs and NTPs, which, as we have seen, can transphosphorylate freely with ATP. Thanks to these reactions, any NDP or NMP that forms is immediately reactivated to NTP, ready for use in a new biosynthetic process (see pp. 130–131). Conversely, in times of acute demand for ATP, as at the onset of muscular contraction, cells may call on their NTPs and on their NDPs (including ADP) to help restore the consumed ATP by reversal of the transphosphorylation reactions. The brunt of this responsibility, however, falls on another group of high-energy compounds, called phosphagens, characterized by an amidophosphate linkage. The phosphagen of vertebrates is creatine phosphate, which arises from creatine by a γ_d type of phosphoryl transfer from ATP:

$$\text{ATP} + \text{Creatine} \rightleftharpoons \text{ADP} + \text{Creatine}{\sim}\text{P}$$

The equilibrium of this reaction is such as to favor ATP formation. Only when the ratio of ATP to ADP concentration is sufficiently high, as it is in cells that are not subjected to an energy stress, is the left-to-right direction favored: the creatine phosphate reservoir is restored. As soon as ATP starts being consumed and the ADP level rises, the right-to-left direction becomes the favored one and creatine phosphate serves to regenerate ATP from ADP. This tides the cell over the period needed to get oxphos units in full action (see Chapter 14).

The enzymes that catalyze γ_d transphosphorylation reactions from ATP are called phosphokinases or, more simply, kinases (Greek *kinein*, to move). In addition to serving in many biosynthetic processes and in energy metabolism, phosphokinases also play an important role as primers of catabolic reactions. This is how their existence was discovered, after it was found that glucose needs to be activated by a "hexokinase" to enter the glycolytic chain (Chapter 7). In Chapters 13 and 18, we will encounter a special group of phosphokinases acting on proteins. They control a number of central regulatory processes, including those that govern cell division, and could thereby be implicated in the anarchic multiplication of cancer cells.

Reactions dependent on β_d transfer. The NTP involved in this rare type of reaction acts as an NMP carrier bearing a pyrophosphoryl group (previously assembled from P_i and activated by the central repair machinery):

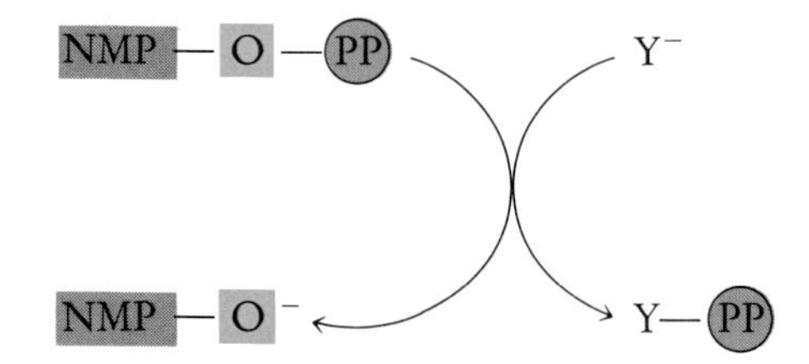

The most important acceptor of the pyrophosphoryl group is thiamine, or vitamin B_1, which happens to be the first vitamin discovered. Thiamine pyrophosphate (TPP) is an important coenzyme of decarboxylation reactions. It costs all of 28 kcal to make a pyrophosphoryl group from P_i. Part of this energy is conserved in TPP, to be dissipated only when the PP group is hydrolyzed off. There is, however, a considerable difference, of the order of -8 to -10 kcal per gram-molecule, between the free energy of hydrolysis of the pyrophosphate bond linking PP to AMP in ATP and that of its ester attachment to thiamine in TPP. This is more than enough to make the transpyrophosphorylation entirely irreversible.

Reactions dependent on β_p transfer. Here, the NTP is to be seen as an activated NMP-yl group offered by a pyrophosphate carrier, as shown at the top of the left-hand column on the facing page. The energetic contribution of pyrophosphatase is made at the initial transfer step, which therefore has available close to 28 kcal per gram-molecule for making the NMP—Y bond.

Most processes that involve the incorporation of nucleotidyl groups into stable biosynthetic products take

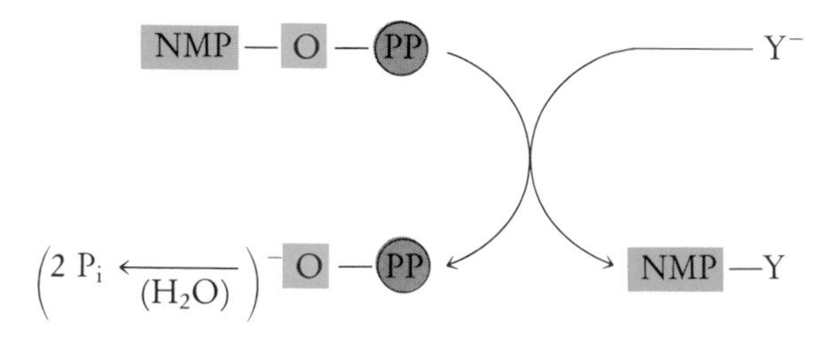

place by β_p transnucleotidylation. They include the fundamental processes whereby RNA (see Chapter 16) and DNA (see Chapter 17) are assembled, as well as reactions involved in the synthesis of NAD and of other coenzymes, such as NADP, FAD, and coenzyme A (see pp. 142–143), which likewise contain AMP. Similarly to phosphorylation, adenylylation plays an important role in the regulation of some enzymic proteins.

In a special, self-attacking variant of β_p transfer, the attacking agent is the internal 3′-hydroxyl group of the transferred nucleotidyl group itself. The most important such reaction, catalyzed by adenylate cyclase, leads to the formation of 3′,5′-cyclic AMP (cAMP), an important intracellular mediator of hormone action (see Chapter 13):

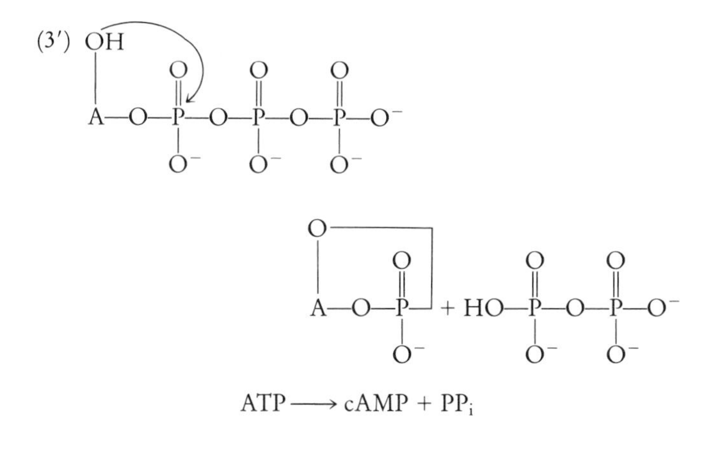

Many of the bonds made by β_p transfer are high-energy bonds. Nevertheless the transfer is completely irreversible in vivo, thanks to pyrophosphatase.

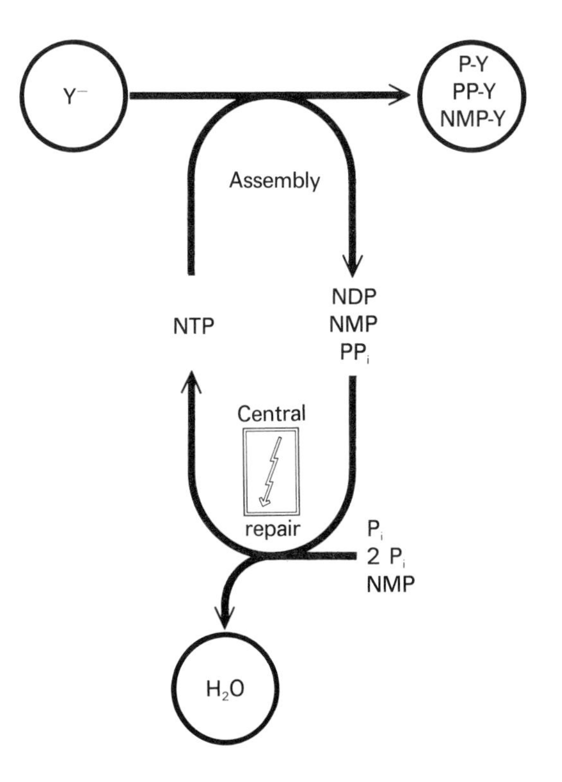

Summary of single-step biosynthetic reactions. Inorganic phosphate or mononucleotides are the building blocks. They are activated by the central repair machinery to NTP, from which they are transferred to their final acceptor in the form of a phosphoryl (γ_d), pyrophosphoryl (β_d), or nucleotidyl (β_p) group.

Two-Step Processes

Reactions in this class conform to the basic pattern of sequential group transfer. Either they are catalyzed by a single enzyme of ligase type, and then proceed by way of an enzyme-bound Janus intermediate, or they are carried out by two distinct enzymes, often physically separated from each other. The Janus intermediate then transports the X-yl group from the activation to the assembly site, with, as carrier, the group donated by the activating NTP, in combination with the oxygen atom it has appropriated from the X—O^- building block.

Reactions dependent on γ_d transfer. They proceed as follows, with NDP as carrier of the X-yl group in the Janus intermediate:

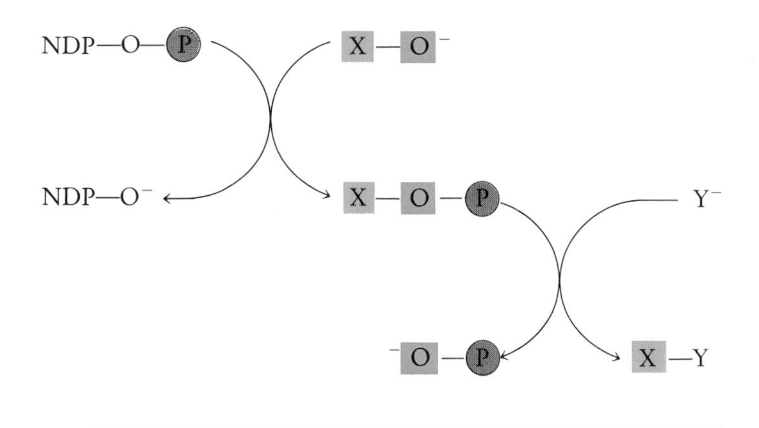

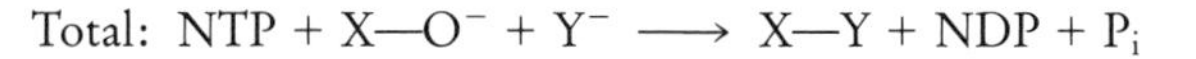

Total: NTP + X—O^- + Y^- ⟶ X—Y + NDP + P_i

In a number of such reactions, the X—O^- building block is a carboxylic acid, and the double-headed intermediate is the corresponding acyl phosphate:

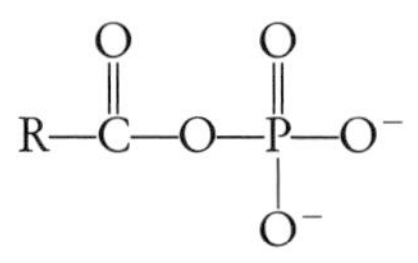

Such compounds are often unstable. In addition, like many anhydrides—substances that arise by the dehydrating condensation of two acids—they are high-energy compounds, with group potentials comparable to that of the γ pyrophosphate (anhydride of phosphoric acid) bond of NTPs. These drawbacks are obviated by the participation of ligase-type enzymes. In such cases, ATP is the standard energy purveyor, and the ligases are known as ADP-forming, to distinguish them from the AMP-forming ligases, which we will meet when we look at reactions powered by the β bond.

The final acceptor of the activated acyl group is often ammonia (NH_3) or a primary amino group (R—NH_2). The resulting amide linkage (—CO—NH—) has a relatively low "physiological" free energy of hydrolysis, of the order of −6 to −8 kcal per gram-molecule. The overall process is thus sufficiently exergonic to be irreversible. The synthesis of asparagine and glutamine from aspartic and glutamic acids, respectively (Chapter 2), and that of the tripeptide glutathione from glutamic acid, cysteine, and glycine are examples of γ_d-powered, concerted two-step mechanisms. So is the synthesis of carbamoylated derivatives (R—CO—NH_2), which include intermediates in the formation of the amino acid arginine, of urea, and of pyrimidine bases. But this reaction presents us with an interesting difference: activation and assembly are catalyzed by two distinct enzymes, linked by a freely circulating Janus intermediate—carbamoyl phosphate. The thermodynamic obstacle to such a mechanism is overcome thanks to a concerted process whereby carbamate, the substrate of the activation step, arises itself as the enzyme-bound product of the condensation of bicarbonate with ammonia, also catalyzed, as it happens, by a γ_d two-step mechanism. Thus, we are dealing with a chain of two consecutive γ_d two-step mechanisms. The first three reactions in it are catalyzed by a single, trifunctional enzyme—carbamoyl phosphate synthetase—by way of two unstable, enzyme-bound intermediates (shown between brackets in the sequence of reactions at the top of the facing page): carboxyl phosphate, the Janus product of the first activation, and carbamate, the product of the first assembly step, which becomes the substrate of the second activation. The second assembly step (dashed arrows) is catalyzed by a separate enzyme.

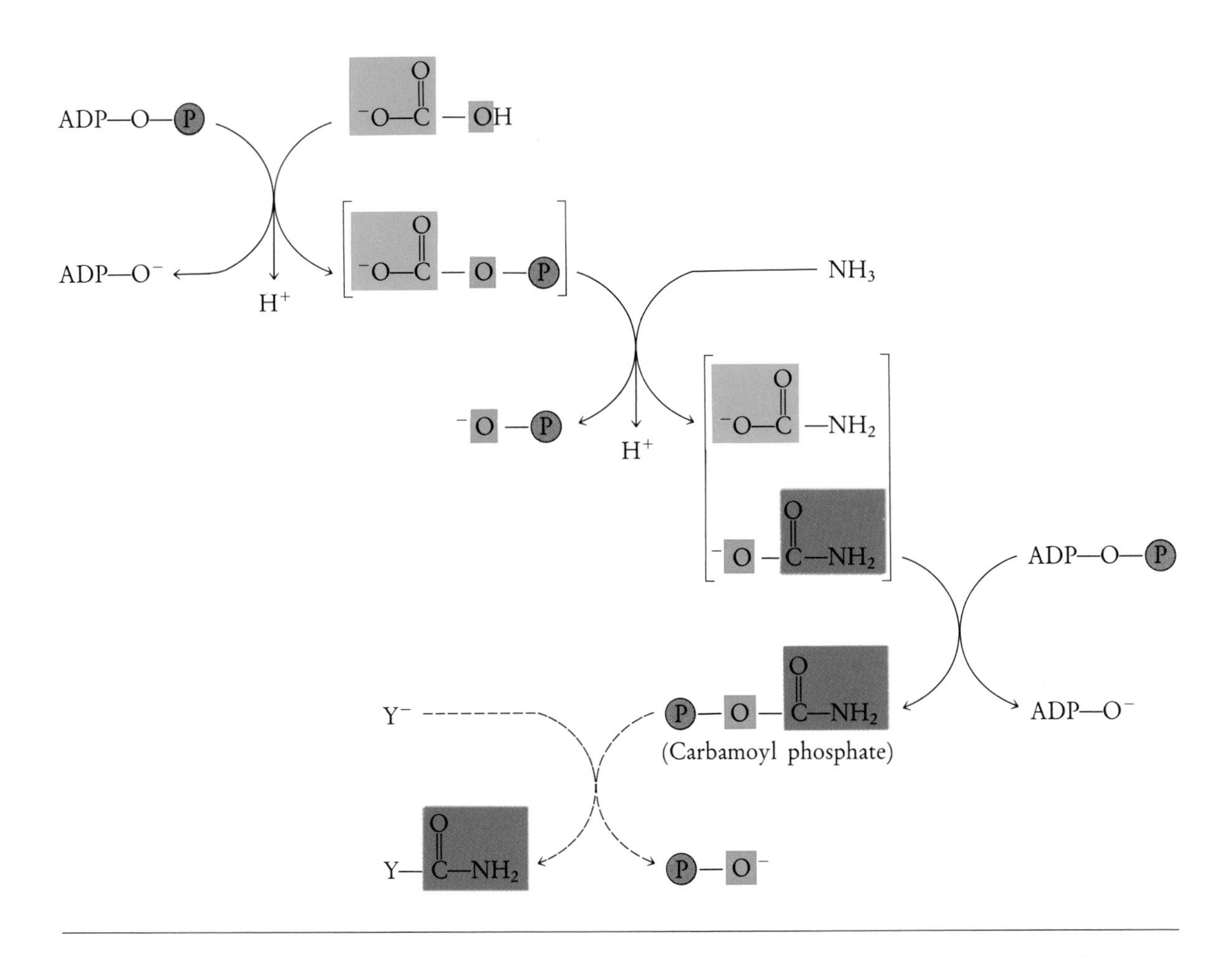

Total: 2 ATP + Bicarbonate + Ammonia + Y^- ⟶ Carbamoyl—Y + 2 ADP + 2 P_i + 2 H^+

Two γ bonds, or 28 kcal per gram-molecule, are consumed in the making of the low-energy amide bond of carbamate and the high-energy bond of its anhydride with phosphoric acid, for a total gain of about 20 to 22 kcal per gram-molecule. The overall process is strongly exergonic, thanks to the conservation of energy in the enzyme-bound carbamate. So is the subsequent transfer of the carbamoyl group to its biosynthetic acceptor, which brings it from a high to a low group potential.

It sometimes happens that the final product, X—Y, of a γ_d two-step reaction is itself a high-energy compound—for instance, a thioester (characterized by a —CO—S— linkage). Then the overall process is freely reversible and can also serve for the assembly of an NTP from the corresponding NDP and P_i, at the expense of the splitting of the X—Y bond:

$$\mathrm{R{-}\overset{O}{\overset{\|}{C}}{-}S{-}R'} + \mathrm{NDP} + \mathrm{P_i} \rightleftharpoons \mathrm{R{-}\overset{O}{\overset{\|}{C}}{-}O^-} + \mathrm{R'{-}SH} + \mathrm{NTP}$$

Reactions of this type play an important role in the operation of some substrate-level oxphos units, in which the thioester bond is made by the oxidative condensation of the R′—SH thiol with an aldehyde (R—CH=O):

$$\mathrm{R{-}\overset{O}{\overset{\|}{C}}{-}H} + \mathrm{R'{-}SH} \rightleftharpoons \mathrm{R{-}\overset{O}{\overset{\|}{C}}{-}S{-}R'} + 2\,e^- + 2\,\mathrm{H^+}$$

Adding the two reactions, we observe the oxidation of the aldehyde to the corresponding acid, with the coupled assembly of an NTP. This is exactly what happens in the oxphos unit of the glycolytic chain (Chapter 7):

$$\mathrm{R{-}\overset{\overset{\displaystyle O}{\|}}{C}{-}H + NDP + P_i} \rightleftharpoons \mathrm{R{-}\overset{\overset{\displaystyle O}{\|}}{C}{-}O^- + NTP} + 2\ e^- + 2\ \mathrm{H^+}$$

In other substrate-level oxphos units, the substrate of the oxidation is an α-keto acid that similarly combines oxidatively with a thiol to form a thioester, with, in this case, concomitant decarboxylation.

$$\mathrm{R{-}\overset{\overset{\displaystyle O}{\|}}{C}{-}COO^- + R'{-}SH} \rightleftharpoons \mathrm{R{-}\overset{\overset{\displaystyle O}{\|}}{C}{-}S{-}R' + CO_2} + 2\ e^- + \mathrm{H^+}$$

In these reactions, the R′—SH thiol plays a catalytic role in the coupling. The thioester is an intermediate of a very rare kind, capable of acting as transducer between electron-linked and group-linked energy: it can be made at the expense of either.

Reactions dependent on pseudo-γ_p transfer. In the synthesis of glycogen, a treelike polymer made of thousands of molecules of glucose (Chapters 2 and 7), activated glucosyl units are transferred to the ends of growing branches ("tail growth," see p. 139), with UDP as carrier:

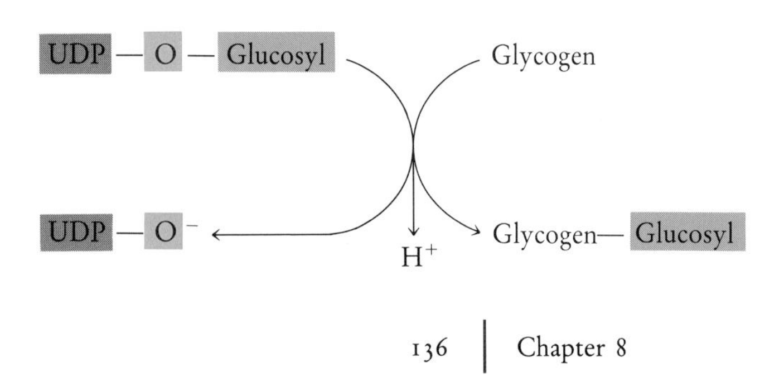

This example may serve as a paradigm of saccharide synthesis, whether they be disaccharides, oligosaccharide side chains of glycoproteins and glycolipids, or polysaccharides. Invariably, the activated sugar molecule is presented to the acceptor by an NDP, which may be UDP, ADP, GDP, or CDP, depending on the nature of the sugar. The transfer is made either directly to the final biosynthetic acceptor, as in glycogen synthesis, or by way of a fat-soluble carrier, dolichyl mono- or diphosphate, as in some of the glycosylation reactions that take place in the ER (Chapter 6). The NDP carriers also act as handles. For example, glucose molecules undergo a variety of metabolic transformations while attached to UDP.

It is interesting to note that the NDP-sugars have exactly the structure that would be expected for Janus intermediates arising by γ_p transfer (trans-NDP-ylation) on the free sugar molecule:

$$\mathrm{NTP + Sugar \longrightarrow NDP{-}Sugar + P_i}$$

This, however, is not Nature's way. In reality, the NDP-sugars are made by β_p transfer with a glycosyl phosphate as acceptor. The glycosyl phosphate itself arises, directly or indirectly, by a γ_d transphosphorylation from ATP, as shown in the first sequence of reactions on the facing page.

Note the "doubly double-headed" character of the Janus intermediate. As it arises, it has an NMP-yl and a glycosyl-phosphoryl face. For assembly, it metamorphoses, so to speak, into a molecule with an NDP-yl and a glycosyl face. Because of the hydrolysis of PP_i, each glycosidic bond (from 6 to 8 kcal per gram-molecule) costs two γ bonds, or 28 kcal per gram-molecule, which is twice the price that would have been paid for a simple two-step γ_p mechanism. Why the latter was not selected for may never be known. Perhaps chance never gave it the opportunity. Or perhaps it has drawbacks that we do not perceive.

Reactions dependent on β_d transfer. In these very rare reactions, pyrophosphate is the carrier of the X-yl group in the Janus intermediate, as shown in the second sequence of reactions on the facing page.

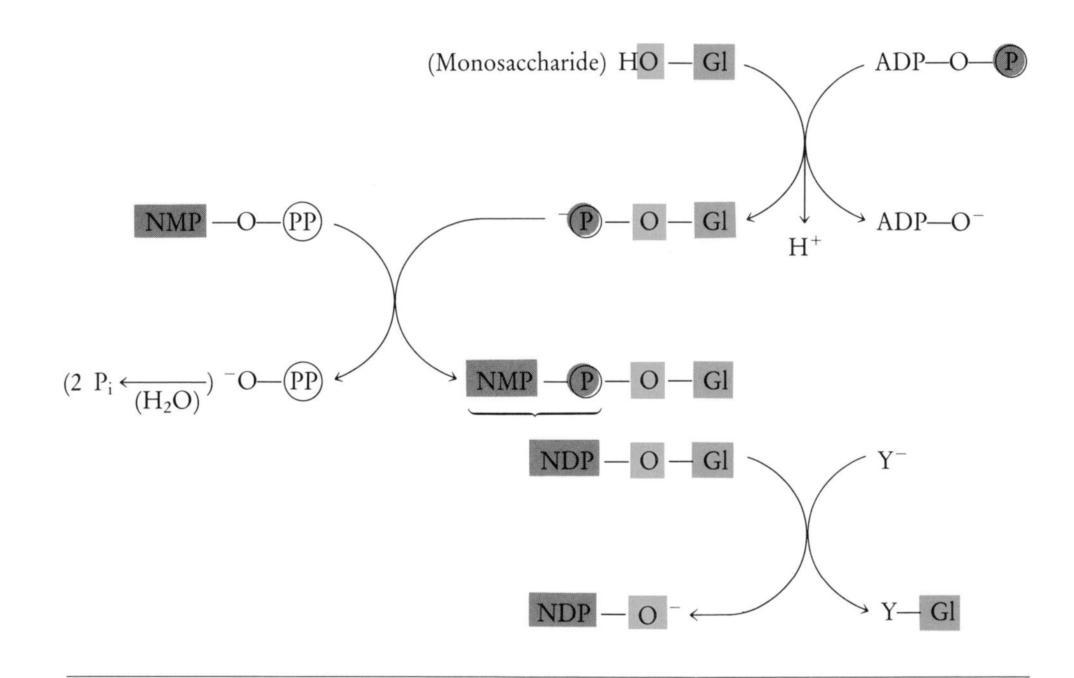

Total: ATP + NTP + Monosaccharide + Y^- ⟶ Glycosyl—Y + ADP + NDP + PP_i (2 P_i)

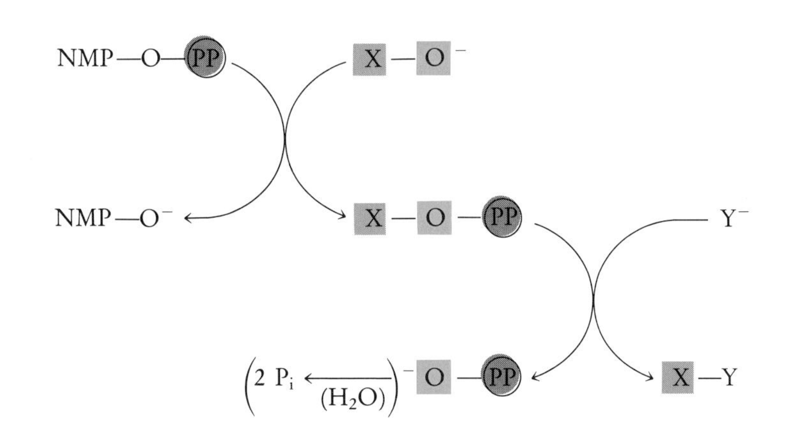

Total: NTP + X—O^- + Y^- ⟶ X—Y + NMP + PP_i (2 P_i)

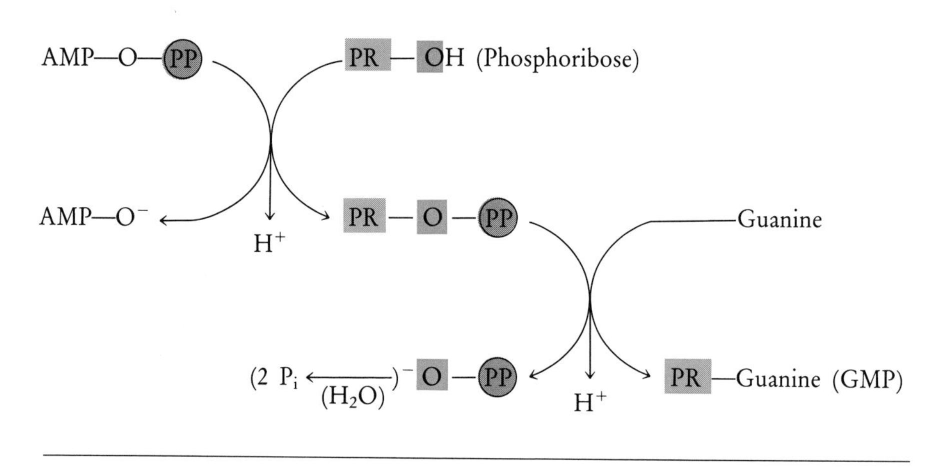

Total: ATP + PR + Guanine $\longrightarrow$ GMP + AMP + PP_i (2 P_i)

The total amount of energy available for the synthesis of X—Y is 28 kcal per gram-molecule. Note, however, that much of this energy may be conserved in the double-headed intermediate, as it is the assembly step that benefits from the low PP_i concentration maintained by pyrophosphatase action. This is in contrast with β_p mechanisms (see pp. 139–140).

The most important X—O^- building block of β_d two-step reactions is 5-phosphoribose, which is the molecule that is left when the purine or pyrimidine base is hydrolyzed off from a mononucleotide. Reconstitution of some mononucleotides can then take place by a typical β_d two-step mechanism, with phosphoribosyl pyrophosphate (PRPP) as Janus intermediate. For example, guanine can thereby be rejoined with 5-phosphoribose to form GMP, as shown above.

This salvaging of bases is only one of the functions of PRPP. It is an intermediate in the synthesis of the amino acids histidine and tryptophan and in that of the purine ring.

Another key Janus intermediate with the structure X—O—PP is isopentenyl pyrophosphate, a precursor of a host of important fat-soluble molecules, including quinonic electron carriers (see Chapters 9 and 10); vitamins A, D, E, and K; sterols and steroids; carotenoids; terpenoids; latex (rubber); essential oils; and many other substances that belong to what is known as the isoprene group:

$$CH_3\text{—}\underset{}{\overset{\overset{\displaystyle CH_2}{\|}}{C}}\text{—}CH_2\text{—}CH_2\text{—}O\text{—}\overset{\overset{\displaystyle O}{\|}}{\underset{\underset{\displaystyle O^-}{|}}{P}}\text{—}O\text{—}\overset{\overset{\displaystyle O}{\|}}{\underset{\underset{\displaystyle O^-}{|}}{P}}\text{—}O^-$$

Isopentenyl pyrophosphate

This important substance does not arise by β_d pyrophosphoryl transfer, but rather by a complex mechanism that includes two γ_d phosphoryl transfers. We are dealing here, therefore, with a pseudo-β_d mechanism. It shares with the authentic mechanism the extra boost provided to the assembly step by pyrophosphatase action and deserves to be considered here.

As a rule, two or more 5-carbon units assemble together, with release of inorganic pyrophosphate, by an iterative transfer mechanism in which the terminal carbon of isopentenyl pyrophosphate is the acceptor and the growing chain is the transferred group. This is shown schematically in the sequence of reactions at the top of the facing page. It is a typical example of "head growth" of a

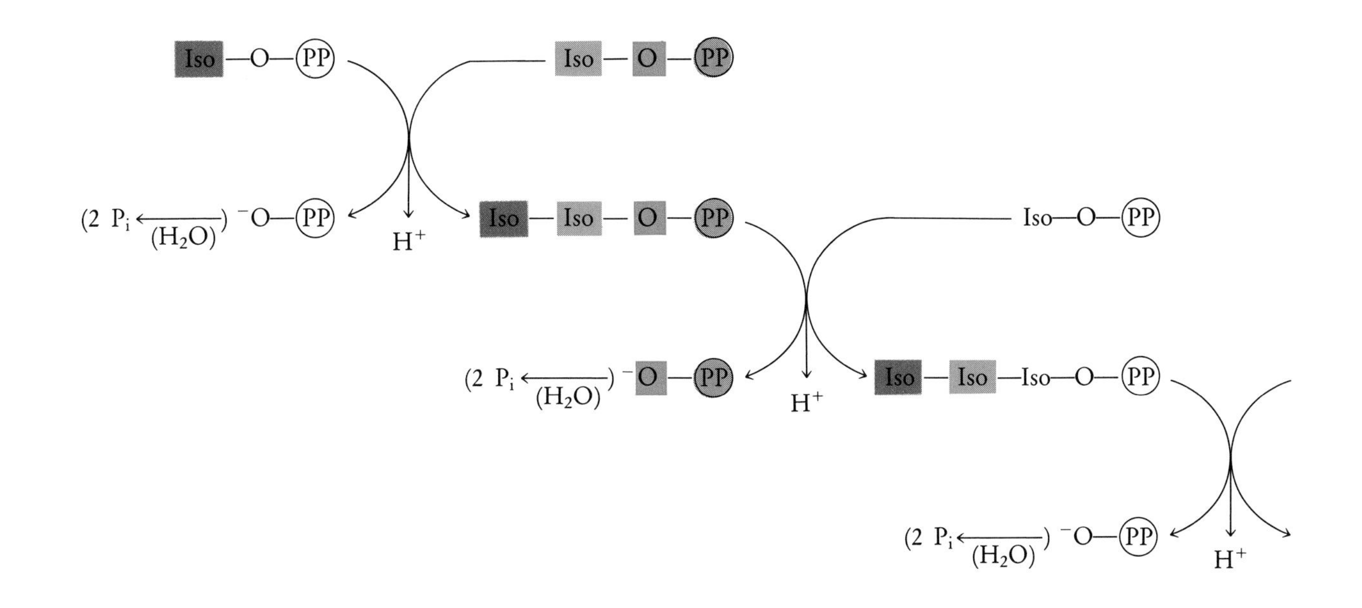

lengthening chain. Fatty acid synthesis (see pp. 145–146) and protein synthesis (see Chapter 15) are others. The characteristic of such mechanisms is that the double-headed product of the activation reaction first acts as acceptor of the part of the chain already completed, before doing its duty as group donor. In tail growth, on the other hand, the double-headed intermediate donates its group immediately to the growing chain. We saw an example of it with polysaccharide synthesis.

Reactions dependent on β_p transfer. This is probably the most widely used biosynthetic mechanism. It takes place according to the following scheme, with NMP—O—X as Janus intermediate:

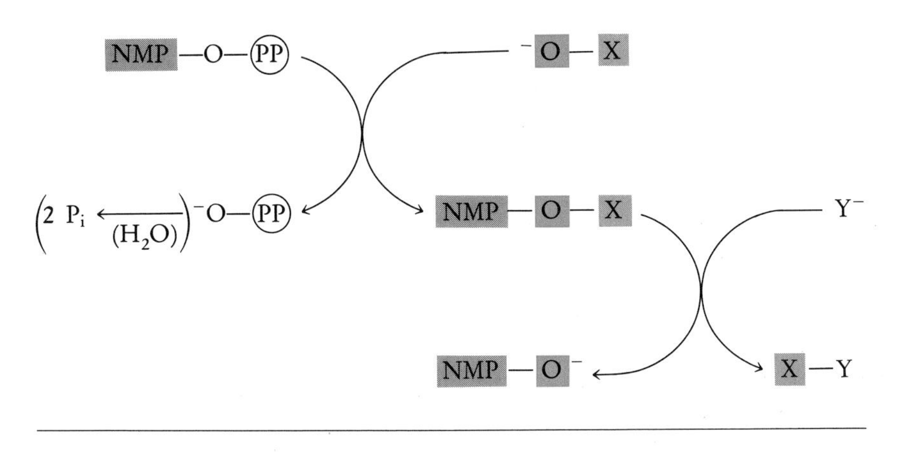

Total: $NTP + X\text{—}O^- + Y^- \longrightarrow X\text{—}Y + NMP + PP_i$ (2 P_i)

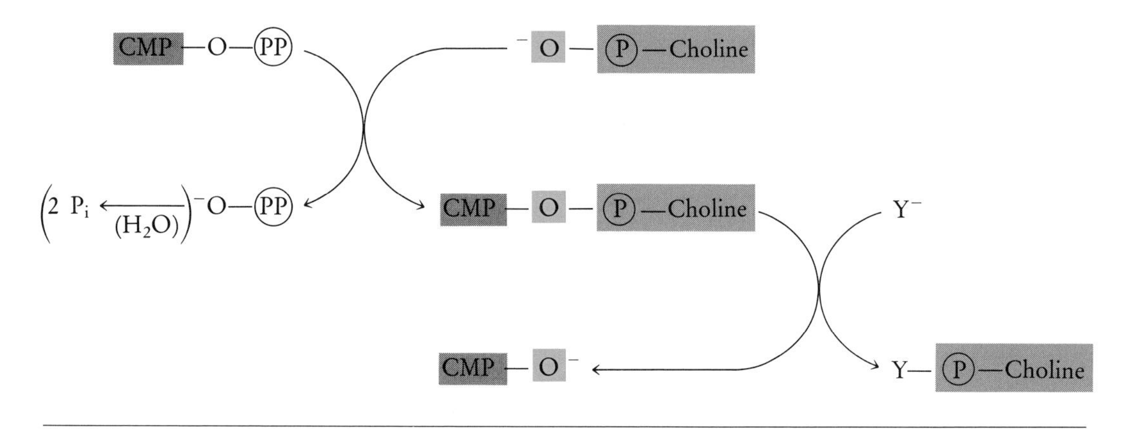

Total: CTP + Phosphorylcholine + Y^- ⟶ Y—Phosphorylcholine + CMP + PP_i (2 P_i)

The advantage of this mechanism is that it has the full strength of the β bond—up to 28 kcal per gram-molecule—available for activation. The resulting Janus intermediates are often unstable molecular arrangements that remain enzyme-bound. Activation and assembly are catalyzed in concerted fashion by a single enzyme. These ligases use ATP as energy donor, as do the ADP-forming ligases that catalyze similar concerted reactions powered by the γ bond (see p. 134). An important enzyme of this group is DNA ligase, one of the main agents participating in DNA synthesis (see Chapter 17). Many other biosynthetic processes are carried out by AMP-forming ligases.

A few two-step processes dependent on β_p transfer involve the participation of two distinct enzymes, linked by a stable Janus intermediate in which the activated X-yl group is transported by an NMP carrier. NDP-sugars, it may be remembered, arise in this form, even though they behave somewhat differently upon assembly. More-orthodox Janus intermediates of the same class are generated by β_p transfer between CTP and various phosphorylated building blocks used in the synthesis of phospholipids, including phosphatidic acids, phosphorylcholine, and phosphorylethanolamine. These compounds have the structure CDP—R and are designated as such (CDP-choline, for example). Unlike the NDP-sugars, however, they do not act as R donors in the subsequent assembly but continue to behave as intermediates with a CMP-yl and an R-phosphoryl face, as shown at the top of this page.

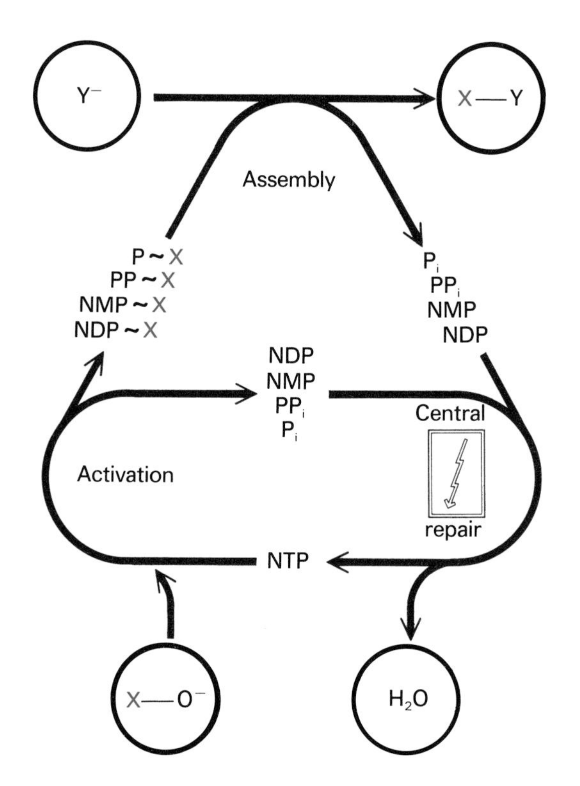

Summary of two-step biosynthetic reactions. An oxygen-containing building block ($X{-}O^-$) is activated by an NTP to a Janus intermediate in which it is linked to a phosphoryl (γ_d), pyrophosphoryl (β_d), nucleotidyl (β_p), or nucleoside-diphosphoryl (pseudo-γ_p) group. After transfer of the activated building block to its final acceptor, the split products of the NTP used are reunited by the central repair machinery.

Three-Step Processes

These processes start systematically by a two-step concerted mechanism catalyzed by a ligase. They differ from the other processes of this kind already considered in the fact that the activated group is attached to a carrier instead of to a biosynthetic building block. Final assembly occurs in a third step by transfer of the group from the carrier to its biosynthetic acceptor.

Reactions dependent on γ_d transfer. Two important group-carrier complexes are assembled by ADP-forming ligases according to the general scheme of γ_d two-step sequential group transfer. One is carboxy-biotin, which is made very much like carbamate, with enzyme-bound carboxyl phosphate as Janus intermediate, and then serves as donor of the activated carboxyl group in a number of carboxylation reactions, as shown in the sequence of reactions below.

Biotin, or vitamin H, which plays a catalytic role in the overall three-step process, is a molecule of historical interest. It celebrates by its name an early discovery in the field of nutrition—that of a yeast growth factor—by Eugène Wildiers, a Belgian biochemist. Wildiers was so impressed by the life-giving power of his factor that he called it Bios. The name biotin was subsequently given to one of the vitamins to commemorate this event. In the performance of its carrier function, biotin is bound covalently to a flexible polypeptide arm included in a multienzyme complex that contains the ligase and the carboxylase. Its transport role is limited to the shuttling of carboxyl groups between the active centers of the two enzymes.

A similar three-step mechanism, but with enzyme-bound formyl phosphate as Janus intermediate, serves to attach formate (H—CO—O$^-$) to a carrier called tetrahydrofolate (THF), a derivative of the vitamin folic acid (Latin *folium*, leaf), one of the secret ingredients of Popeye's spinach, known for its antianemia properties.

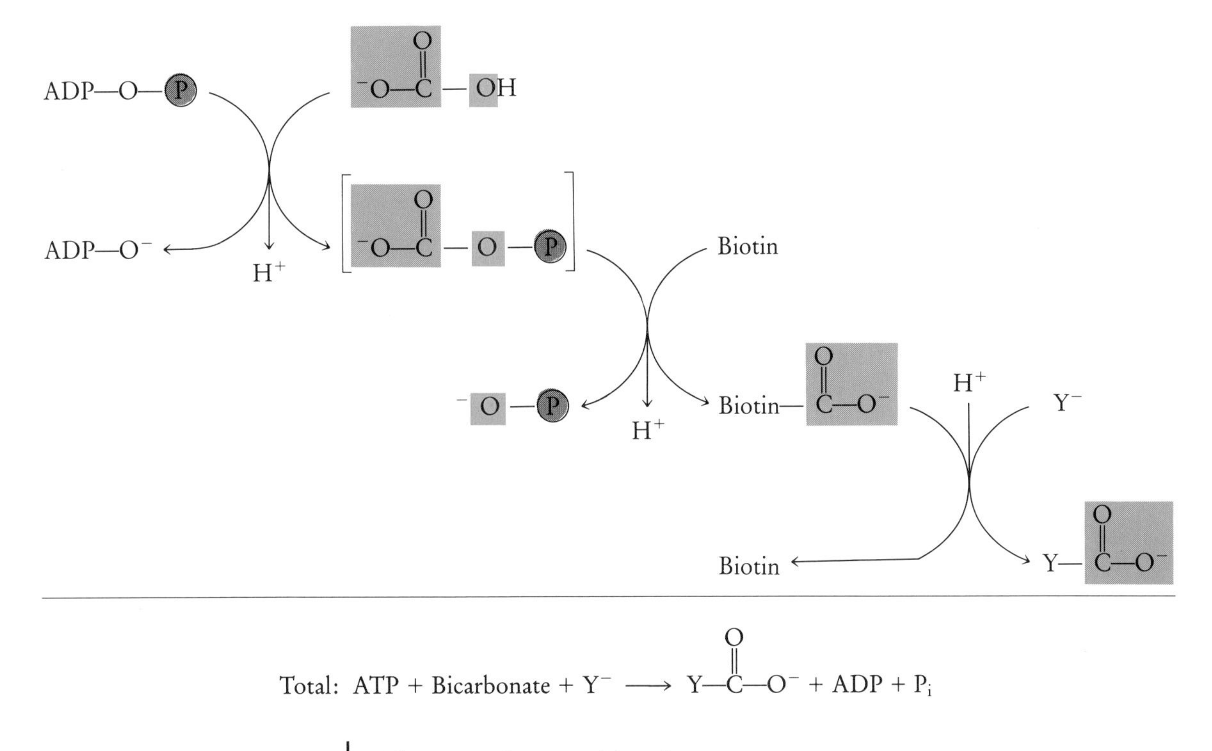

THF not only carries the activated formyl group to various formyl transferases, but also offers this group to a number of modifying enzymes, which may convert it into a methenyl (—CH═), methylene ($—CH_2—$), hydroxymethyl ($—CH_2OH$), methyl ($—CH_3$), or formimino (—CH═NH) group. Each group participates in a number of transfer reactions. Thus we are dealing with a very versatile coenzyme, which acts both as carrier and as handle for activated groups. Among the many substances that depend on THF for their formation are the amino acid methionine (see pp. 146–147), the purine bases, and the pyrimidine base thymine, a constituent of DNA (see Chapter 15).

Reactions dependent on β_p transfer. Two major biological processes follow this mechanism. Both have a carboxylic acid ($R—CO—O^-$) as $X—O^-$ building block, use ATP as energy donor, and depend on an AMP-forming ligase to generate a stable, soluble acyl-carrier complex by way of an enzyme-bound acyl-AMP Janus intermediate:

One such process is protein synthesis, in which the amino acids are the $X—O^-$ building blocks, and the corresponding transfer RNAs (tRNAs) are the carriers. The whole of Chapter 15 will be devoted to this matter, and we will not consider it further here, except for pointing out that polypeptide chains grow by head growth (see pp. 138–139), which means that, between the intermediary and the final assembly steps in the scheme below, there is intercalated a step in which the carrier—CO—R complex (aminoacyl-tRNA) serves as acceptor of the growing chain.

The other process that follows the three-step β_p mechanism uses a variety of organic acids, among them the fatty acids found in lipids, as $X—O^-$ building blocks and coenzyme A as carrier. This coenzyme owes its name to the fact that it was first discovered as a cofactor of acetylation reactions. It is a derivative of vitamin F, or pantothenic acid, a ubiquitous substance (*pantothen* means everywhere in Greek), which at one time enjoyed a dubious notoriety as a hair restorer because its deficiency

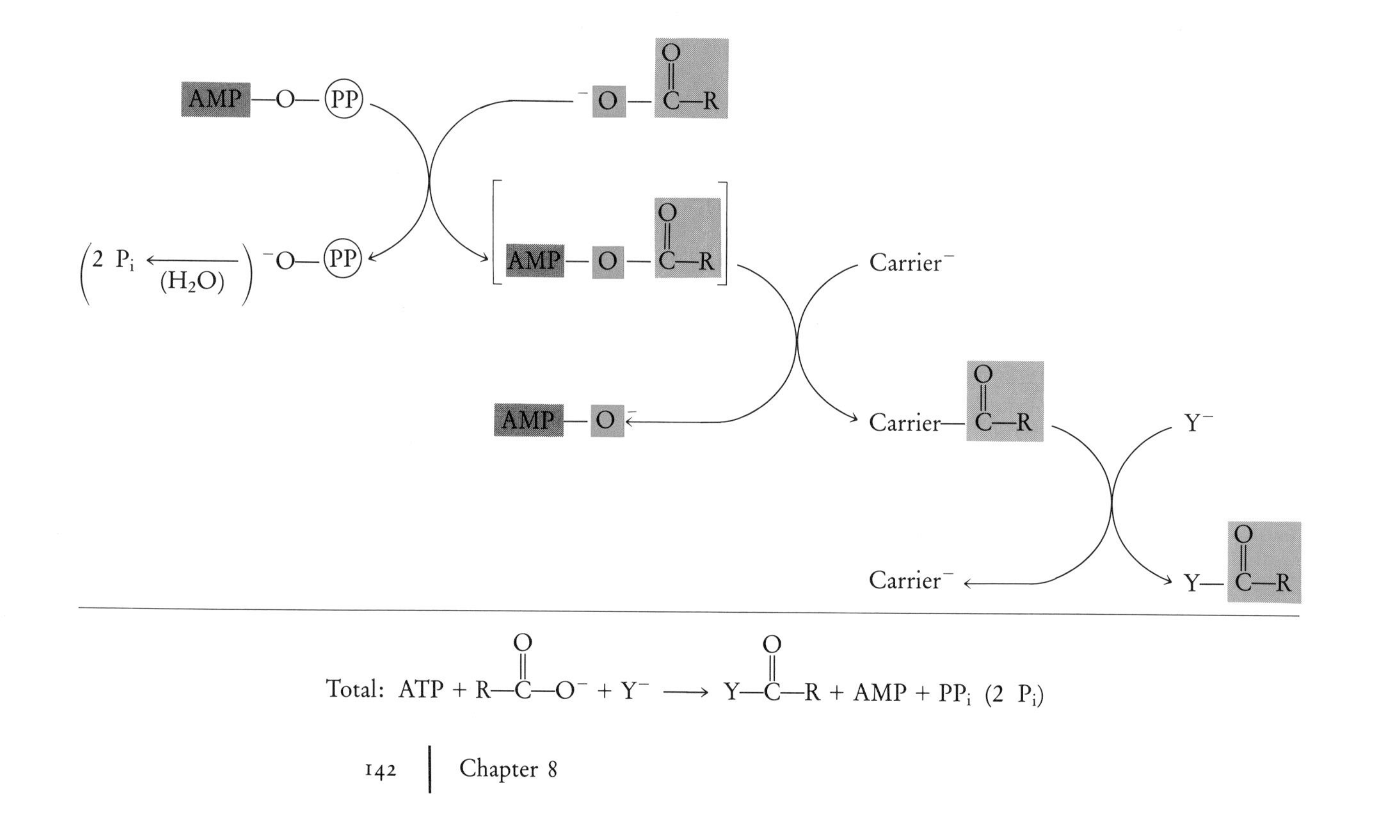

$$\text{Total: ATP} + R—\overset{O}{\overset{\|}{C}}—O^- + Y^- \longrightarrow Y—\overset{O}{\overset{\|}{C}}—R + \text{AMP} + \text{PP}_i \ (2\ \text{P}_i)$$

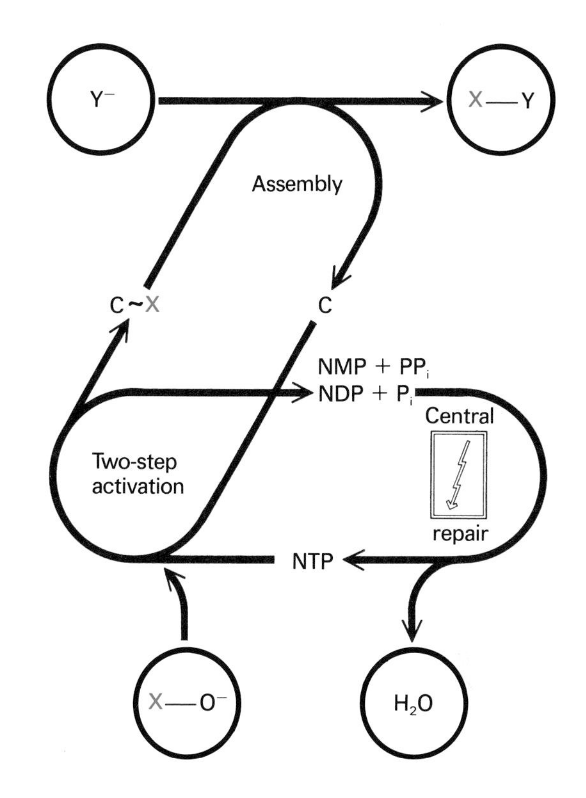

Summary of three-step biosynthetic reactions. Two-step mechanisms (see the illustration on p. 140) serve to attach the activated building block to a carrier C, from which it is delivered to its final acceptor.

causes premature graying in rats. Coenzyme A is a complex molecule that contains several other constituents, including a molecule of AMP, in addition to pantothenic acid. The molecule has a thiol group (—SH) as reactive end and is accordingly abbreviated CoA—SH. Acyl-CoA derivatives are thioesters, which, as we have seen, are high-energy compounds, with "physiological" group potentials of the order of −14 kcal per gram-molecule. Their formation, nevertheless, is highly exergonic, as it consumes 28 kcal per gram-molecule.

The main Y—H acceptors of the acyl groups borne by coenzyme A are alcohols:

$$\mathrm{CoA{-}S{-}\overset{\displaystyle O}{\overset{\|}{C}}{-}R + R'{-}OH \longrightarrow R'{-}O{-}\overset{\displaystyle O}{\overset{\|}{C}}{-}R + CoA{-}SH}$$

The resulting esters are low-energy compounds, making the final transfer essentially irreversible. Among the molecules made in this way are the neurotransmitter acetylcholine (acetic acid plus choline, see Chapter 13), the various esters of fatty acids and glycerol found in neutral lipids and phospholipids, and many others. In addition, coenzyme A is involved in numerous other reactions of central importance—as mediator of activated acyl groups, as metabolic handle, and as a participant in some substrate-level oxphos units. (Remember the importance of thioester bonds in the operation of such units, mentioned on pp. 135–136). It is a key piece of the cell's machinery.

A Few Exceptions That Confirm the Rule

Proteins, nucleic acids, neutral lipids, phospholipids, polysaccharides, steroids, terpenoids, nucleotides, coenzymes, amino acids, purines, pyrimidines: the list of substances that we have met in our brief excursion reads like a biochemical *Who's Who*. Obviously, our simple scheme covers a lot of ground. Not surprisingly, it does not cover everything. Some of the exceptions are worth mentioning.

First, some transfer reactions do not depend on a straightforward nucleophilic attack. Among them are transamination and transthiolation, in which an exchange of groups actually takes place. Then there are the various instances where the bond to be formed has a higher free energy of hydrolysis than the NTP bond used up in the biosynthetic process. It is interesting to see where the required energy supplement comes from, or rather how it is delivered. Almost invariably, its source is ATP hydrolysis, as might be expected. As to the means, they may be classified as either "boosting the donor" or "boosting the acceptor."

An example of the former is sulfurylation, a reaction whereby the sulfate ester groups of sulfated mucopolysaccharides and sulfolipids are constructed from inorganic sulfate. The reaction is initiated by a typical β_p attack of sulfate on ATP, giving rise to the double-headed adenylyl sulfate, which is released in freely soluble form. However, the "physiological" free energy of hydrolysis of this anhydride is so high, especially at the very low concentration of inorganic sulfate prevailing in most living cells, that even the full complement of 28 kcal per gram-molecule made available by the splitting of the β bond of ATP does not suffice to lift its concentration to the level needed for an adequate rate of diffusion to distant assembly sites. For this reason, it cannot serve as donor of the sulfuryl group (reaction shown by dashed arrows in the diagram below). Nature's solution of this problem is phosphorylation of the 3′-hydroxyl group of adenylyl sulfate, a highly exergonic reaction capable of raising the concentration of its phosphorylated product at least four orders of magnitude above that of its substrate. The resulting 3′-phospho-adenylyl sulfate can now fulfill the role of transport form and act as donor of sulfuryl groups in a variety of assembly reactions. After it has done its duty, the 3′-phosphate group of AMP is cleaved off. The complete process adds up as follows:

$$2\ ATP + SO_4^{2-} + Y^- \longrightarrow Y{-}SO_3^- + AMP + ADP + PP_i\ (2\ P_i) + P_i$$

The total cost amounts to 42 kcal per gram-molecule.

Examples of acceptor activation are seen in the synthesis of fatty acids, in that of the porphyrin ring, and in a number of amination reactions. In each case, the acceptor in the final assembly reaction is fitted with a carboxyl

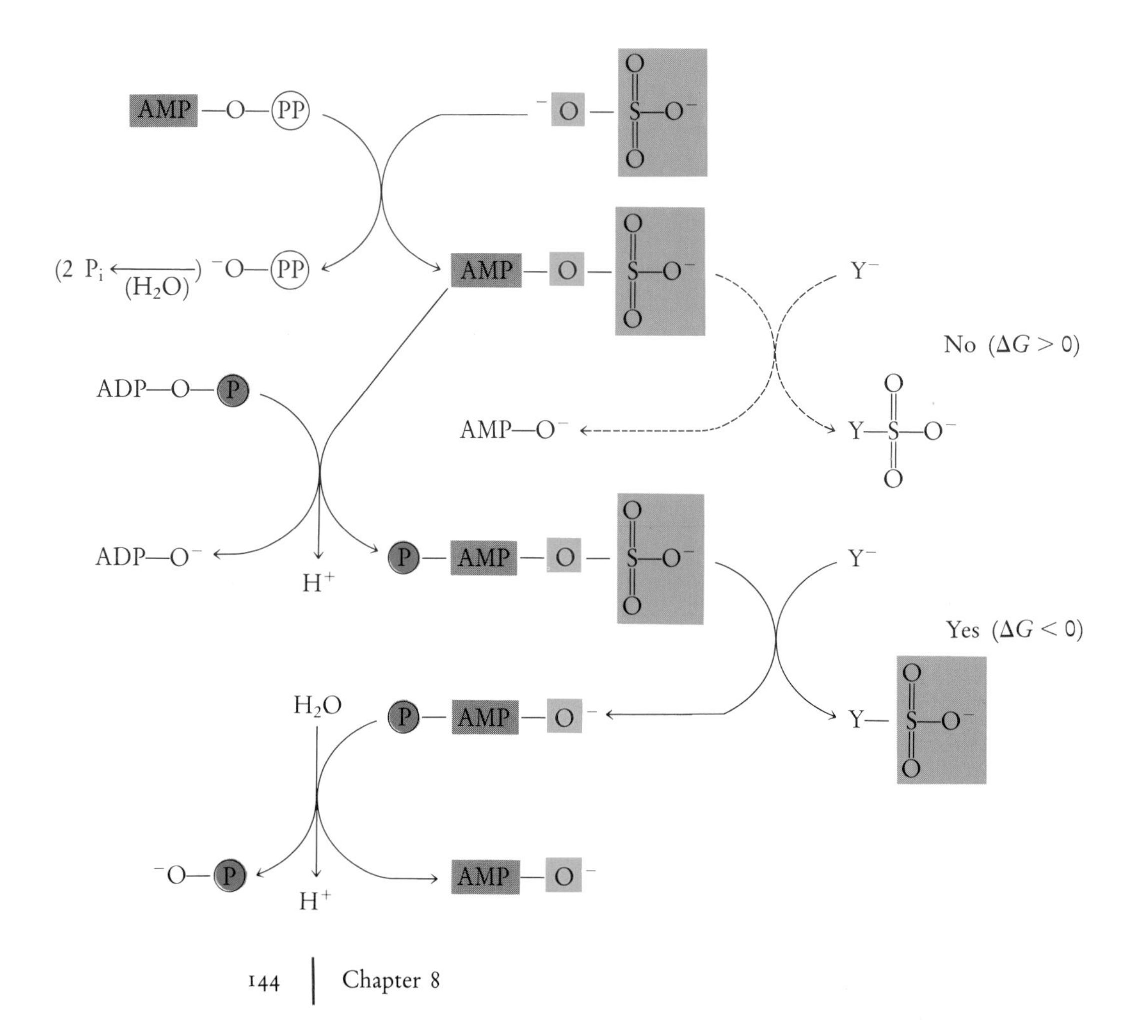

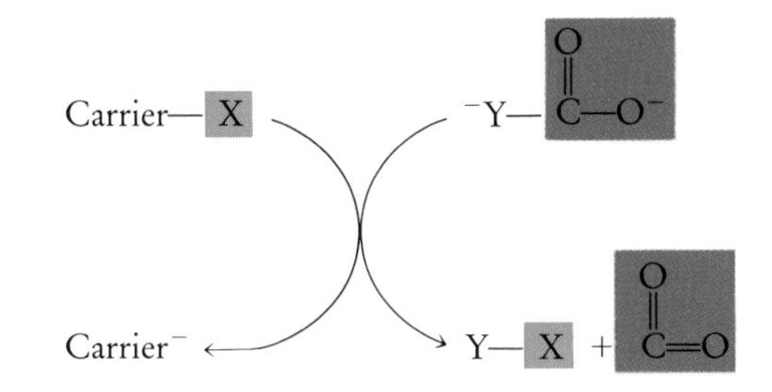

(—CO—O^{-}) or acyl (—CO—R) group that falls off upon assembly, thus adding the free energy of decarboxylation or deacylation to the potential of the donated group for supporting the cost of making the new bond, as shown at the right.

or:

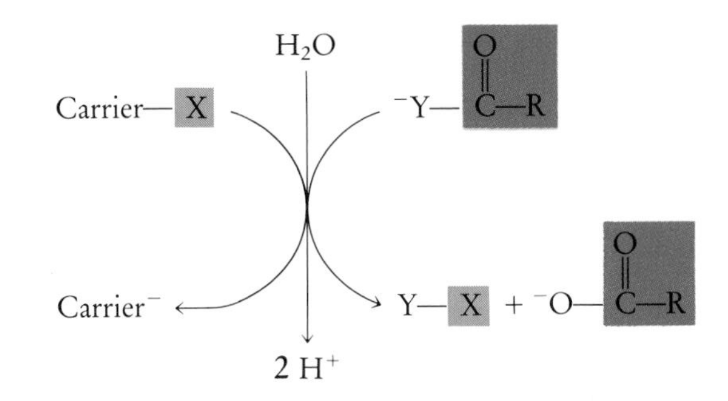

In fatty acid synthesis, the donor is a growing acyl chain borne by a protein carrier (acyl-carrier protein, ACP), which includes a piece of coenzyme A as active group. The acceptor is malonyl-ACP, derived from malonyl-CoA, which is acetyl-CoA activated by carboxylation (by a biotin-dependent γ_d mechanism of the kind described on p. 141), as shown below:

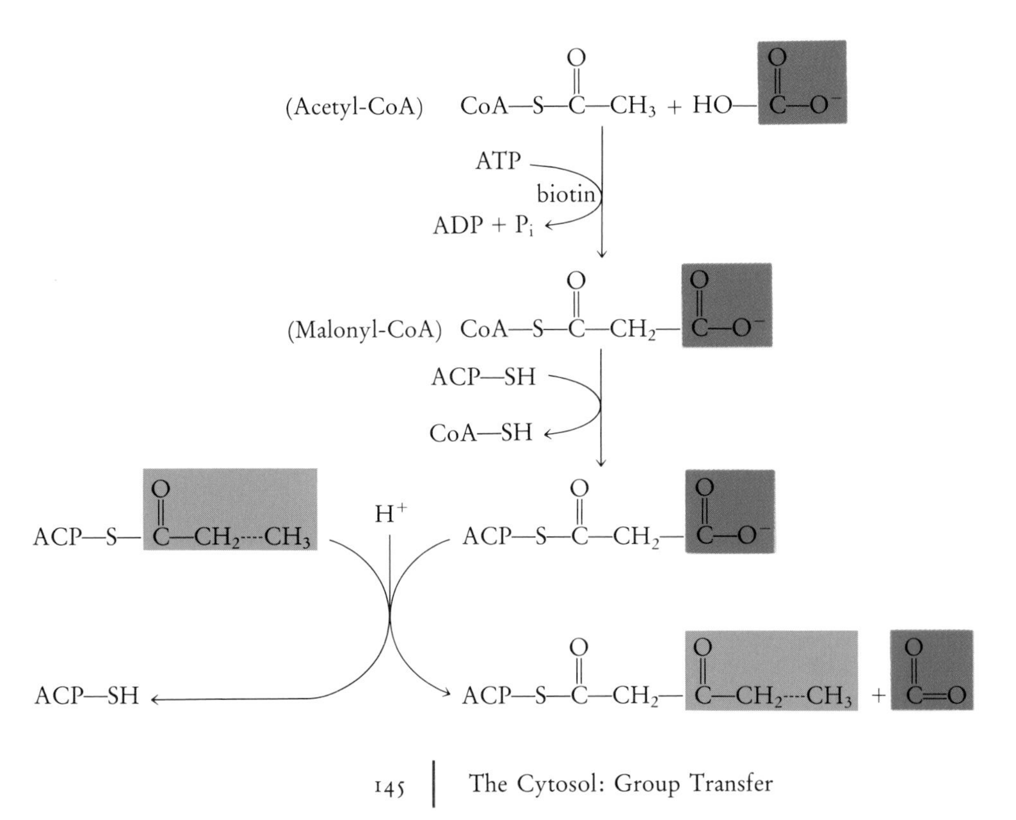

Note that the final product of the assembly reaction (a β-ketoacyl-ACP two carbons longer than the acyl-ACP donor) is exactly what would be obtained by a direct transfer of the acyl group onto acetyl-ACP. However, such a reaction cannot normally occur, because the potential of the transferred acyl group is distinctly higher in the β-ketoacyl product than in the thioester donor. This reaction does take place in living cells (with coenzyme A as carrier), but in the reverse direction (thiolysis of the β-ketoacyl CoA derivative by coenzyme A), as part of fatty acid degradation. By carboxylating the acetyl-ACP acceptor, at the expense of an additional 14 kcal per gram-molecule, the cell provides it with about 6 to 8 kcal per gram-molecule extra energy, enough to drive the reaction in the direction of assembly. It will be noted further that the reaction consists in the addition of a growing chain to a two-carbon building block. After reduction of the β-keto group, this process will be repeated, alternating with reductive steps until the chain is completed. It is another example of head growth (pp. 138–139).

There is an analogous step in the construction of the porphyrin ring, which enters into the composition of such important molecules as hemoglobin, cytochromes (see Chapter 9), and chlorophylls (see Chapter 10). Here the donor is succinyl-CoA, and the acceptor is the amino acid glycine, which, in this reaction, behaves as methylamine (CH_3—NH_3^+) activated by carboxylation, as shown at the top of the facing page. The product is δ-aminolevulinic acid, a precursor of the porphyrin ring.

In amination reactions, which, with rare exceptions, are rendered thermodynamically unfavorable because of the low concentration of ammonia in cells, ammonia is replaced as acceptor by glutamine, the amide of glutamic acid, which is assembled by a typical γ_d two-step process, as was seen earlier. In those amination reactions, glutamine has the character of a molecule of ammonia activated by acylation with glutamic acid (see facing page, middle).

In some amination reactions, the activated form of ammonia is aspartic acid. The mechanism is different from that considered here.

To conclude this brief tour of biosynthetic oddities, we will take a look at the most important representative of the very rare processes that depend on an α_p attack on ATP (transadenosylation). The attacking agent is the amino acid methionine, which is characterized by a thiomethyl group (—S—CH_3). Exceptionally, in this case, the attack is made by a sulfur atom, not by an oxygen atom. In addition, the subsequent assembly reaction does not involve transfer of the whole activated molecule but only of its terminal methyl group, which is rendered easily transferable as a result of the positive charge acquired by the sulfur atom upon adenosylation. Inorganic triphosphate, the other product of the α_p attack, does not appear as such but is hydrolyzed to inorganic phosphate and pyrophosphate, which is itself split by pyrophosphatase action (see facing page, bottom).

This process is responsible for many important methylation reactions, including those affecting nucleic acids (see Chapters 16–18). It is, you will notice, a very expensive business: all of 35 kcal per gram-molecule. And it does not even start from scratch, with methanol (CH_3OH) as building block, but uses methionine as source of the methyl group. Methionine itself can be reconstituted from homocysteine, with, in this case, methyltetrahydrofolate as methyl donor (see pp. 141–142):

$$\text{Homocysteine} + \text{THF-CH}_3 \longrightarrow \text{Methionine} + \text{THF}$$

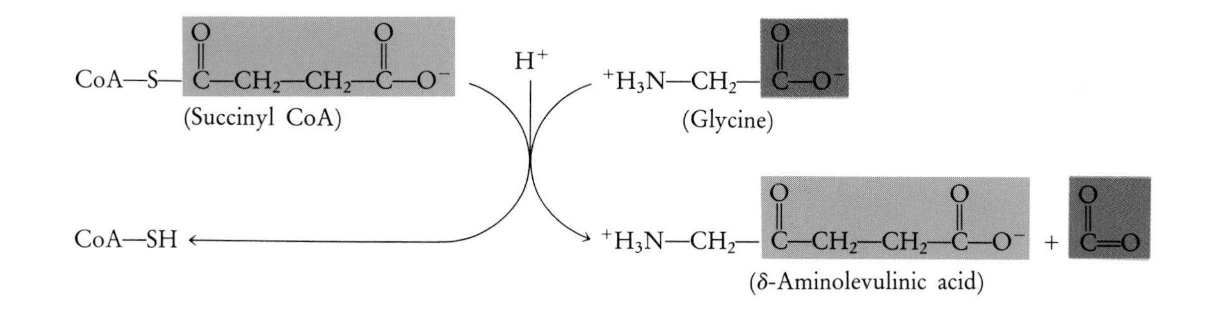

CoA—S—C—CH₂—CH₂—C—O⁻
(Succinyl CoA)
H⁺
⁺H₃N—CH₂—C—O⁻
(Glycine)
CoA—SH
⁺H₃N—CH₂—C—CH₂—CH₂—C—O⁻ + C=O
(δ-Aminolevulinic acid)

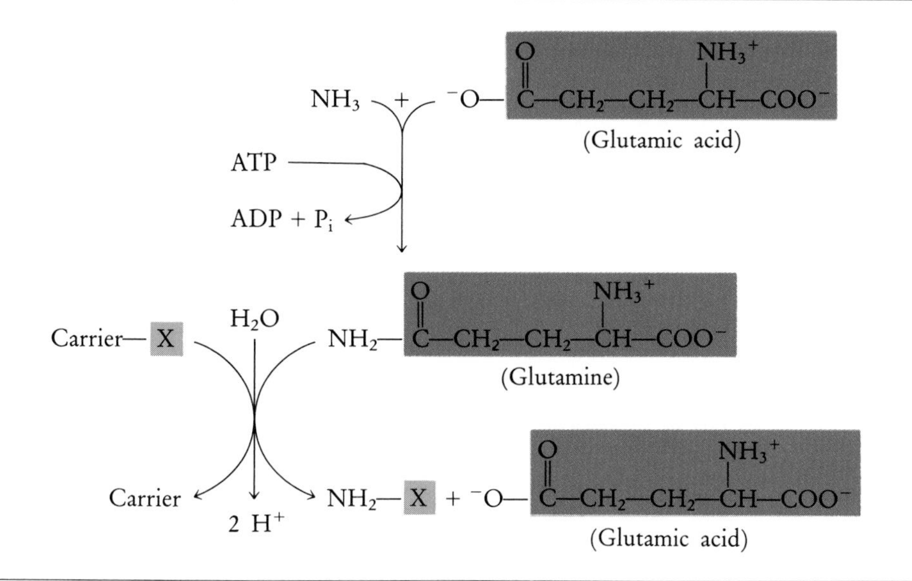

NH₃ + ⁻O—C—CH₂—CH₂—CH—COO⁻
(Glutamic acid)
ATP
ADP + Pᵢ
Carrier—X
H₂O
NH₂—C—CH₂—CH₂—CH—COO⁻
(Glutamine)
Carrier
2 H⁺
NH₂—X + ⁻O—C—CH₂—CH₂—CH—COO⁻
(Glutamic acid)

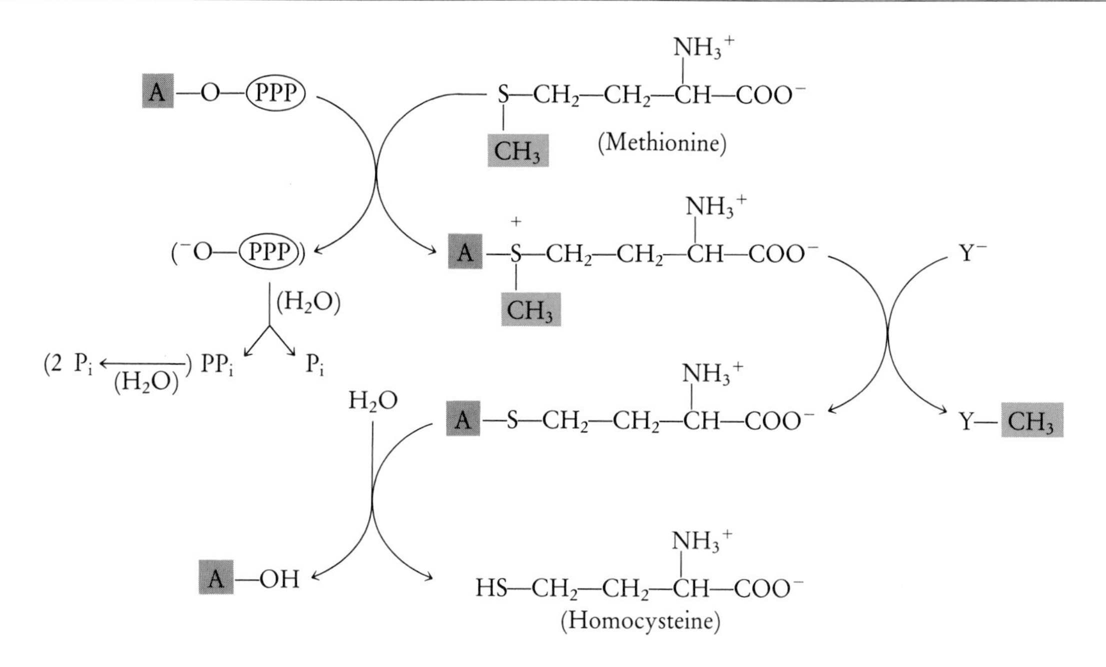

A—O—PPP
S—CH₂—CH₂—CH—COO⁻
CH₃
(Methionine)
(⁻O—PPP)
A—S—CH₂—CH₂—CH—COO⁻
CH₃
Y⁻
(H₂O)
(2 Pᵢ ←) PPᵢ
(H₂O)
Pᵢ
H₂O
A—S—CH₂—CH₂—CH—COO⁻
Y—CH₃
A—OH
HS—CH₂—CH₂—CH—COO⁻
(Homocysteine)

9 The Mitochondria: Respiration and Aerobic Energy Retrieval

Air is so vital to us and to most living beings that we are hard put to imagine life without it. Yet, as was mentioned in Chapter 7, life almost certainly began, and thrived for a long time, in an oxygenless world. Only after the emergence of the advanced photosynthetic mechanism known as photosystem II, about three billion years ago (see Chapter 10), did oxygen start appearing in the atmosphere in significant quantity, slowly rising to its present level of 20 per cent.

This phenomenon must have posed a major threat to life and brought about the extinction of numerous species comparable to today's obligatory anaerobes. The reason for this lies in the ability of molecular oxygen to interact in a variety of ways with reducing biological molecules, giving rise to such highly toxic products as the superoxide ion, O_2^-, and hydrogen peroxide, H_2O_2. The survivors, except for those that found refuge in an oxygen-free niche, were those that developed protective enzymes, such as superoxide dismutase and catalase (see Chapter 11). Some went further and actually succeeded in taming the enemy and turning it into life's greatest ally, thanks to an adaptation of their ATP-yielding oxphos units to the use of oxygen as final electron acceptor. Their progeny now fill most of the living world.

The Taming of Oxygen

Judging from known aerobic bacteria, adaptation to oxygen must have been a gradual process, which eventually culminated in one of Nature's greatest achievements, the phosphorylating respiratory chain: a string of fifteen or more electron carriers spanning the whole 1,070-mV difference between NADH and oxygen, and arranged in such a manner as to include up to three successive oxphos units in series. A system of this sort is found in the plasma membrane of a number of contemporary bacteria, which presumably have inherited it from those remote ancestors that first acquired it well over one billion years ago. Essentially the same system is present in the inner of the two membranes that surround mitochondria, those discrete, membrane-wrapped bodies about the size of bacteria that are found scattered in large numbers throughout the cytoplasm of the vast majority of eukaryotic cells in both plants and animals, where they serve as the main centers of respiration and of oxidative energy retrieval. The link between the bacterial and the mitochondrial systems—assuming there is one, which is highly probable—makes fascinating speculation.

In the most popular version of the story, the hero is the primitive phagocyte, that hypothetical, giant, voracious, bacteria-gobbling cell that is postulated to be an intermediate between prokaryotes and eukaryotes (Chapter 6). Among its daily catch—so the story goes—were some aerobic bacteria that failed to be killed and broken down for food. They did not kill their captor either, as do many pathogenic bacteria that escape destruction. Rather, they established a permanent, mutually advantageous, symbiotic relationship with it. Their descendants have survived unto this day as the mitochondria of eukaryotic cells. Fully integrated with their host cell, as might be expected after more than a billion years of living together, these organelles have nevertheless retained the remnants of a genetic system of typical bacterial kind, together with some other vestigial properties that attest to their ancestry. As we shall see in the next chapter, the chloroplasts of plant cells may have arisen in the same way from symbiotically adopted photosynthetic bacteria.

Known as the endosymbiont hypothesis, this theory has much to commend it, including some of the phylogenetic trees deduced from molecular sequencing (see Chapter 18). Scientists, however, are a disbelieving lot, especially when it comes to reconstructing events of the

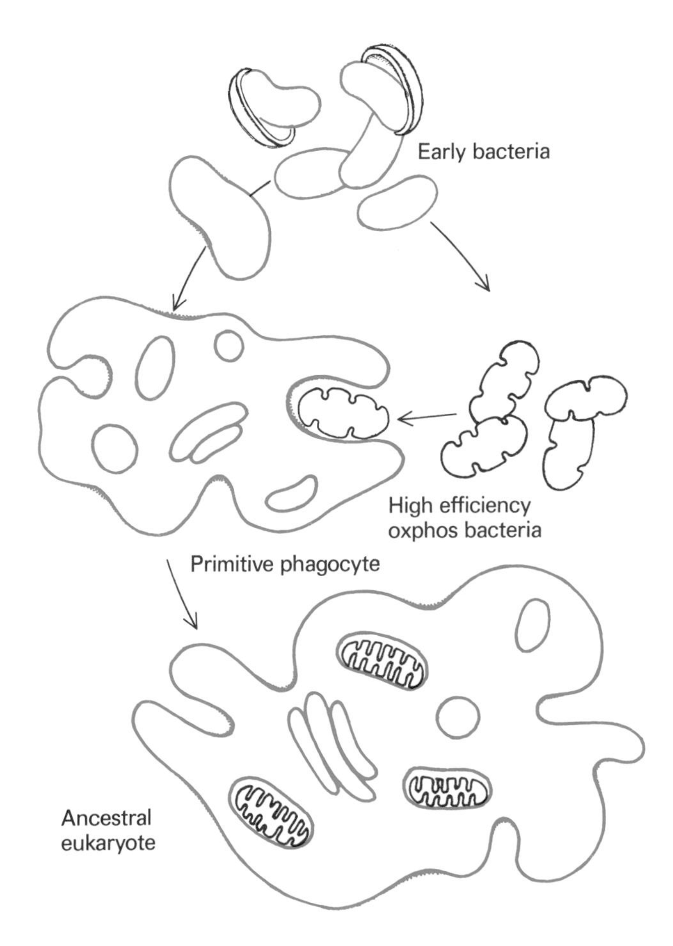

Endosymbiont hypothesis of origin of mitochondria. A primitive phagocyte (Chapter 6) engulfs and adopts symbiotically aerobic bacteria having a highly elaborate system of oxidative phosphorylation in their plasma membrane. These bacteria have developed from photosynthetic bacteria (see Chapter 10) in response to the progressive appearance of oxygen in the atmosphere. After their endosymbiotic adoption, they evolved into the mitochondria of a primitive eukaryote ancestral to all animal and plant cells.

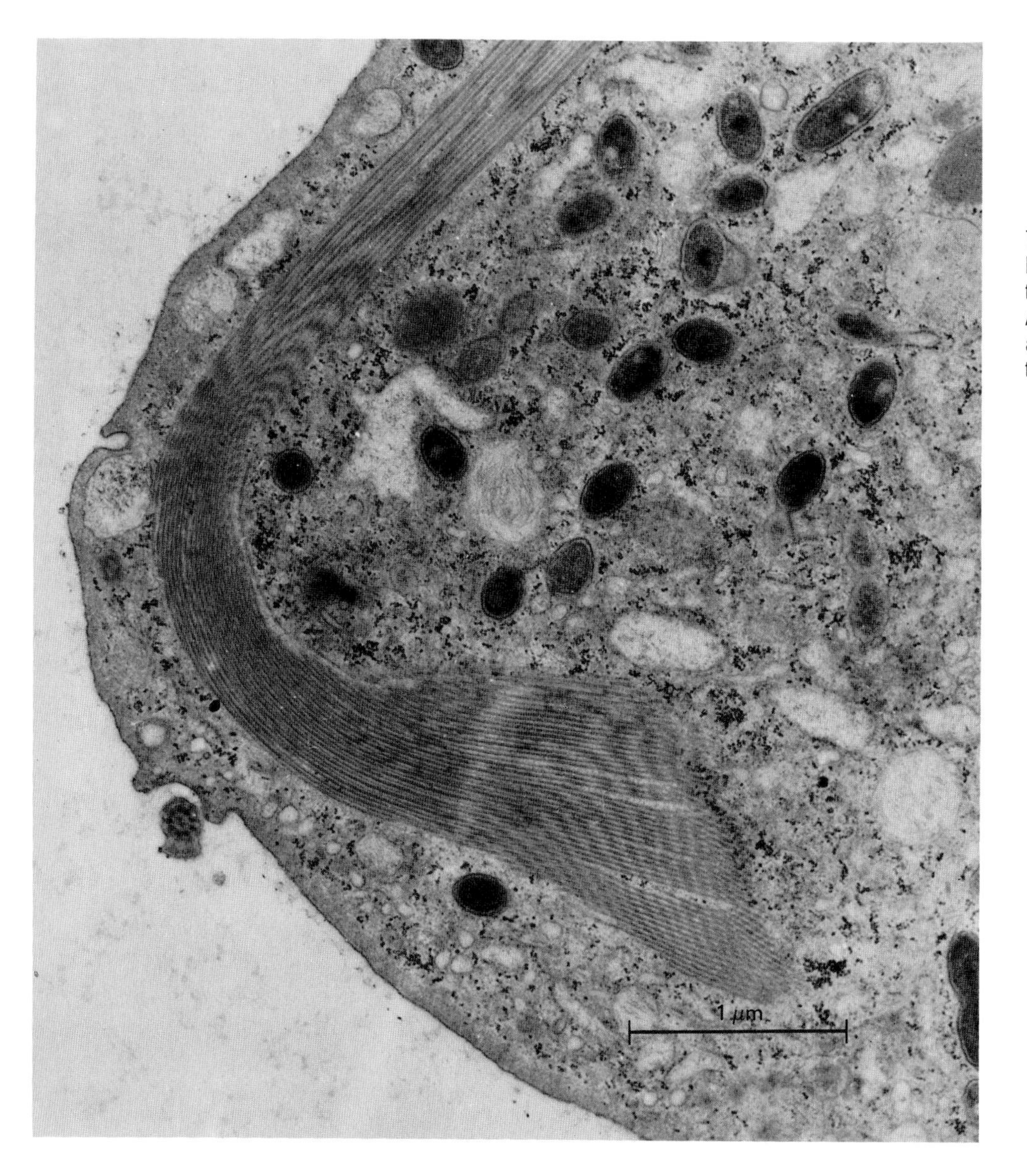

This electron micrograph shows bacterial endosymbionts living in the cytoplasm of the protozoon *Pyrsonympha*, which itself lives as a symbiont in the hindgut of the termite *Reticulitermes*.

distant past. Mitochondria, some counter, could have originated just as well from infoldings of the plasma membrane of an enlarging aerobic bacterium, much as the vacuome is supposed to have done (Chapter 6), but with a different kind of differentiation leading to segregation of the phosphorylating respiratory chain in the resulting vesicles. A number of other possibilities can be evoked, including that of an independent origin of the bacterial and mitochondrial systems through a phenomenon of convergent evolution.

It does not befit a mere guide to try to settle an argument that is still dividing experts. However, some of the features of mitochondria (and of chloroplasts) that support their origin from bacterial endosymbionts will be pointed out for the benefit of those who are seduced by the romantic appeal of the endosymbiont hypothesis.

Model reconstructed from serial sections showing a single giant mitochondrion of a yeast cell.

Reclining Figure, by Henry Moore.

Some Anatomical Details

The term mitochondrion is derived from the Greek *mitos*, thread, and *chondros*, grain. It refers to the filamentous appearance of the mitochondria in certain cells. In other cells, however, the mitochondria may be rodlike, egg shaped, or almost spherical. Sometimes several are joined together in bizarre concatenations, which, it has been claimed, may include the entire mitochondrial population of a cell, converting it into a strange, convoluted sort of open structure that could have been designed by the sculptor Henry Moore. The sizes of mitochondria are variable but tend to be large, as cellular dimensions go: 1 μm or more. Multiply this figure by one million, and you approach our own size. With a little luck, especially if we choose our target judiciously—the

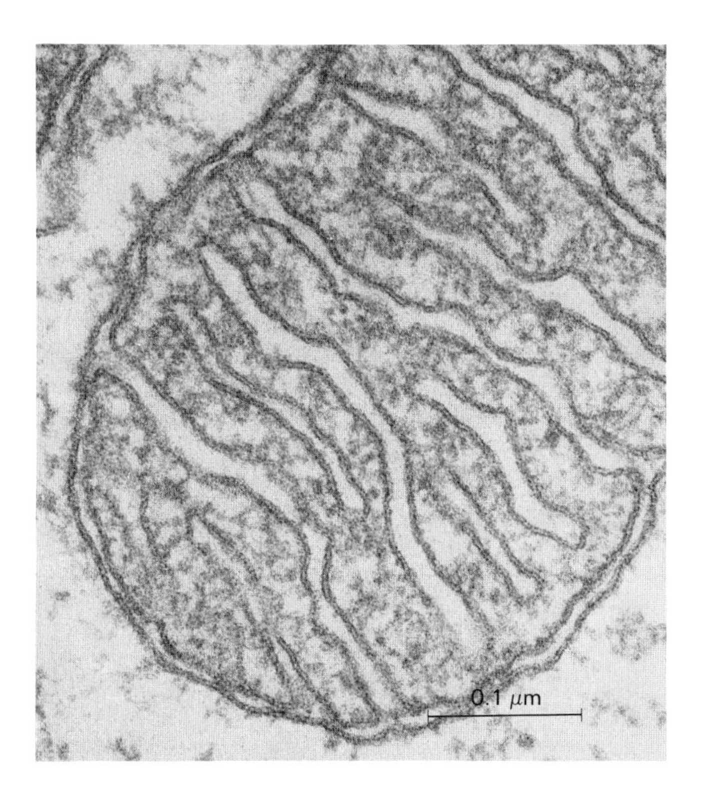

Thin section through a mitochondrion of a guinea-pig acinar pancreatic cell shows a continuous outer membrane and an inner membrane folded inward into numerous incomplete ridgelike partitions, the cristae mitochondriales.

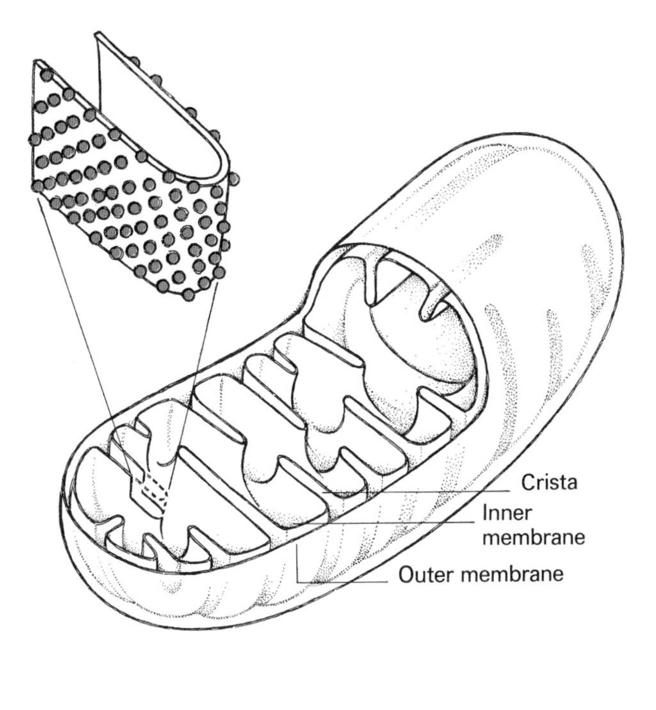

Three-dimensional reconstruction of the anatomy of a mitochondrion illustrates disposition of cristae. Enlargement shows that the inner face of the inner membrane bears small knobs (see the electron micrograph on p. 157).

bluebottle fly has particularly roomy mitochondria in its wing muscles—we may just be able to fit inside. There is no question, however, of entering without damage. We will have to resort to vivisection.

Seen from the outside, mitochondria appear encased in a thin, smooth, semitransparent envelope. Through this outer membrane can be discerned an inner one, bright pink and ravined by numerous deep clefts. Viewed from the inside, the structures corresponding to these clefts appear as ridges, or cristae (*crista* = ridge, in Latin), which form a number of incomplete partitions in the interior of the mitochondrial body. Actually, the cristae are no more than infoldings of the inner membrane, which is continuous and entirely closed. We can verify this by ripping the outer membrane and allowing the mitochondrion to take up water by applying some osmotic imbalance. It will swell to a considerable size without bursting, unfolding its inner membrane. The formation of cristae—replaced in certain cell types by tubular infoldings—represents a device for increasing the surface area of the membrane without enlarging the volume of the body. The usefulness of such an increase will become obvious, once we find out that the inner membrane is the very heart of the cellular power plant, its actual energy generator; it contains the whole respiratory chain and its associated phosphorylating systems.

From the evolutionary point of view, this inner membrane is believed to be derived from the plasma membrane of the putative ancestral bacterial endosymbiont. The outer membrane, on the other hand, presumably originates from the vacuolar system of the ancestral host-phagocyte. Indeed, it shows some kinship with the endoplasmic reticulum, with which it shares, for instance, the possession of a special pigment known as cytochrome b_5.

Energetics of Mitochondrial Oxidations

The cavity delimited by the ridged, inner mitochondrial membrane is filled with a structureless fluid, the mitochondrial matrix. It is a protein-rich sap made up largely of catabolic enzymes involved in the oxidative breakdown of all major foodstuffs, including the amino acids, which compose the proteins; the fatty acids, which are the main constituents of lipids; and pyruvic acid, which, we have seen, is formed in the cytosol as the product of aerobic glycolysis. All these pathways join in a central metabolic vortex known as the Krebs cycle, named for Sir Hans Krebs, the British-German biochemist who discovered it in the late 1930s.

A motley collection of exotic molecules participate in these transformations, and we will make no attempt to identify them, noting only the end-products, which are very simple substances: water, carbon dioxide, ammonia or urea, inorganic sulfate—the same substances, or almost, as would be formed by combustion in a furnace. An astonishingly cool furnace, though, in which matter is burned, but little heat is produced.

The paradox, you may remember from Chapter 7, is readily explained: foodstuffs do not combine with oxygen; they combine with water and give off hydrogen atoms or electrons to appropriate acceptors. These transactions are conducted at cell temperature, and in such a way as to release little energy. On the rare occasions when they do release an appreciable amount of energy, they are directed to take place through a substrate-level oxphos unit such as that encountered in glycolysis.

Take glucose, for example. Altogether, twelve pairs of electrons are released for every molecule of this sugar oxidized in a living cell:

$$C_6H_{12}O_6 + 6\ H_2O \longrightarrow 6\ CO_2 + 24\ e^- + 24\ H^+$$

Two of these pairs arise through oxidative glycolysis (amputated snake, Chapter 7):

$$C_6H_{12}O_6 \longrightarrow 2\ CH_3\text{—}CO\text{—}COOH + 4\ e^- + 4\ H^+$$

The other ten are yielded by the further oxidation, through the Krebs cycle, of the pyruvic acid molecules produced by glycolysis:

$$2\ CH_3\text{—}CO\text{—}COOH + 6\ H_2O \longrightarrow 6\ CO_2 + 20\ e^- + 20\ H^+$$

In glycolysis, as we have seen, the electrons are transferred to NAD^+ across a substrate-level oxphos unit:

$$4\ e^- + 2\ H^+ + 2\ NAD^+ + 2\ ADP + 2\ P_i \longrightarrow 2\ NADH + 2\ ATP + 2\ H_2O$$

In the Krebs cycle, four of the five electron pairs released by the oxidation of one molecule of pyruvic acid are transferred to NAD^+, in one case through a substrate-level oxphos unit. The fifth pair is transferred to a flavin coenzyme called FAD (see below). Thus, for two molecules of pyruvic acid:

$$20\ e^- + 12\ H^+ + 8\ NAD^+ + 2\ FAD + 2\ ADP + 2\ P_i \longrightarrow 8\ NADH + 2\ FADH_2 + 2\ ATP + 2\ H_2O$$

Adding up the two processes we find:

$$24\ e^- + 14\ H^+ + 10\ NAD^+ + 2\ FAD + 4\ ADP + 4\ P_i \longrightarrow 10\ NADH + 2\ FADH_2 + 4\ ATP + 4\ H_2O$$

which gives, for the anaerobic oxidation of glucose, as it actually occurs in living cells with the help of the immediately involved electron acceptors:

$$C_6H_{12}O_6 + 2\ H_2O + 10\ NAD^+ + 2\ FAD + 4\ ADP + 4\ P_i \longrightarrow 6\ CO_2 + 10\ NADH + 10\ H^+ + 2\ FADH_2 + 4\ ATP$$

The aerobic part of the process takes place through the transfer of the electrons from the reduced coenzymes to oxygen:

$$10\ NADH + 10\ H^+ + 2\ FADH_2 + 6\ O_2 \longrightarrow 10\ NAD^+ + 2\ FAD + 12\ H_2O$$

Foodstuff	Free energy of oxidation (kcal per gram-molecule)				
		Recovered in			
	Total	NADH	$FADH_2$	ATP*	Heat
Glucose	686	490	74	56	66
Palmitic acid	2338	1519	555	84	180
Glutamic acid	478	343	74	28	33

*ATP made by substrate-level phosphorylation minus ATP used for substrate activation.

The energy balances of the anaerobic and aerobic parts are easily computed. We have already seen (Chapter 7) that the "physiological" free energy of oxidation (electron potential) of NADH is −49 kcal per pair of electron-equivalents transferred to oxygen. Considering that the "physiological" electron potential of $FADH_2$ is of the order of −37 kcal per pair of electron-equivalents transferred to oxygen, we find for the aerobic part a total of −564 (10 × 49 + 2 × 37) kcal per gram-molecule of glucose oxidized.

On the other hand, we know from calorimetric measurements that the "physiological" free energy of oxidation of glucose, with oxygen as electron acceptor, equals −686 kcal per gram-molecule. This leaves an anaerobic balance of −122 (686 − 564) kcal per gram-molecule, of which 56 (4 × 14) are retrieved through the operation of the substrate-level oxphos units and 66 are lost as heat.

In other words (see the table above), when living cells "burn" glucose with their own electron snatchers instead of oxygen, they conserve more than 90 per cent of the free energy that would be released with oxygen as electron acceptor. Most of this energy is stored in reduced coenzymes, and a small part in ATP. What this means, in terms of the hydrodynamic analogy offered at the end of Chapter 7, is that the electrons released by the biological oxidation of glucose are first transferred to energy-rich reservoirs situated high above the water/oxygen level. In eight out of twelve cases, these transfers take place across small differences in altitude, with little loss of energy. In the four others, the difference is substantial, but the electron fall is harnessed to a substrate-level ATP-yielding oxphos unit.

Fatty acids and amino acids suffer an essentially similar kind of cold combustion, as illustrated by the examples of palmitic acid and glutamic acid shown in the table. Their complete oxidation occurs with more than 90 per cent of the free energy of combustion stored in energy-rich cofactors.

Knowing these values, we can better appreciate the plight of our early anaerobic ancestors. Since NAD, FAD, and the other electron-carrying coenzymes are present in cells in catalytic amounts, the advantages of cold combustion can be enjoyed only if an outlet exists for the electrons stored in these coenzymes. Fermenters

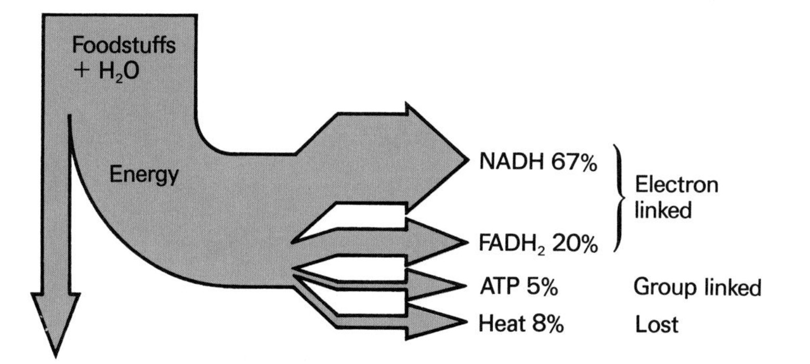

"Cold combustion," with water as oxygen supplier, brings about the complete oxidation of foodstuffs. Less than 10 per cent of the free energy released is dissipated as heat. Most is stored as high-potential electrons in NADH and, to a lesser extent, in reduced flavoproteins. A small fraction supports substrate-level oxphos units making ATP.

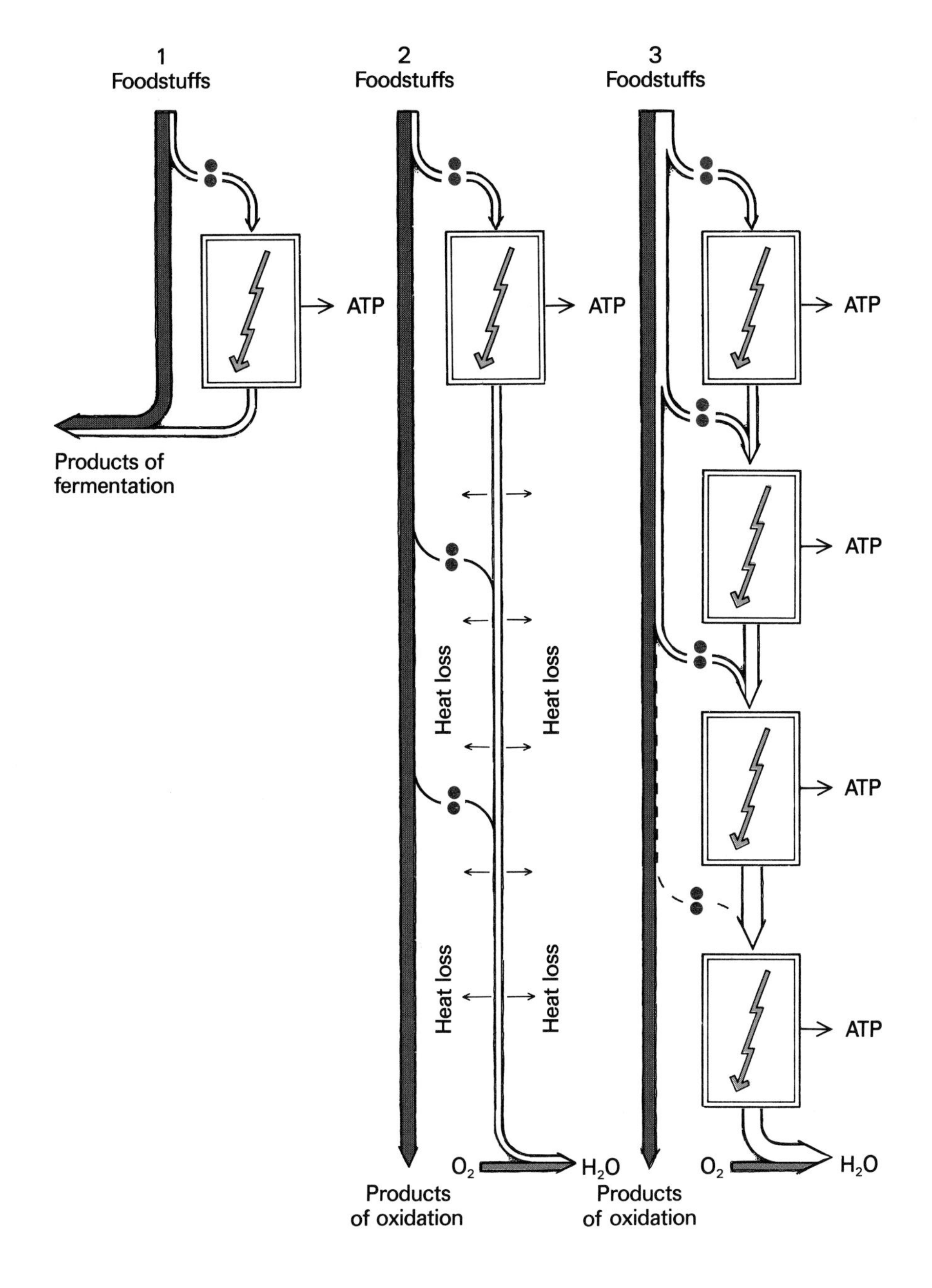

The problem of electron disposal.

Solution 1: Electrons return to metabolic flow and are excreted together with products of anaerobic fermentation. The drawbacks are poor utilization of foodstuffs and strict stoichiometric constraints.

Solution 2: Electrons are transferred freely to oxygen (or other exogenous acceptor). The advantages are greater flexibility and complete oxidation of foodstuffs, but the yield of utilizable energy (ATP) is very low.

Solution 3: Electrons are transferred to oxygen through phosphorylating respiratory chain. Foodstuffs are utilized completely, with a high energy yield.

satisfy this requirement by returning the electrons to the substrate flow. This procedure works, but it has a very low energy yield. For instance, anaerobic glycolysis, as we have seen, releases only 47 kcal per gram-molecule of glucose utilized, or less than 7 per cent of the energy that would be available by complete oxidation. It provides the cell with only two ATPs, or 28 kcal of utilizable energy.

Another drawback of fermentation is that it imposes rigid stoichiometric constraints on metabolic pathways. Only as many electrons may be released from the substrates as can be collected back into the substrate flow. These constraints are lifted if the electrons stored in coenzymes can be unloaded onto an exogenous acceptor. Then any kind of oxidizable foodstuff can be used, and used completely. The result is greater flexibility and often a better energy yield, at least from the cell's point of view.

From the point of view of world resources, however, it is shockingly wasteful if the electrons are simply unloaded without energy retrieval.

Consider again the example of glucose. As we have seen, substrate-level phosphorylation provides the anaerobic fermenter with two ATPs, or 28 kcal, for every gram-molecule of glucose utilized. The table shows that this yield is doubled in complete oxidation. But at what cost to total resources! Whereas, in fermentation, only 47 kcal are released out of the 686 that are available and the remainder is returned to the environment in the form of energy-rich molecules of lactic acid or ethanol, in unharnessed oxidation, the whole 686 kcal would be dissipated for a paltry additional gain of 28 kcal per gram-molecule of glucose used. Tolerable in a world of plenty, such profligacy becomes prohibitive when food is scarce. Yet, it is possible that some primitive aerobes operated in this way, and a number of such wasteful reactions still occur in the most advanced eukaryotes (Chapter 11).

Obviously, the only way to minimize oxidative energy waste is by making use of the energy that is released when the electrons are transferred from the reduced coenzymes to the final acceptor. This requires the successful insertion of oxphos units into the pathway of the electrons. Many different devices of this sort were developed in the course of evolution and harnessed to a variety of electron acceptors. The best results were obtained with oxygen as final acceptor, especially once the phosphorylating respiratory chain was perfected to its ultimate form, as is found today in mitochondria and some aerobic bacteria. Cells equipped with these super energy-savers can make as many as 38 ATPs with a single molecule of glucose, retrieving in usable form almost 80 per cent of the free energy released by the oxidative process.

The Phosphorylating Respiratory Chain

Physically, the mitochondrial energy transducers are situated entirely within the pink, pleated sheath that makes up the inner membrane of the body. It is a thin casing, 7 nm thick, hardly more than one-quarter of an inch at our millionfold magnification, smooth on the outside, but covered with small knobs on the inside. These knobs are about 9 nm in diameter (one-third of an inch if enlarged a millionfold) and are attached to the membrane by short, narrow stems, making the inner face of the membrane look as though it is covered with small mushrooms.

For further inspection, we will need our molecular magnifying glass. The sight it discloses is truly amazing, one of the greatest masterpieces of molecular engineering ever assembled. The whole expanse of the membrane is covered with tiny microcircuits, each made up of some fifteen to twenty different species of electron carriers beautifully put together to ensure efficient electron transfer. As many as 100,000 such microcircuits may be present on a single mitochondrial body.

Usually referred to as the respiratory chain, these microcircuits include a variety of different molecules (see Appendix 1). Prominent among them are the flavins and the hemes. The flavins (Latin *flavus,* yellow) are a remarkable class of greenish-yellow pigments derived from riboflavin, or vitamin B_2. The most important flavin derivatives are flavin mononucleotide (FMN) and flavin adenine dinucleotide (FAD). Their key property lies in their ability to act as hydrogen carriers. As such, they serve as coenzymes for numerous flavoprotein dehydrogenases.

Hemes (Greek *haima,* blood) are close relatives of the chlorophylls, the green photosynthetic pigments with which they share the characteristic porphyrin nucleus—a planar, disklike molecule made of four pyrrole rings linked by methene bridges (—CH=). In the center of this disk is a hole ringed by the electrons of four resonating nitrogen atoms. Insert a magnesium ion into this hole and you have the makings of a chlorophyll molecule. Substitute iron for magnesium and you have a heme. At the same time, thanks to the ability of iron to assume either the divalent ferrous (Fe^{2+}) or the trivalent ferric (Fe^{3+}) state, you have an electron carrier:

$$\text{Heme—Fe}^{2+} \rightleftharpoons \text{Heme—Fe}^{3+} + e^-$$

In combination with proteins, hemes make up a whole

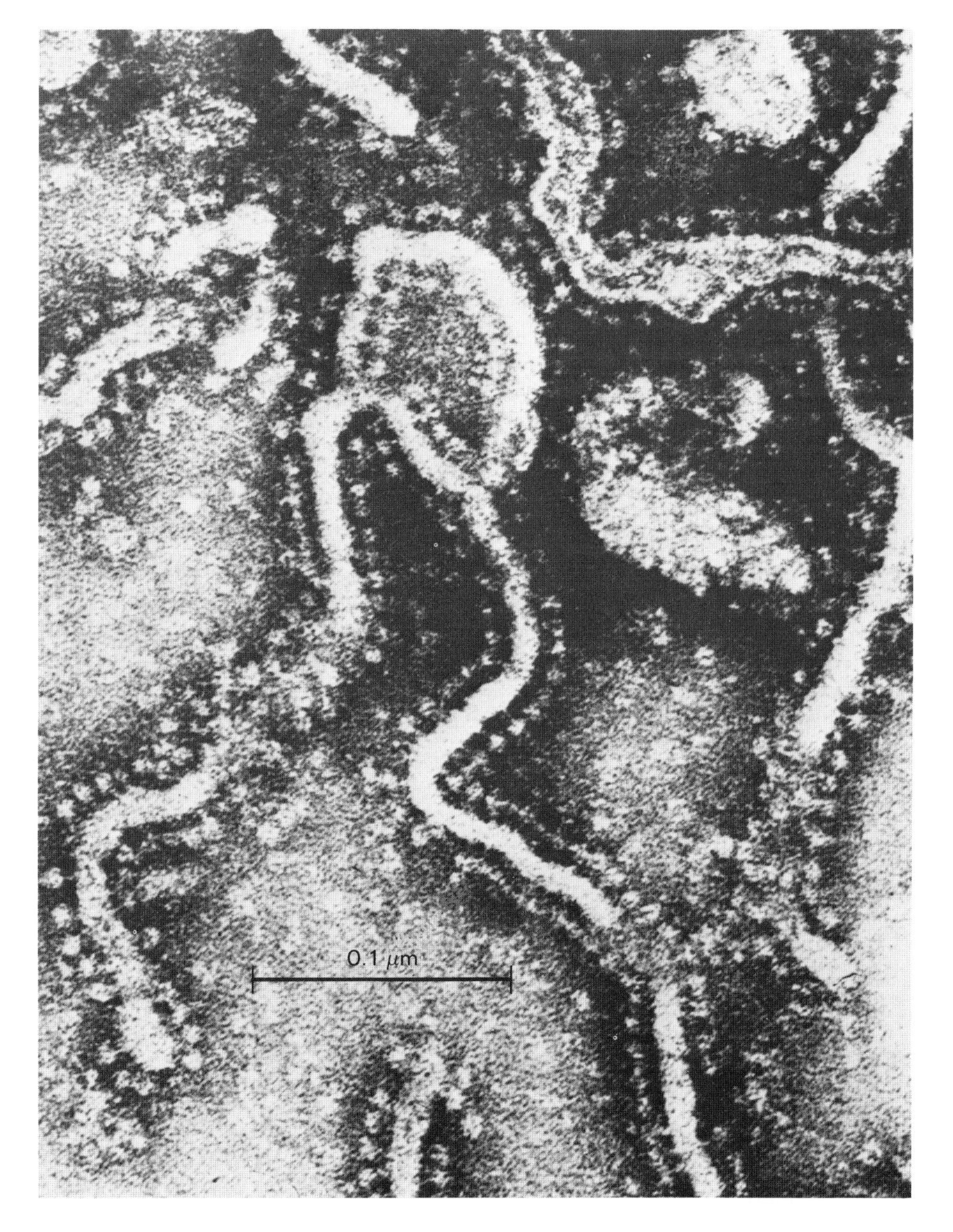

High-magnification electron micrograph of mitochondrial cristae. The preparation was negatively stained with electron-dense potassium phosphotungstate, which clearly delineates the knobs and their stalks protruding into the mitochondrial matrix. The knobs contain ATP-driven proton pumps.

panoply of colored molecules, with shades ranging from blood-red to pea-green. They serve in a variety of ways in the utilization of oxygen. Their best-known function is oxygen transport, as carried out by the red blood pigment hemoglobin. In the respiratory chain, hemoproteins are represented by a number of cytochromes, so named because they were among the earliest pigments discovered as regular cell components.

In addition to these two main classes of electron carriers, the respiratory chain also contains diphenols, which are oxidized to the corresponding quinones; iron-sulfur proteins, in which electron-carrying iron ions are encased within a shell of sulfhydryl groups; protein-bound copper ions; and possibly other components. It is noteworthy, and relevant to subsequent discussions, that some of these molecules act strictly as electron carriers (the hemoproteins and the metalloproteins), whereas the others transport hydrogen atoms (i.e., electrons in combination with protons).

Within each respiratory microcircuit, these various molecules are organized in a manner that allows them to act as a highly efficient "bucket brigade" for electrons. This means, first, that they are arranged in descending order of electron potential, so that the electrons can cascade down smoothly from one component to the next. In addition, they must be so oriented to each other that their active centers can readily exchange electrons with their partners on either side, relying only on the small displacements afforded by thermal vibrations and rotations to provide the necessary contacts. Finally, they must be fitted with appropriate inlets and outlets. As we shall see, the main inlet is provided by a flavoprotein enzyme that transfers electrons from matrix NADH to the top of the respiratory chain. A few secondary inlets also exist at lower energy levels. The main outlet is to oxygen, through an elaborate complex of cytochromes and copper known today as cytochrome oxidase, but originally designated by its discoverer, the German biochemist Otto Warburg, as *Atmungsferment,* respiratory enzyme. A most appropriate name, since most of the oxygen respired in the biosphere is utilized through the action of this enzyme.

To assemble such an electron-transport chain is certainly no mean feat. Yet these aspects of its architecture

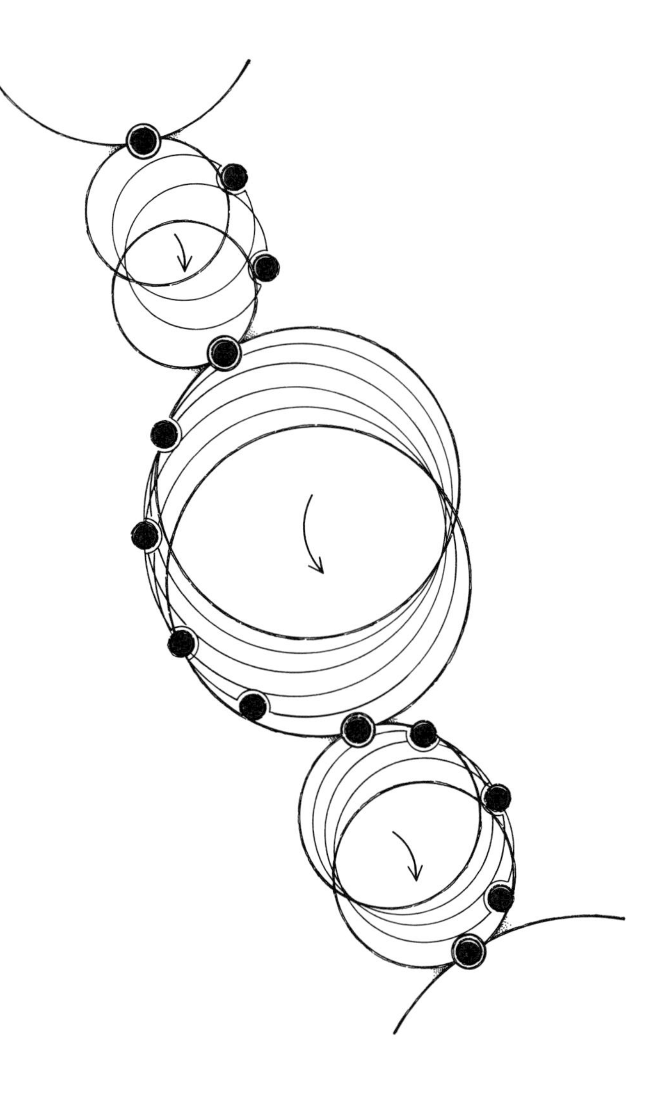

A molecular "bucket-brigade" for electrons. Members of the respiratory chain are organized so that thermal vibrations and rotations can bring their electron-carrying centers alternatively in contact with those of *two* of their neighbors.

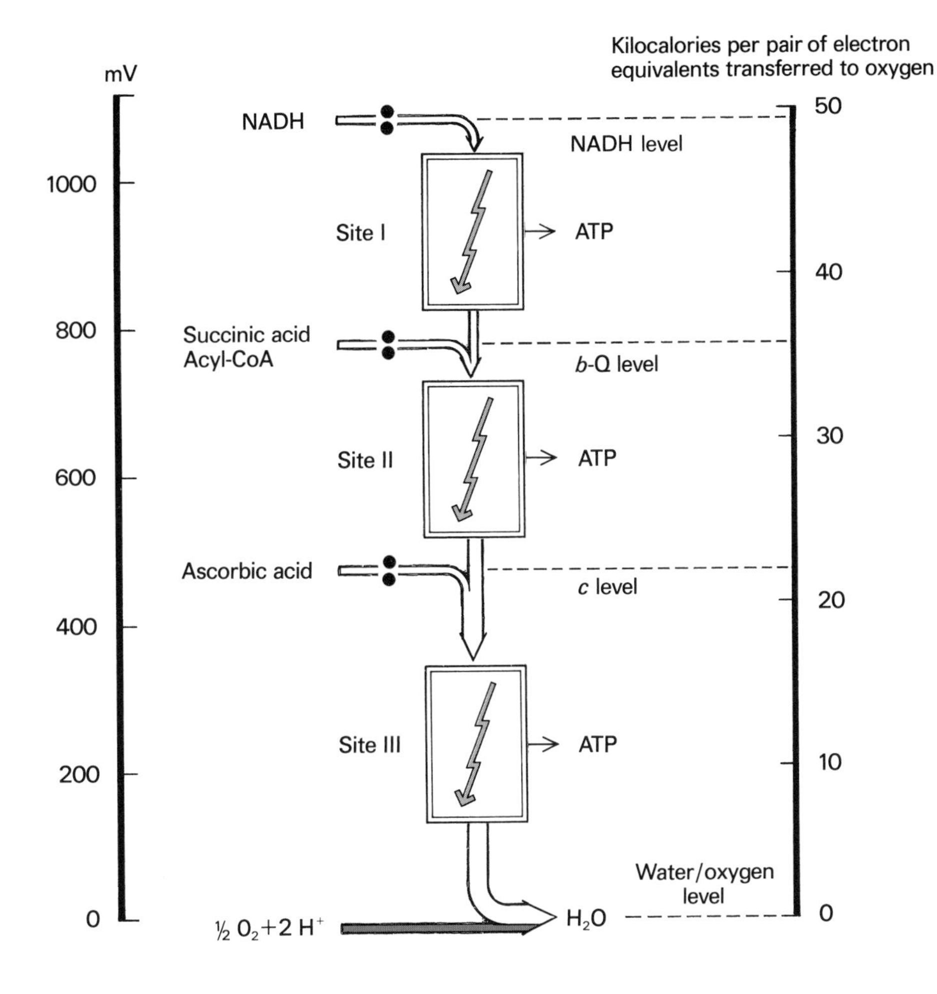

Respiratory chain phosphorylation. Electron pairs falling down to oxygen traverse between one and three oxphos units, depending on the level at which they enter.

are almost trivial in comparison with some of its other properties, for the microcircuit is not just an electron conductor. It is an energy transducer, organized so as to trap the energy dissipated by the electrons as they tumble down to oxygen and to render it usable for biological work through the mediation of ATP. In other words, the microcircuits contain incorporated oxphos units.

This introduces two additional conditions into the design of the microcircuits. First, the potentials of the carriers must be so poised as to provide appropriate electron falls. Remember, for a pair of electrons to yield enough energy to sustain the synthesis of an ATP molecule, it must fall down a potential difference of at least 300 mV. The second condition is that the electrons going down such falls must be subjected to some sort of constraint that couples their flow obligatorily to the synthesis of ATP. This is similar in principle to the kind of constraint that deviation through a hydroelectric turbine imposes on a waterfall, coupling the flow of water obligatorily to the generation of electricity.

The first condition is satisfied very efficiently. As many as three oxphos units, or phosphorylation sites, are intercalated in series in the pathway of the electrons. They are called sites I, II, and III. They are separated by platforms, which we will call the NADH level, the *b*-Q level, and the *c* level, each after some characteristic electron carrier. Each of these platforms is accessible to electrons supplied from the outside by appropriate donors. Electrons can be collected back from them with certain artificial acceptors, such as ferricyanide or methylene blue. Normally, however, the only electron outlet is to oxygen, at the zero level of energy.

Let us consider the *c* level first. It derives its name from cytochrome *c*, its principal component. Among the substances capable of donating electrons to the chain at this level is ascorbic acid, or vitamin C, the substance in fresh fruits and vegetables that prevents the nutritional disease called scurvy (scorbut), a common plight of navigators in the old days (ascorbic comes from the Greek *a*, not, plus scorbut). The antiscurvy property of vitamin C is not known to be directly linked to its ability to donate electrons, although it could be in some specialized way. In laboratory experiments, however, ascorbic acid will interact with mitochondria at the *c* level, and this has allowed the identification of the oxphos unit at site III through the finding that one molecule of ATP is synthesized for every pair of electrons transferred from ascorbic acid to oxygen.

Access to site II is through the *b*-Q level, so named because it is occupied jointly by cytochrome *b* and coenzyme Q, or ubiquinone, a quinone type of electron carrier. Several FAD-dependent flavoproteins are connected with this level, to serve as port of entry for electrons provided by certain metabolic substrates, such as succinic acid, an intermediate in the Krebs cycle, and fatty acyl-coenzyme A derivatives, the activated forms of fatty acids. When a pair of electrons enters at this level, it falls down two consecutive oxphos units: two ATP molecules are made.

The NADH level is occupied by a special flavoprotein that transfers electrons from NADH into phosphorylation site I. Altogether, three ATPs are synthesized for each electron pair that enters at this level and is collected by oxygen at the bottom. NAD^+ itself serves as electron acceptor in countless reactions. So this pathway channels most of the catabolic electrons in aerobic organisms.

The efficiency of these mitochondrial transducers is quite remarkable. With NADH as electron donor, more than 85 per cent of the free energy of oxidation is recovered as ATP: $3 \times 14 = 42$ kcal per pair of electron-equivalents out of a total of 49. With $FADH_2$, the efficiency is a little lower, but still considerable: $2 \times 14 = 28$ kcal per pair of electron-equivalents out of 37, or 76 per cent.

Combined with the high efficiency of the cold combustion reactions themselves, these values lead to overall efficiencies approaching 80 per cent, as can readily be calculated from the data of the table shown on page 154. These are staggering yields, to which no man-made combustion power plant has ever come near. They reflect the remarkable quality of the energy transducers incorporated into the microcircuits of the inner mitochondrial membrane.

How these transducers work is a question that has occupied some of the best brains in the business for more than 30 years. The answer, according to the British scientist Peter Mitchell, the father of the widely accepted "chemiosmotic theory," is: through protonmotive power. It is a somewhat offbeat topic for tourists, but let us have a go at it anyway. For, as will be even clearer after Chapter 10, we are looking at one of life's most central mechanisms. It is worth the extra effort.

Protonmotive Power, the Secret of Oxphos

The inner mitochondrial membrane is "proton tight": it is impermeable to hydrogen ions. This means that it is also impermeable to hydroxyl ions, OH^-, which otherwise could act as carriers for H^+ ions, taking them through as H_2O and returning as OH^-. In fact, the membrane is completely "ion tight": no ion, whether positive or negative, can get through it at an appreciable rate by passive diffusion. The whole commerce of ions between mitochondria and the surrounding cytosol is strictly regulated by special gates or pumps.

There is such a pump for protons—or rather, there are two types of such pumps. They are both built to pump protons forcibly out of the mitochondria with the help of a supply of energy. They are both reversible, by which is meant that they can run backward and generate power at the expense of an inward flow of protons. But their power supplies are different: one is electron driven, the other ATP driven. Such, in a nutshell, is the secret of oxphos: electron flow and ATP synthesis are interlinked by protonmotive power.

Let us look first at the ATP-driven pump. For each molecule of ATP split into ADP and inorganic phosphate,

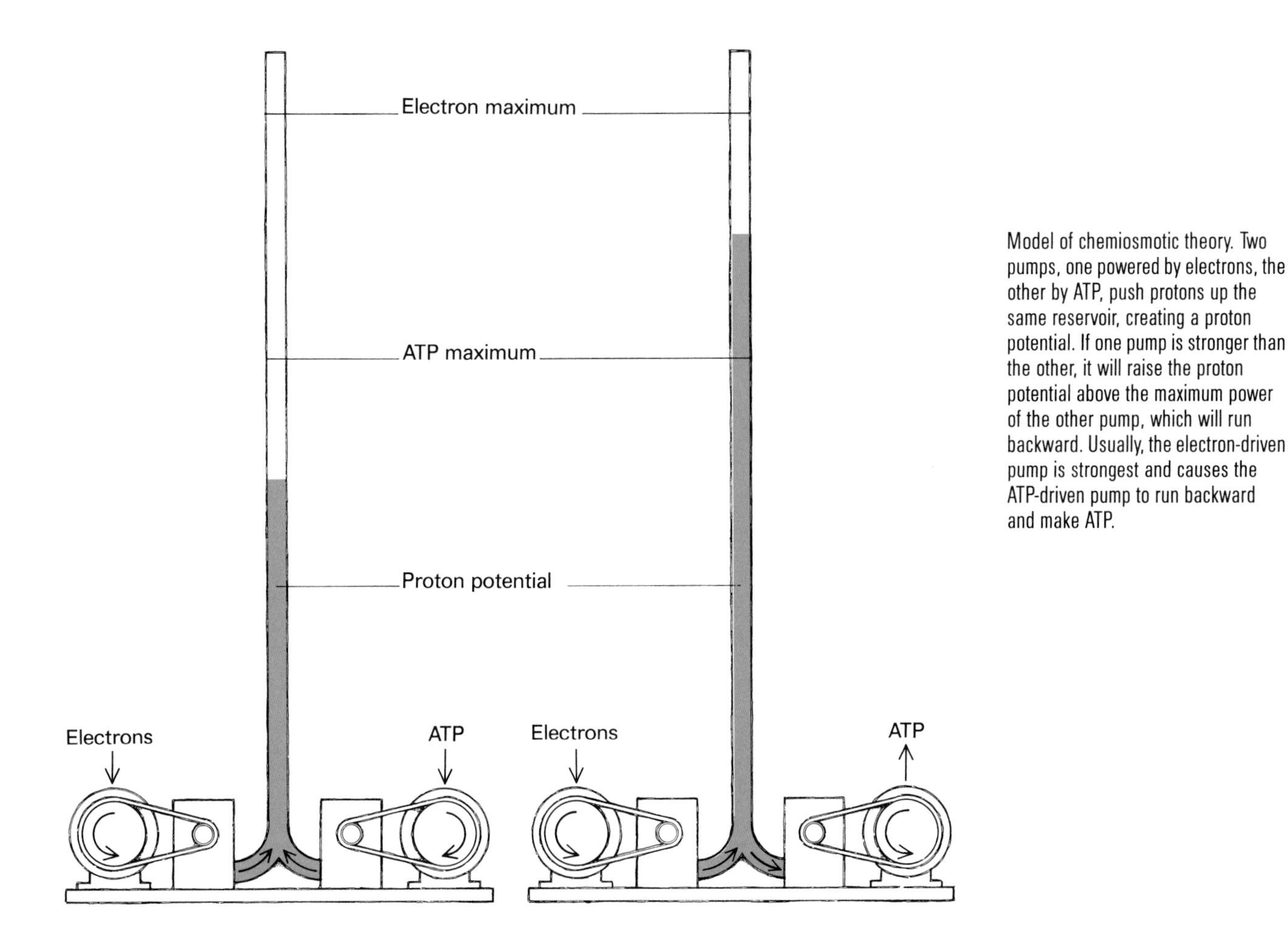

Model of chemiosmotic theory. Two pumps, one powered by electrons, the other by ATP, push protons up the same reservoir, creating a proton potential. If one pump is stronger than the other, it will raise the proton potential above the maximum power of the other pump, which will run backward. Usually, the electron-driven pump is strongest and causes the ATP-driven pump to run backward and make ATP.

it forces n protons out of the mitochondrial body. It can go on doing so as long as it is able to overcome the proton potential that builds up outside. When this potential becomes equivalent to the power of the pump, the pump will stop, just like an electric pump that can lift water only to a certain level, depending on the power of the motor. We know the power the ATP-driven pump has available; it is of the order of 14 kcal per gram-molecule of ATP split. Therefore, the maximum level to which it can lift protons is about $14/n$ kcal per proton-equivalent.

The same reasoning applies to the electron-driven pump, except that it is powered by the energy that is released by a pair of electrons falling down a potential difference. If this difference is 300 mV, the energy available will be 14 kcal per pair of electron-equivalents. Such a pump will stop, like the ATP-driven pump, when the proton potential reaches $14/n$ kcal per proton-equivalent, in which n represents the number of protons that the pump translocates per pair of electrons downgraded. As it happens—the coupled system described here would not work otherwise—n is the same for the two pumps.

If you now have the two pumps working together against the same proton potential, and one happens to be slightly stronger than the other, the more powerful pump will lift the proton potential above the limit level of the weaker pump, which, if reversible, will run backward. Just so with two electric pumps pushing water up into the same reservoir; if one pump is stronger than the other, it will bring the water level sufficiently high to reverse the flow through the weaker pump and cause it to act as a generator. In a reversible system of this sort, the electricity generated by the second pump will be equal to the electricity consumed by the first pump, barring the inevitable small losses due to the imperfections of the system.

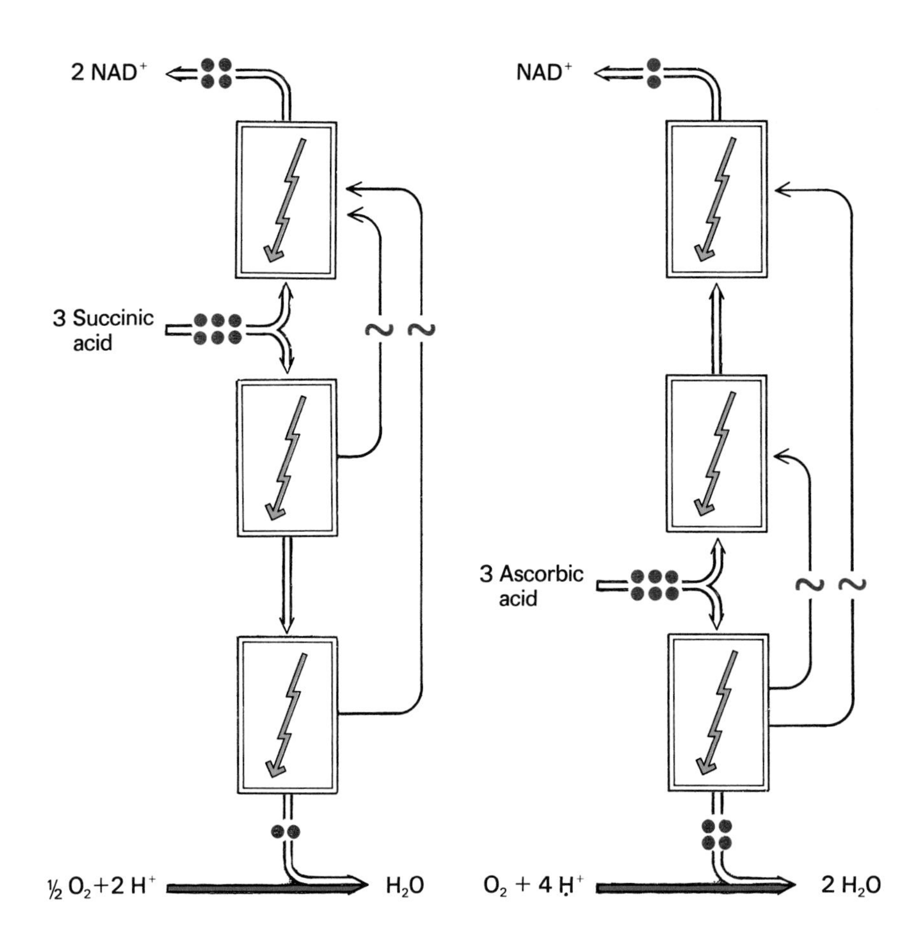

Examples of reverse electron transfer. The fall of one electron pair from succinic acid to oxygen produces two "squiggles" of protonmotive force at sites II and III. With a supply of NAD$^+$ as acceptor, two additional electron pairs are delivered from succinic acid and *lifted* through site I, acting in reverse with the help of the protonmotive force produced in sites II and III. Similarly, two electron pairs are transferred from ascorbic acid to oxygen through site III, producing two "squiggles" that help a third electron pair climb from ascorbic acid to NAD$^+$ through sites II and I acting in reverse.

Electric energy is thus being transferred by way of a hydraulic transducer.

An analogous situation obtains in the mitochondrial membrane, but with different kinds of pumps and energy. Under most circumstances, the more powerful of the two mitochondrial pumps is the electron-driven one, since it is constantly supplied with fuel by metabolism, whereas ATP is continuously consumed by the various forms of work accomplished by the cells. And so the ATP-driven pump runs backward, making ATP with the help of the protonmotive power supplied by electron flow.

Normally, this system does not run in the reverse direction, for there is no separate source of ATP to drive it. But what can happen is that one electron-driven pump has more power than the others. Remember that there are three distinct oxphos units in the respiratory chain, each connected to different electron carriers and therefore hooked onto a different proton pump. Of the three, that at site III is stronger than the other two. It has available a potential drop close to 500 mV and is essentially irreversible for this reason, whereas those at sites I and II are freely reversible, with potential differences of the order of 300 mV. Thanks to the extra drop down site III, electrons fed into the system at the *c* level or at the *b*-Q level may generate enough of a proton potential to reverse the flow of electrons through the site or sites situated above their port of entry. For such reverse electron transfer to manifest itself, a suitable acceptor—NAD^+, for instance—must be supplied to collect the uplifted electrons at the higher level. Such a phenomenon has been produced artificially with mitochondria. As we shall see, it functions physiologically in many autotrophic organisms, where it is of immense biological importance.

In trying to visualize this system, you have to realize that the mitochondrial proton pumps do not actually secrete acid in the way the proton pump in the lining of the stomach does. The reason is that the extruded protons are neither accompanied by a negative ion nor exchanged for another positive ion. As protons are forced out, an electric imbalance is created by the loss of positive charges from inside the mitochondrion; an electric potential, positive outside, is built up across the mitochondrial membrane. As the potential increases, it becomes increasingly difficult to push protons out against this potential—remember, charges of equal sign repel each other according to Coulomb's law—and eventually the pump will grind to a halt when the work required to force an additional positive charge against the electric potential becomes equal to the power of the pump. In other words, the proton potential in this particular system is expressed mainly in the form of an electric potential difference across the membrane. There is a small difference in acidity between the inside and the outside of mitochondria, but it is of minor importance.

We can calculate the mitochondrial membrane potential from the known power of the pump: $14/n$ kcal per proton-equivalent ($58{,}600/n$ joules per proton-equivalent), or:

$$\frac{58{,}600}{96{,}500 \times n} = 0.6/n \text{ V}$$

According to Mitchell, $n = 2$, which makes the membrane potential 300 mV. But not all workers agree with Mitchell. Some claim that $n = 3$ or 4, in which case the potential would be 200 or 150 mV. Others believe in an even lower value of membrane potential, not so much because they disagree with Mitchell's stoichiometry, but because they see the proton potential as being partly manifested in conformational changes in the membrane. Unfortunately, we do not have a voltmeter to settle the question. Nor is our eyesight good enough to find out whether the molecules are twisted out of shape when the microcircuits become energized.

What, now, do we know about the pumps themselves? Very little, unfortunately. Several ATP-driven proton pumps are known—remember, there is one in the membranes of endosomes and lysosomes—but the molecular mechanism linking proton transfer to ATP hydrolysis has not been elucidated for any pump so far. What is known is that the mitochondrial ATP-driven proton pump is situated in the little knobs that can be seen protruding from the inner face of the membrane and that it can be selectively jammed by a substance called oligomycin. As the name indicates, this substance is made by a mold (Greek *mykês*). We will encounter a number of other "mycins" on our tour. They have been invaluable tools to biochemists and molecular biologists in helping to unravel some of the most complex biological mechanisms. It is interesting that all these substances are in some way offshoots of penicillin. Before the discovery of that substance, few scientists showed an interest in molds. But, after it was found that a mold (*Penicillium notatum*) manufactures a substance (named penicillin for obvious reasons) capable of blocking the development of certain pathogenic microbes, a vast search was instituted for other mold products with antibiotic properties. Hundreds of thousands of strains of molds from all over the world were screened for this purpose, yielding a rich crop of active substances. A few of these were adopted for clinical use; streptomycin is an example. But most had to be rejected because they proved highly toxic to human organisms; they inhibit some essential biological process, such as DNA transcription, protein synthesis, or oxidative phosphorylation. In biochemistry, we can often learn a great deal about the working of a machine from the way it is blocked by inhibitors. And so these dropouts of drug research became instruments of key scientific advances, illustrating, as does much of the history of science, the unpredictability of discovery and the folly of those bureaucrats, unfortu-

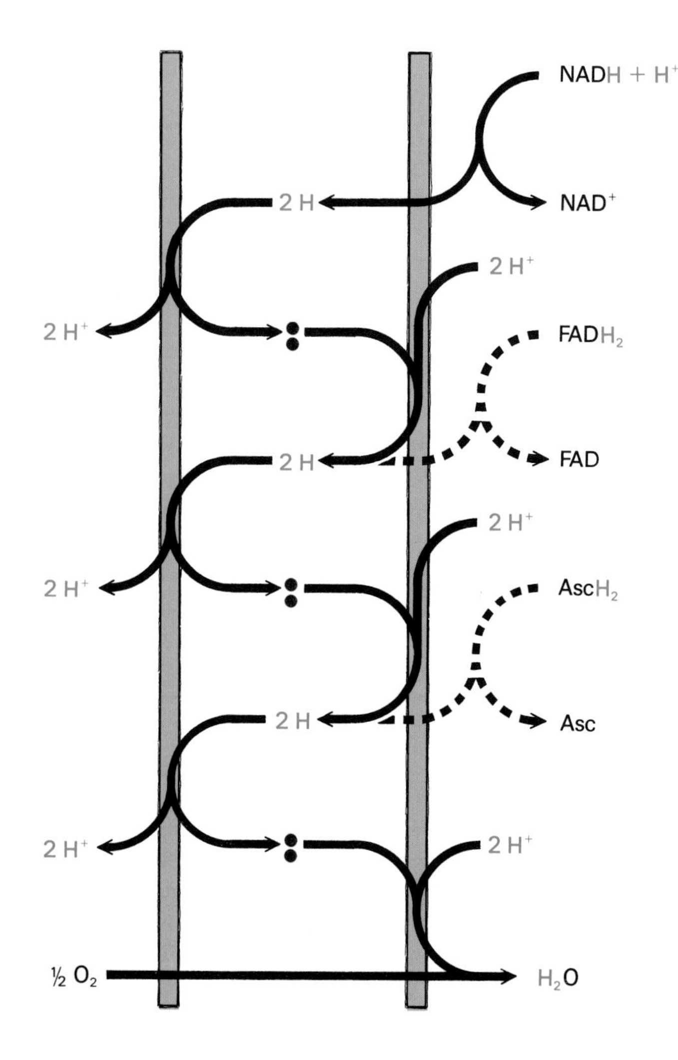

Loop hypothesis proposed by Mitchell to explain coupling between electron transfer and proton extrusion. At each prosphorylation site, electrons are assumed to carry protons from inside to outside, and to return naked.

nately more and more numerous, who would tie scientists down to rigidly predetermined programs.

Turning now to the electron-driven proton pumps, we must look for them in the molecular architecture of the microcircuits themselves, as their functioning depends on an obligatory coupling between electron transfer and proton extrusion. Mitchell has proposed an ingenious hypothesis to account for this coupling. Expressed in simple terms, it assumes that electrons enter oxphos units as hydrogen atoms supplied from inside the mitochondrion and that they leave the units naked, after having shed the protons outside. This model implies that the microcircuits are made of alternating hydrogen-carrying and electron-carrying segments and that they are organized in loops in the thickness of the membrane. This, then, would be the answer to the question, What sort of constraint does the molecular anatomy of the microcircuits impose on the flow of electrons? The constraint is that the electrons are forced along a sinuous path that takes them alternately from one face of the membrane to the other, together with protons in one direction and without protons in the other. In reality, the mechanisms are considerably more complex than might be gathered from this simplistic description, and involve subtle physicochemical and conformational changes associated with the passage of the carriers from the oxidized to the reduced state. Add these requirements to all those that were enumerated above and you have a pretty refined piece of machinery. And mind you, it is all compressed in a little "biochip" about 0.3-millionth of an inch thick and one-millionth of an inch wide!

In consequence of their built-in constraints, the mitochondrial oxphos units never spend energy unnecessarily. Electron flow and ATP synthesis are obligatorily coupled, which means that the rate of electron flow (and of substrate and oxygen consumption) is automatically adjusted to the rate of ATP consumption. At rest, little ATP is used, and consequently food consumption and respiration are low. Let there be a sudden call on ATP, as in an athlete starting a race, and electron flow immediately increases manifold, causing catabolic oxidations and oxygen uptake to increase simultaneously in the same proportion. This regulatory mechanism is called respiratory control.

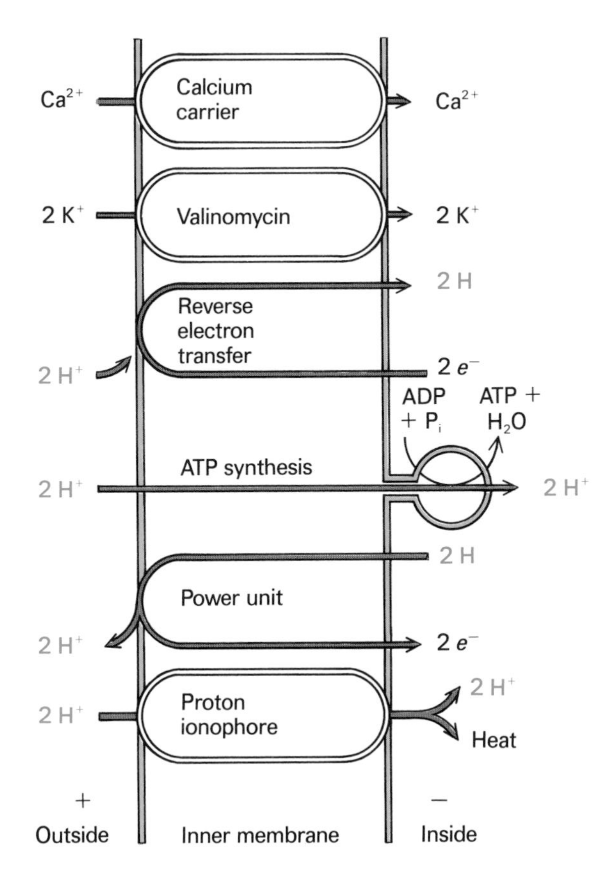

Some of the forms of work that can be powered by mitochondrial protonmotive power. Protons can *return* inside, with the coupled synthesis of ATP; or they can *return* with low-level electrons and bring them to a higher level; or they can be *exchanged* for calcium, thanks to a calcium carrier present naturally in the membrane; or they can be *exchanged* for potassium, if valinomycin (potassium ionophore of mold origin) is added; or they can *return*, with the help of an added proton ionophore (uncoupler), with collapse of the potential and dissipation of the energy as heat. Note that all these forms of work can also be powered by the ATP-synthesizing machinery, acting in reverse as a proton-extruding ATP hydrolase.

This control depends on the strength of the coupling, which is itself determined by the proton-tightness of the membrane. Any substance capable of transporting protons across the lipid bilayer of the membrane (a proton ionophore) will immediately cause the proton potential to collapse and so uncouple the system. Many such uncouplers are known. Their prototype is 2,4-dinitrophenol.

Another way of discharging the membrane potential is by providing a positive ion capable of getting across the membrane into the mitochondrion. For this to happen, there must be an appropriate carrier in the membrane. One exists naturally for calcium ions, with the consequence that mitochondria tend to keep their surroundings free of calcium ions—an important regulatory mechanism (see Chapter 12). A passage for potassium ions can be created artificially by the addition of a mold toxin called valinomycin, which has the property of a potassium ionophore. Note that, in these cases, work is still performed. Electron flow now supports the active inward transport of the added cation in exchange for the outward transport of protons.

As a last comment, it should be pointed out that substrate-level oxphos units, such as are present in glycolysis or in the Krebs cycle, do not operate by protonmotive force. They are purely chemical transducers. In fact, early understanding of the mechanism of substrate-level phosphorylations impeded for a long time the elucidation of the type of membrane-linked, mediator-level phosphorylation that is found in mitochondria, because most investigators were searching for nonexistent chemical intermediates.

Growth and Multiplication of Mitochondria

Mitochondria grow and divide, as would be expected of a symbiotic microorganism. Unlike their putative bacterial ancestors, however, they manufacture hardly any of their constituents. Most mitochondrial proteins are synthesized in the extramitochondrial cytoplasm according to instructions sent out from the nucleus. How these proteins are directed to the mitochondria and pass across two membrane layers is far from clear. The question is considered in Chapter 15.

This definitely is not how bacteria acquire their proteins. Bacteria make their proteins themselves, with their own ribosomes, reading from their own genes. One would expect their symbiotic descendants to do the same. Since they do not, should we not reject the symbiont theory? Not necessarily, for mitochondria are not entirely controlled by nuclear genes; they do, in fact, have a genetic system of their own. It is a very rudimentary system, coding for only a small number of proteins, but it is essentially complete, including DNA as well as all the pieces that are needed for the replication, transcription, and translation of the genetic information it contains. And it is operative, expressing itself in certain specific mitochondrial proteins, such as part of the cytochrome-oxidase complex. It can even suffer mutations, which are therefore transmitted by the cytoplasm and not by the usual Mendelian chromosomal mechanism. The so-called "petite" mutation of yeast is a typical example of such a hereditary change affecting the mitochondrial DNA.

The fascinating thing about this mitochondrial genetic machinery is that it has typical bacterial characteristics. The DNA is circular, resembling bacterial DNA. The ribosomes are smaller than the cytoplasmic ribosomes and about the same size as bacterial ribosomes. Their capacity to make proteins is blocked by the antibiotic chloramphenicol, an inhibitor of bacterial protein synthesis but not of eukaryotic protein synthesis, and it is not affected by the eukaryotic inhibitor cycloheximide, which also is inactive on bacteria. Most impressive of all, mitochondria do not even speak the same language as the rest of the cell; they employ a partly different genetic code.

These are some of the reasons why many scientists believe that mitochondria are descended from some distant prokaryotic ancestors that were adopted endosymbiotically and that, in the course of their prolonged process of integration, lost greater and greater control over their own fate. But a few vestiges of their erstwhile autonomy appear to have survived this attrition process to give us an inkling of historical events of immense importance that may have happened more than one billion years ago. Those events would have remained forever buried in the darkness of the past but for these telltale relics. The revelation, however, is not unambiguous and has been interpreted differently by others.

10 Chloroplasts: Autotrophy and Photosynthesis

Although our tour is concerned mostly with animal cells, it would hardly be complete without at least a passing excursion into the plant world. Plants belong to the general group of autotrophs—literally self-feeding organisms—better, though less commonly, designated lithotrophs (Greek *lithos*, stone). Indeed, the "foodstuffs" of autotrophs come entirely from the mineral world in the form of carbon dioxide (CO_2), water (H_2O), nitrate (NO_3^-), sulfate (SO_4^{2-}), and related inorganic components. The required elements are there, but in strictly "no cal" form, valueless from the energetic point of view. Autotrophs, therefore, need a separate source of energy, since they cannot, like heterotrophs (organotrophs), derive power from their food.

How to Survive in a Lifeless World

In what form this energy? There are several answers to that question, but they all boil down to a single word: electrons. That electrons are necessary for autotrophic life is obvious from the nature of its building blocks. To convert CO_2, H_2O, NO_3^-, SO_4^{2-}, and the like into carbohydrates, proteins, lipids, and other biological constituents, a large quantity of electrons must be supplied—as many, in fact, as are released in heterotrophs by the oxidation of these constituents. That electrons may also be sufficient is suggested by our knowledge of heterotrophs. Provided an organism has at least one functional oxphos unit available, it can use electrons to manufacture ATP and thereby cover all its energy needs.

The electrons must be of good quality: they must be delivered at a high enough level of potential energy. As far as ATP synthesis is concerned, this requirement is flexible and depends on the level at which the electrons leave the oxphos unit. All that is needed is that they enter at least 300 mV above exit level (if, as is usual, they travel in pairs). The requirement is, however, more stringent for reductive syntheses, which most commonly use electrons supplied at a level at least equivalent to that of NADH. The reason is obvious: that is the level from which most of the stored electrons tumble back down during catabolism.

In primitive systems, autotrophic reductions are carried out via NADH—that is, with the same coenzyme as catabolic oxidations. Animal glycolysis also functions in this way. As we saw at the end of Chapter 7, the oxphos unit of the glycolytic chain operates near thermodynamic equilibrium, so that relatively small changes in the concentrations of the participating substances may suffice to reverse the flow of electrons through the unit and to cause the cell to switch from glycolysis to gluconeogenesis.

By and large, however, Nature has assigned a special coenzyme to biosynthetic reductions, thus separating anabolism from catabolism. This coenzyme is a phosphorylated derivative of NAD and is represented as NADP, which stands for nicotinamide adenine dinucleotide phosphate. Despite its close kinship with NAD, NADP is treated differently. With very rare exceptions, dehydrogenases use either one or the other coenzyme, not both. Thanks to this distinct mode of handling, helped, if need be, by the expenditure of energy, the NADPH reservoir is maintained physiologically at a significantly higher energy level than is NADH: about 1,200 mV above water/oxygen level, or 55 kcal per pair of electron-equivalents transferred to oxygen. This is consistent with the fact that the only direction electrons can take when they move under their own power is downward. Thus in synthetic reductions, the electrons fall down from the NADPH level to that of substrate. And in catabolism they fall further down, from substrate to NAD^+, from NADH to the next acceptor, and so on all the way down to oxygen. In addition to an adequate supply of high-grade electrons, the autotrophic way of life requires an appropriate catalytic machinery. Much of this could be

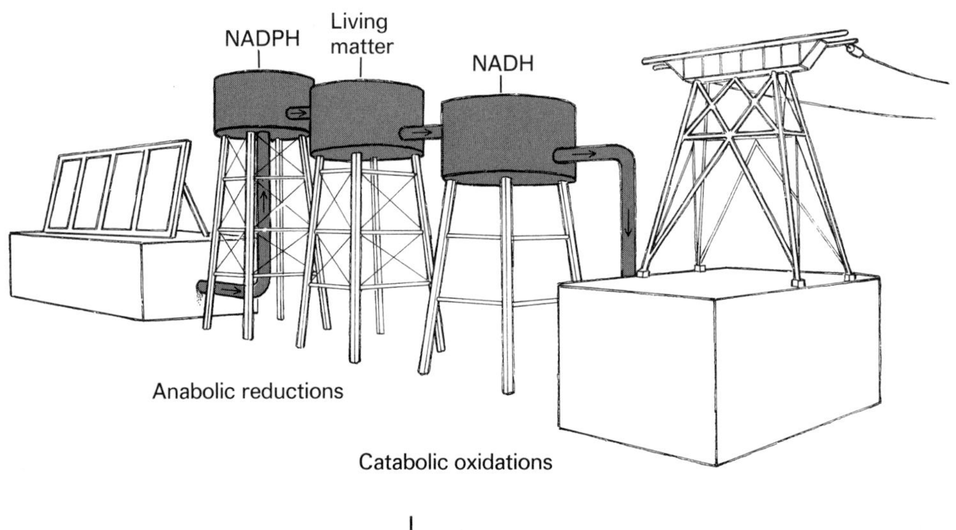

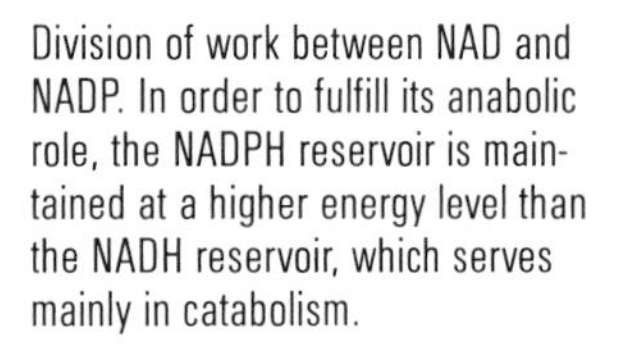
Division of work between NAD and NADP. In order to fulfill its anabolic role, the NADPH reservoir is maintained at a higher energy level than the NADH reservoir, which serves mainly in catabolism.

The dawn of life.

common to heterotrophs thanks to the ready reversibility of many of the reactions going on in these organisms. But some processes are unique to autotrophs.

To try to understand autotrophic mechanisms, we must go back in imagination to those very early, prebiotic times, some 4.5 billion years ago, when, according to most scientists, our young, cooling, and as yet lifeless planet gradually turned into a hotbed of organic syntheses. What happened in those distant days will probably never be known in any detail. But it is fair to assume that organic molecules of various kinds—among them many of the building blocks of present-day living organisms—began to be constructed spontaneously, thanks to a combination of favorable chemical and physical factors. These may have included a reducing atmosphere rich in moisture and in carbon-nitrogen gases; energy in the form of heat, ultraviolet radiation, and the discharges of electrical storms; catalytic clays and other minerals; and, finally, shallow puddles of water in which the thickening "primeval soup" incubated. It all would sound pretty fanciful, except that some of these events have already been reproduced in the laboratory through the simulation of the conditions believed by geophysicists to have prevailed in prebiotic times.

What happened next is even more difficult to imagine. Somehow, polymers of various kinds were formed, joining in all sorts of random assemblies, out of which a rare few turned out to contain the seeds of self-reproduction and self-regulation. Henceforth, natural selection had something to work on; "chance and necessity" (see Chapter 18) could consort to bring forth associations of increasing complexity and organizational stability. Eventually, structures that we would recognize today as living cells appeared. Incredible as it may seem, this whole process took no more than a billion years at most. By that time, organisms resembling bacteria existed in places as far apart as Greenland and South Africa. Life had emerged.

If you find it difficult to believe that something as complex as a bacterial cell could have arisen in this span of time, you may choose to go along with Francis Crick, of double helix fame (see Chapter 16), and prefer the alternative that the first bacteria were brought to our planet by a spaceship sent out by some distant civilization (theory of directed panspermia). Even if you accept such a far-fetched ancestry, you have, of course, not solved the problem of the origin of life. You have only shifted it to a more remote point in space and time, presumed to have offered better conditions for life to arise spontaneously than did our own planet. Considering our ignorance of what conditions were on earth those billions of years ago, the reasons for making life a cosmic import product seem anything but compelling. Note further that the modern theory that traces the origin of the universe itself to an initial "Big Bang" that took place about 15 billion years ago leaves our putative forebears hardly more time to reach their own fantastic state of development, supposedly from scratch, than we have had available to attain ours. Nor does it allow for the possibility that life is an eternal feature of the universe, present in germ form throughout outer space, as was proposed about 100 years ago by the Swedish scientist Svante Arrhenius, the father of the original panspermia theory, who is better known for working out an equation defining the influence of temperature on the rate of chemical reactions.

But back to our story. The early bacteria that appeared on earth—or elsewhere—in those distant times could conceivably have started as heterotrophs living, so to speak, on the fat of the land and consuming organic substances produced by abiotic (nonliving) mechanisms. But sooner or later, and at the latest by the time the abiotic supply of nutrients died out, at least one group of organisms must have developed some form of autotrophy. Otherwise, life would have become extinct.

The organisms most likely to have achieved autotrophic survival were those that had available high-grade electrons in their immediate surroundings. Although not very numerous, sources of "good" electrons do exist in the mineral world. Some are even found today, and they must have been more abundant at the time when a highly reducing environment prevailed. Examples are hydrogen gas, carbon monoxide, and especially several of the sulfurous fumes emitted by volcanic springs. It is probably not a mere matter of chance that so many of today's autotrophic bacteria belong to the group of thiobacteria (Greek *theion*, sulfur). They bear witness to the days when life was born in a world that, most unromantically, must have smelled strongly of rotten eggs.

Electrons supplied from the outside at a high enough level of energy—say 50 or more kcal per pair of electron-equivalents above water/oxygen level—could, with the help of suitable catalysts, support all the necessary autotrophic reductions. But ATP supply would still pose a major problem to these primitive heterotrophs if, as speculated in Chapter 7, they were anaerobic fermenters relying on substrate-level phosphorylations. To convert to "electron fuel," they needed a new kind of engine, an oxphos unit of the sort that receives electrons from an outside donor and returns them to an outside acceptor, as is found in the respiratory chain of mitochondria. How soon in the history of life such a device first appeared is beyond our knowledge. But it seems a fair guess, unless early bioenergetic mechanisms were totally different from those we know today, that such an indispensable prerequisite of autotrophic life emerged before or, at the latest, together with autotrophy. Although we have no clue as to how this primitive device operated, it is not unreasonable to assume that it depended on protonmotive force and represented, in fact, the prototype of what is now a universal energy transducer in both autotrophs and heterotrophs, whether prokaryotic or eukaryotic.

Perhaps those early autotrophs, from which all contemporary living forms presumably descend, gained this crucial acquisition originally as an adaptation to extremely acid environments. This adaptation is found today in a type of bacteria known as thermoacidophiles, which belong to the group of archaebacteria (Greek *arkhaios*, ancient), believed to have a particularly ancient history. As their name indicates, thermoacidophiles, which are found in hot sulfur springs, are adapted both to high temperature and to high acidity. They survive in this unkindly environment by actively pumping protons out and use the resulting gradient to perform various kinds of work.

Whatever truth there may be in these speculations, if today's blueprints are at all applicable to the very early forms of life, the most primitive autotroph we can imagine is one that uses high-energy electrons both for biosynthetic reductions and for fueling an oxphos unit hooked to an outside electron acceptor of suitably lower energy level (not oxygen, however, which almost certainly was not available in those days). Make this oxphos unit reversible and link it in series with one or two similar units operating at lower energy levels, as in the respiratory chain, and you have a much more flexible arrangement capable of using a new class of electron donors of moderate energy level. The lower-level oxphos unit(s) then serve(s) to cover all energy needs, including the cost of upgrading the electrons required for biosynthetic reductions. The upper-level oxphos unit does the actual upgrading by running in reverse with the help of energy supplied from below, as we have seen can happen in mitochondria (Chapter 9). Such designs, in one form or another, are characteristic of all present-day chemolithotrophs. Many of these are aerobic and must be evolutionary latecomers. But a few anaerobic chemolitho-

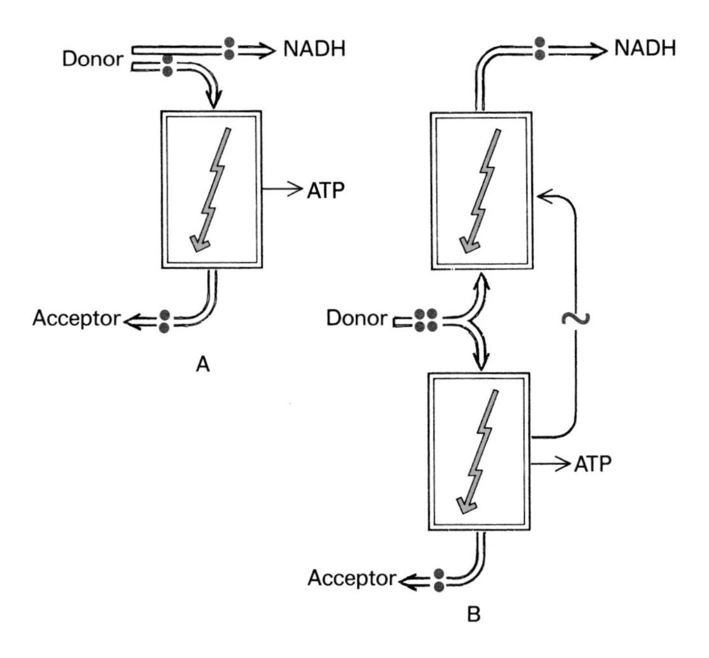

Chemolithotrophy requires: (1) an exogenous electron donor, to support both anabolic reductions and the catabolic supply of energy; (2) an exogenous electron acceptor situated at least 300 mV lower than the donor, to permit ATP formation; and (3) at least one functional oxphos unit inserted between donor and acceptor. If, as in part A, the electron donor is of high enough potential to reduce NAD^+ directly, this suffices. If not, then one or more reverse oxphos units are needed to lift the electrons up to the level of NADH, as in part B.

trophs are known—for instance, *Desulfovibrio desulfuricans* (H_2 reducing SO_4^{2-}) and *Thiobacillus denitrificans* ($S_2O_3^{2-}$ reducing NO_3^-).

Had evolutionary innovation stopped there, life, if subsisting at all, would have remained very primitive and would have clung precariously to those specialized—probably dwindling as time went by—ecological niches capable of sustaining chemolithotrophy. Another key invention was needed before evolution could truly surge forward: the harnessing of sunlight. Actually, we do not know whether this development came before or after that of chemolithotrophy. Both forms of autotrophy share the same essential requirement for an electron-flow oxphos unit. What is added in photosynthetic organisms is a light-powered device for upgrading electrons, a photoelectric unit.

The Conquest of the Sun

The magic component of this unit is chlorophyll (Greek *khlôros*, green; *phyllon*, leaf), a magnesium-containing derivative of the tetrapyrrole (Greek *tetrara*, four) porphyrin ring, which, fitted with iron instead of magnesium, we have already encountered as the active constituent of hemoglobin, the cytochromes (including the respiratory enzyme), and other hemoproteins (Chapter 9). The significance of this molecule for the success of life on earth surely needs no emphasis. But it is interesting that at least one organism has constructed a photoelectric unit with a carotene, a chemical relative of vitamin A, completely unrelated to chlorophyll. Such a device is owned today by the purple *Halobacterium halobium*, a brine-loving microbe, member of the group of archaebacteria, that grows on the surface of evaporating salterns. The carotene family of substances met with only limited success in the capture of solar energy, but it has played a key role in the development of photocommunication. Rhodopsin (Greek *rhodon*, rose; *opsis*, vision), one of the main light-sensitive pigments of the eye, is closely related to bacteriorhodopsin, the principal constituent of the photoelectric energy generator in the membrane of *Halobacterium*.

The basic function of biological photoelectric units is to accept electrons from some low-energy donor, lift them to a higher energy level with the help of light, and then donate them to an appropriate acceptor. This system may have several outlets but must in any event be linked with an oxphos unit through which the photoactivated electrons can fall back to a lower energy level in a manner coupled to the generation of a proton potential that can power the assembly of ATP or the performance of some other kind of work. Such a combination of a photounit with an oxphos unit carries out what is known as photo-

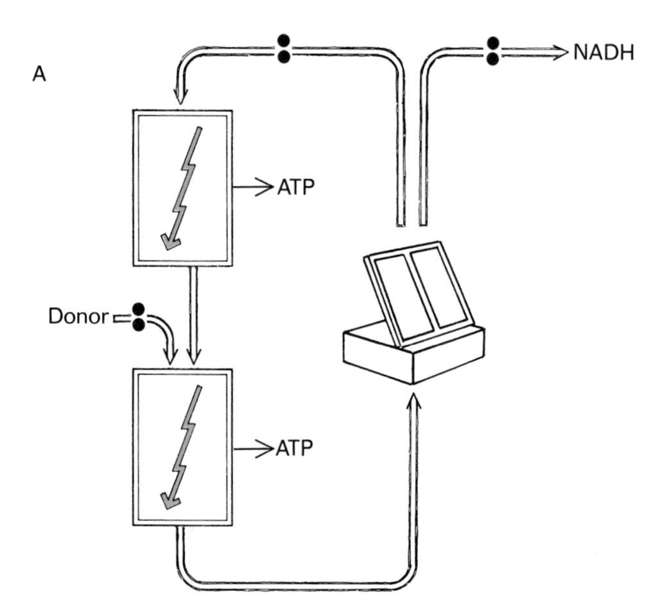

Phototrophy. In bacterial phototrophy, donor is an organic (photoorganotrophy) or inorganic (photolithotrophy) molecule of fairly high electron potential. In example A, characteristic of green photosynthetic bacteria, electrons pass through an oxphos unit and then are lifted by a photounit that either delivers them to NAD^+ or returns them to the oxphos unit (cyclic photophosphorylation), possibly by way of a second oxphos unit. In example B, characteristic of purple photosynthetic bacteria, the phenomena are similar except that the photounit is unable to lift electrons directly to the level of NADH. An additional boost is given by a reverse oxphos unit. Some donors interact directly with the photounit. In plant phototrophy (example C), photosystem II lifts electrons from water up to an endogenous acceptor that donates them to photosystem I. The latter operates like the system in green photosynthetic bacteria (example A) but with $NADP^+$ as final acceptor instead of NAD^+. The number of oxphos units inserted in the electron-transport chain separating the two photounits is unclear. Note that electrons travel singly through photounits.

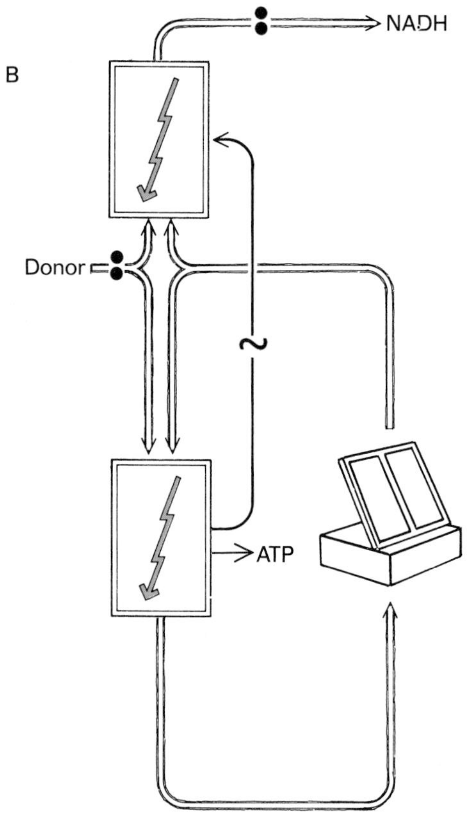

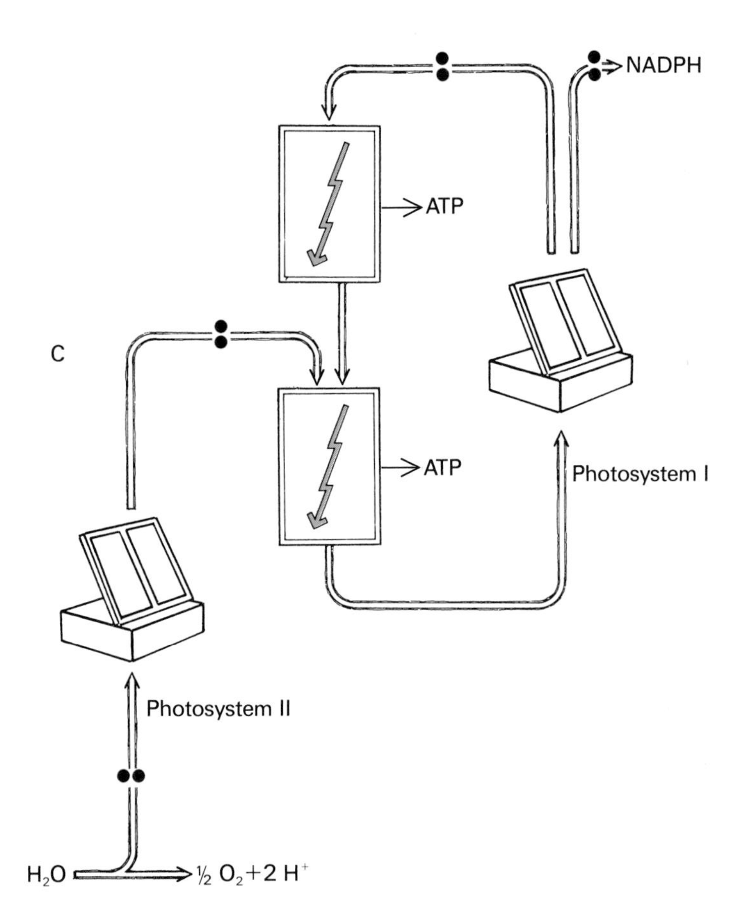

The big electron Ferris wheel of life.

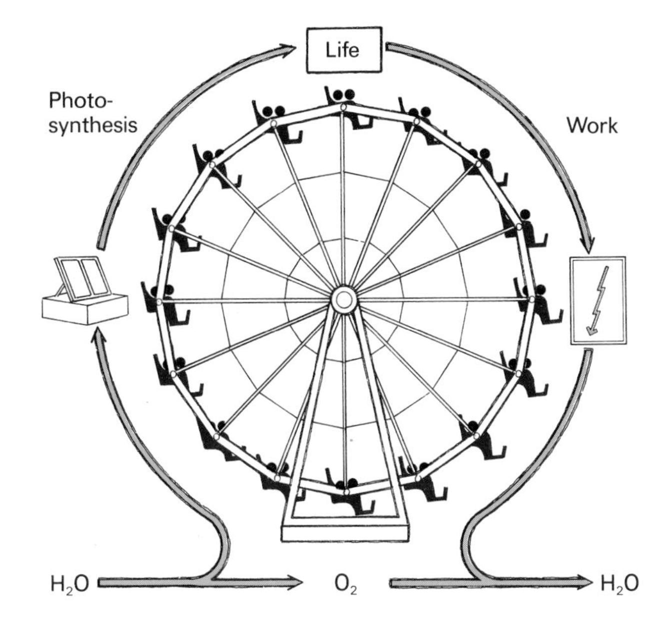

phosphorylation, which can be either cyclic or noncyclic, depending on whether the downgraded electrons are returned to the photounit or transferred to an outside acceptor.

With such a combination, the supply of electrons for biosynthetic reductions can be achieved either directly from the photounit, if its activated level is high enough, or indirectly, after an additional boost provided by another oxphos unit functioning in reverse, as in the second form of chemolithotrophy. As to the electron donors, they belong mostly to the mineral world (photolithotrophy). But cases of photoorganotrophy are also known. They may represent transitional forms between heterotrophy and autotrophy.

The most primitive chlorophyll-dependent complexes are found today in the photosynthetic purple and green sulfur bacteria. These organisms use a variety of sulfur compounds as electron donors and depend on a photochemical machinery known as photosystem I. Most of them are strictly anaerobic. A much more advanced system—actually the crowning innovation in the conquest of autotrophy—is the device designated as photosystem II. Its characteristic is that it uses water as electron donor, oxidizing it to molecular oxygen. The organisms that acquired this system achieved the highest level of energy autonomy, needing only water and sunlight to drive their engines. By itself, photosystem II cannot lift electrons all the way from water to NADPH. It does so by delivering them to photosystem I, which in turn raises them to the level of NADPH. The two photosystems are connected by a phosphorylating electron-transfer chain that allows noncyclic, as well as cyclic, photophosphorylation. The stars of this epoch-making development are the microorganisms called blue-green algae. Although unrelated to the seaweeds (*alga* means seaweed in Latin), which are eukaryotes, these organisms have been called algae because they tend to associate into various types of filaments. Their official name is cyanobacteria (Greek *kyanos*, blue); they are prokaryotes.

When did this all happen? At least 2.5 billion years ago, as indicated by clearly recognizable fossil algae in rocks of that age (Gunflint cherts) found in Canada; perhaps as long as 3.2 billion years ago or earlier, according to some imprints discovered in the South African Pre-Cambrian Fig-Tree rocks. Once they had appeared, the cyanobacteria proved, for understandable reasons, enormously successful. They invaded the entire world, covered it with an extensive mantle of solar-powered chemical factories, and, for some 2 billion years, completely dominated the evolution of life on earth. Through the abundant food they provided, they allowed the heterotrophic mode of life to thrive once again and eventually to branch out into a line of large, highly compartmentalized, voracious phagocytic cells, the putative ancestors of all eukaryotes (Chapter 6). At the same time, the cyanobacteria radically transformed ecological conditions by steadily releasing oxygen into the atmosphere. Adaptation to this change, in turn, produced aerobic life, as we saw in Chapter 9, most likely thanks to the transformation of the original photophosphorylating apparatus into a phosphorylating respiratory chain. Oxygen consumption rose progressively until it reached a steady state with oxygen production, putting in motion the big, sun-driven, electron Ferris wheel of the biosphere: from water to life, through photosynthesis; and back to water again, through respiration. Finally, if holders of the endosymbiont hypothesis are to be believed, the fateful adoption of aerobes by phagocytes, out of which mitochondria and primitive eukaryotic cells were to emerge, was accomplished.

Bacterial photosynthesis has evolved in two steps: first, through the acquisition of carrier-level phosphorylation and photosystem I; next, through the acquisition of photosystem II. According to the endosymbiont hypothesis, cyanobacteria engulfed by an ancestral eukaryote (Chapter 9) became established in their host cell and evolved into the chloroplasts of the unicellular green algae from which all plants are believed to originate.

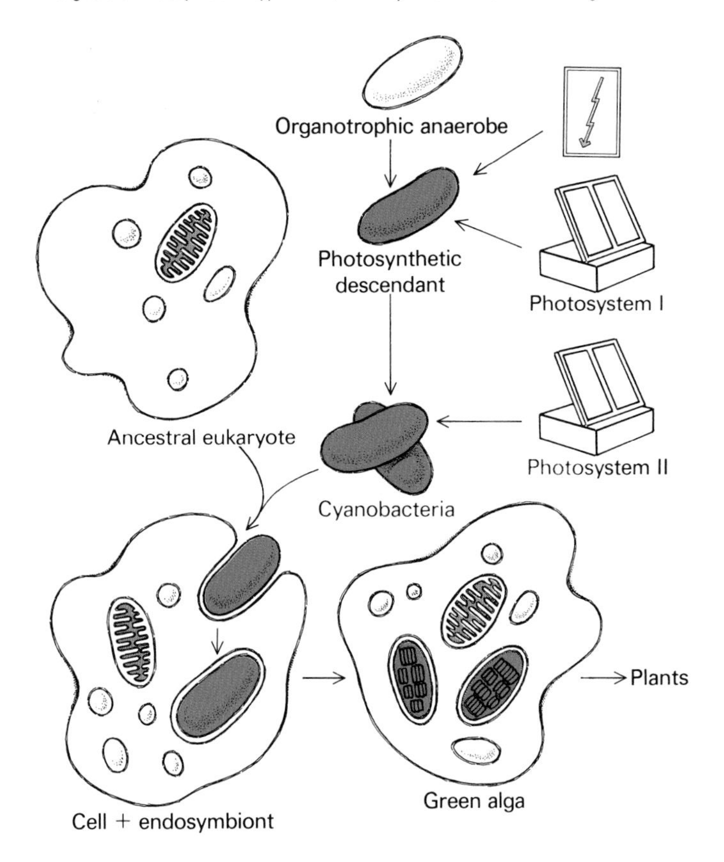

And then, between 1.5 and 1 billion years ago, the last, decisive act in this protracted evolutionary drama—or rather prologue, in view of what followed—was played. Once again, some bacterial victims of phagocytic capture managed to survive and to establish a symbiotic relationship with their captors. The guests, in this case, were cyanobacteria, and the hosts were primitive eukaryotes already equipped with mitochondria. These cells thereby acquired a second type of endosymbiont, which developed into the chloroplasts, the photosynthetic organelles of the unicellular green algae and of their multicellular descendants, the green plants. As described, the scene is, of course, hypothetical. The mechanism it invokes for the origin of chloroplasts does, however, rest on a stronger body of circumstantial evidence than does the endosymbiotic origin of mitochondria and is widely accepted.

So, 3 billion years after life first started seething on the surface of our planet, at the end of some 50 trillion bacterial generations, the stage finally was set for the fantastic development of plants and animals of ever-increasing complexity. We will consider these later parts of the evolutionary saga in Chapter 18.

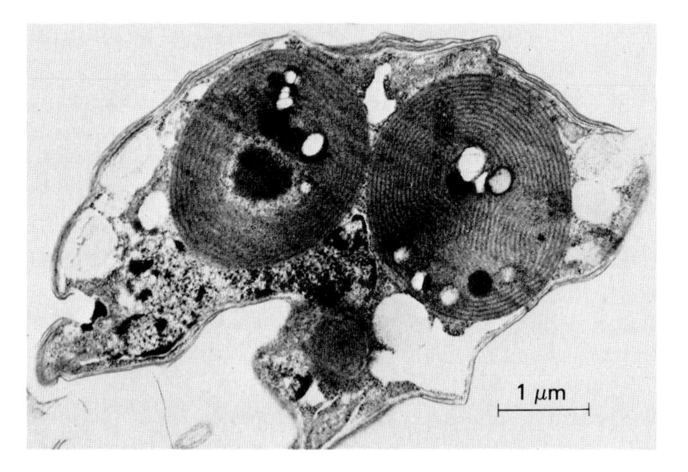

A possible missing link between cyanobacteria and chloroplasts is illustrated by this electron micrograph, which shows two photosynthetic endosymbionts closely related to cyanobacteria (cyanelles) inside a cell of the unicellular, biflagellate protist *Cyanophora paradoxa*.

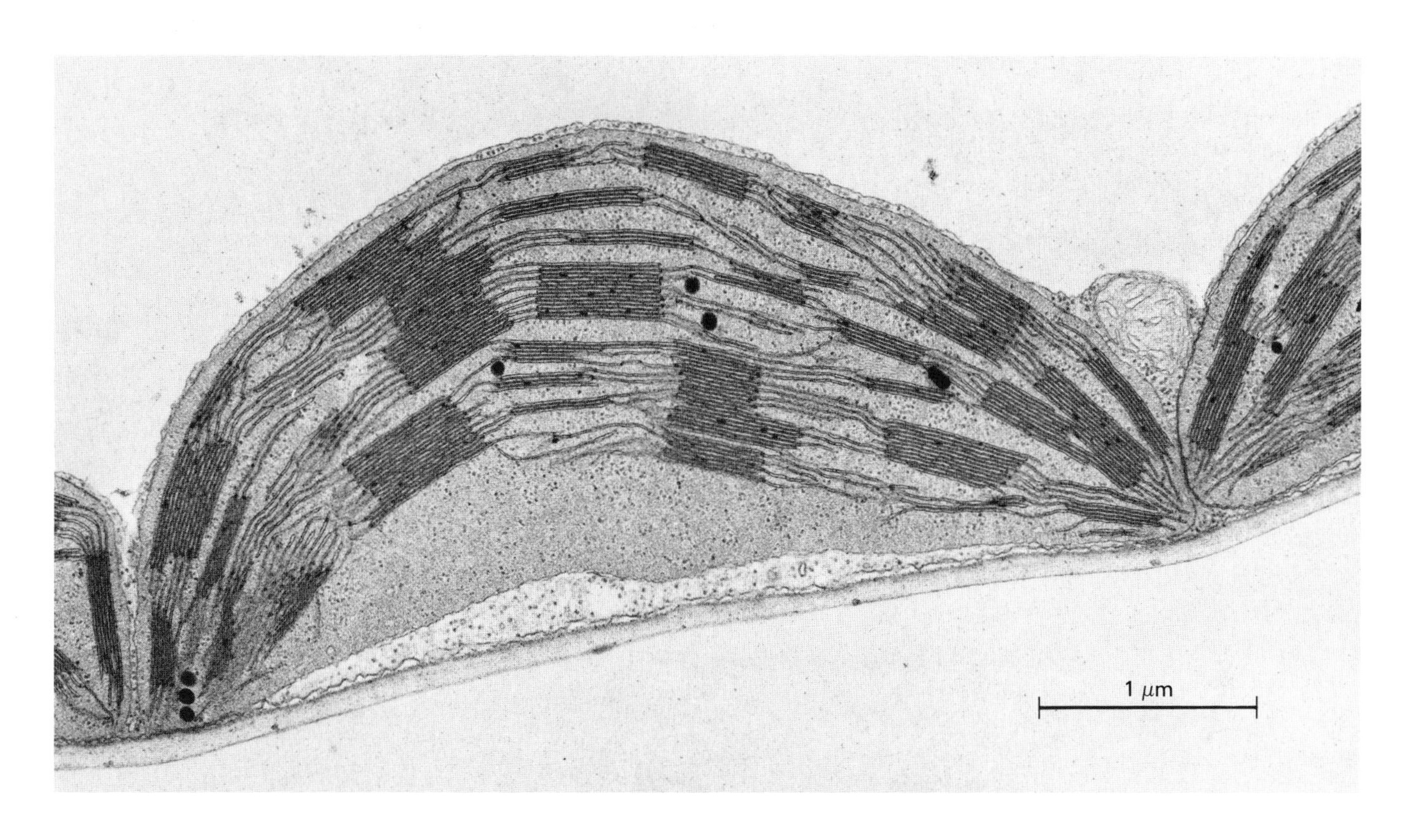

In this electron micrograph of a leaf of timothy grass, *Phleum pratense*, a chloroplast fills almost the entirety of the cytoplasm. Note how piles of thylakoids form grana. The cell is attached to its wall (see the electron micrograph on p. 25). The structure wedged between the chloroplast shown in full view and the one to the right of it is a mitochondrion.

The Green Mansions of Life

Chloroplasts resemble mitochondria in having two surrounding membranes: an outer one, which presumably originates from the vacuolar system of the ancestral phagocyte, and an inner one, supposedly inherited from the plasma membrane of the ancestral blue-green alga endosymbiont. They differ from mitochondria in both their color and their size. As a rule, they are distinctly larger than mitochondria. They measure several micrometers, so that fitting inside would not be too much of a problem for the members of our visiting team were it not for the fact that chloroplasts are generally crammed with stacks of membranes.

These membranes derive, like the mitochondrial cristae, from infoldings of the inner membrane, but with the important difference that the infoldings are severed from that membrane, forming disk-shaped sacs called thylakoids (Greek *thylakos,* pouch). Several thylakoids are stacked together into a cylindrical structure called granum. Each chloroplast contains a number of grana,

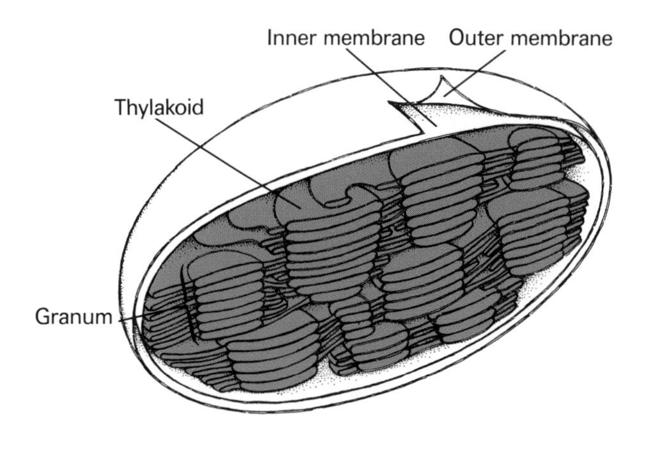

Ultrastructure of a chloroplast.

often linked by tubular connections. This system of inner membranes supports the photosynthetic machinery.

This machinery includes a phosphorylating electron-transport chain flanked by two photoelectric units. The chain resembles the mitochondrial microcircuits. It is similarly made up of a number of structurally associated electron carriers, which include metalloproteins (iron, copper), flavoproteins, quinones, and cytochromes. Electrons circulating through these microcircuits are under the same constraints as those in mitochondria, forcing protons out and building a proton potential. But "out" in a thylakoid is really "in," because the thylakoid is a sealed pouch, unlike the mitochondrial crista, which is an open infolding. And so the impression is given that mitochondria and chloroplasts pump protons in opposite directions. However, this is not really so. In each, the negatively charged side of the membrane is that facing the matrix.

Like mitochondrial inner membranes, thylakoid membranes contain an ATP-driven proton pump situated in knobs protruding into the matrix space. How many sequential oxphos units are associated with the electron-transfer chain is not entirely clear. Certainly one, but possibly two, are on the main noncyclic pathway. There is probably an additional one on the cyclic pathway.

The photosystems themselves are made up mainly of protein-bound chlorophyll, with which are associated variable amounts of accessory pigments, such as the red or blue phycobilins (open-ended tetrapyrroles, resembling bile pigments), and the yellow, orange, or red carotenes and related xanthophylls (Greek *xanthos,* yellow). When leaves start to turn, chlorophyll disappears first, and the blazing colors of the accessory pigments come through to fill our eyes with their autumnal glory.

By definition, a substance is colored because it absorbs some portion of the visible light it receives. Its color corresponds to the light that is not absorbed: it is complementary to the absorbed light. For instance, the chlorophylls are green because they absorb red light (of wavelength 680–700 nm); orange carotenoids absorb blue light (about 450 nm); and so on. In thylakoid membranes, a few hundred such molecules are assembled into

Development of grana. Thylakoids pinch off from infoldings of the inner membrane. Thus, the inner face of the thylakoid membrane corresponds to the outer face of the inner chloroplast membrane.

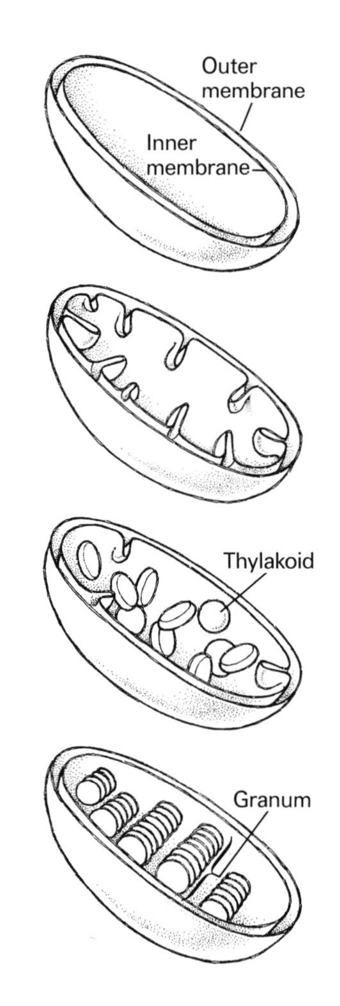

closely organized units centered around a single molecule of a special protein-chlorophyll complex, designated as P700 and P680 in photosystems I and II, respectively (from their wavelengths of maximum light absorption). To find out how such a unit operates, let us leave it in the dark for a while, and then suddenly illuminate it with a flash of light. The pigments in the unit absorb light of appropriate wavelength (depending on their absorption spectrum) and become energized, or excited—a process that involves the destabilization of an electron within the molecular structure by an amount of energy equal to that of the absorbed quantum. With most pigments, energy captured in this way is quickly dissipated as molecular agitation (heat) or sometimes is partly re-emitted in the form of light of higher wavelength (fluorescence), and the destabilized electron falls back to its resting position. Not so in the case of the P complexes, and this explains their biological efficiency. Thanks to their strategic positioning in the thylakoid membrane, they transfer the destabilized electron to an acceptor of high electron potential situated close by and fill in the resulting electron void (positive hole) with an electron taken from a low-potential donor. In this way, they serve to lift an electron from the level of the donor to that of the acceptor with the help of light energy. With P680, the donor is water and the acceptor is a carrier situated somewhere in the upper range of the electron-transport chain. With P700, the donor is a cytochrome at the exit end of the electron-transport chain and the acceptor is ferredoxin, an iron-sulfur protein of high electron potential, which transfers the electron to $NADP^+$ or eventually back into the electron-transport chain.

The P pigments, therefore, constitute the true photoelectric transducers. All the additional molecules of chlorophyll and other colored substances with which they are associated in each photosynthetic unit serve only as light collectors, or solar antennas. Their arrangement is such that they can exchange energy with each other at an extremely rapid rate (exciton transfer), thereby increasing considerably the efficiency of the system. Any quantum of light energy captured by one of the components of the unit can be conveyed to the central P pigment and utilized for photoelectric conversion, provided the connected acceptor molecule is free to receive an electron and there is a donor ready to fill in the hole.

The amount of energy made available by a quantum of light is readily calculated from Planck's formula (Appendix 2). For red light of wavelength 700 nm, the energy is 6.77×10^{-20} calories per quantum, or 41 kcal per einstein. If one quantum moves a single electron through each photosystem, which is most likely, it would take $2 \times 2 \times 41$, or 164, kcal of light energy to move one pair of electron-equivalents from water to $NADP^+$. As we have seen, the energy level in the NADPH reservoir is maintained at about 55 kcal above water/oxygen level.

The photosynthetic machinery of green plants. (Fd stands for ferredoxin.) Note that electrons travel singly through photounits.

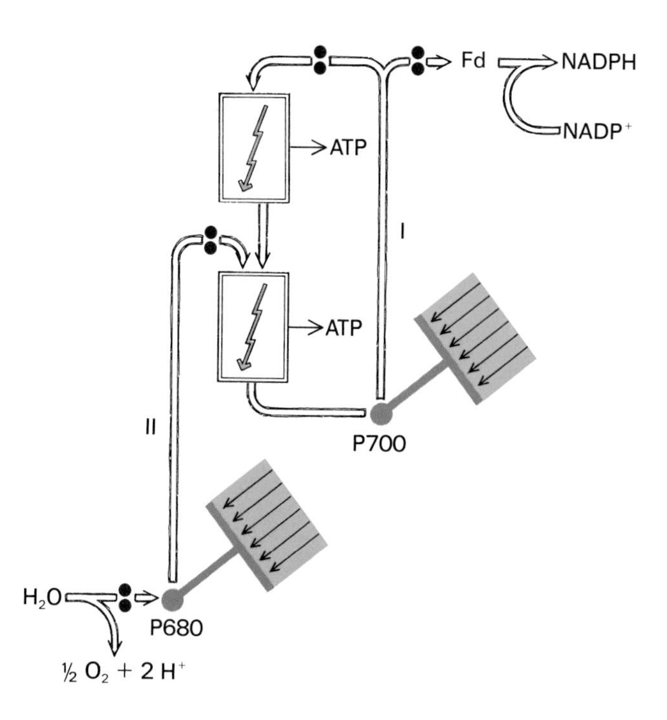

Thus the efficiency of the photoreductive process is of the order of 34 per cent. But this is not counting the gain in ATP. Electrons transferred from photosystem II to photosystem I go through at least one oxphos unit, adding at least 8 per cent to the energy yield in the form of ATP. As to cyclic phosphorylation through photosystem I and the electron-transport chain, it has a minimum efficiency of 17 per cent. But this minimum value could be twice or even three times as high, if the cycling electrons traverse more than one oxphos unit. This point is still unclear, as are many of the finer details of the pathway of photoelectrons. It has even been claimed by one school that photosystem II can by itself lift electrons all the way from water

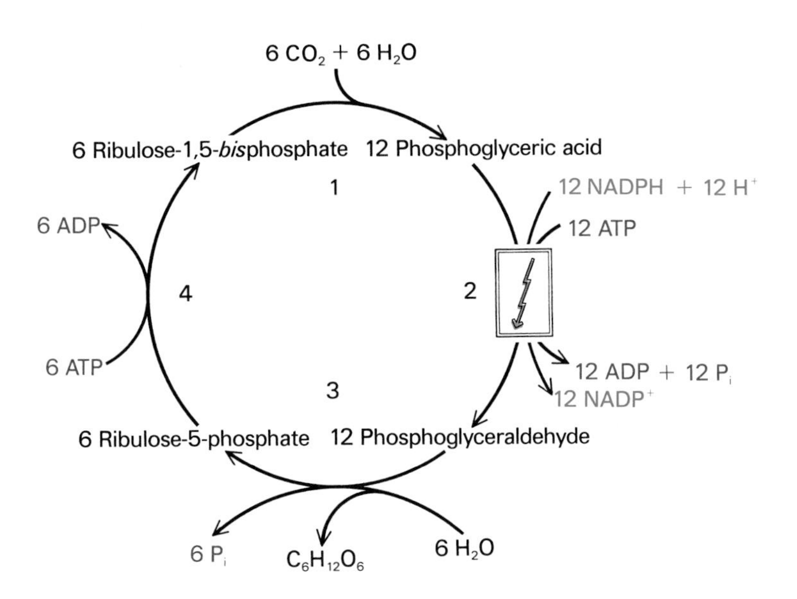

The dark reaction.

1. *Carbon dioxide fixation,* catalyzed by key enzyme ribulose-1,5-*bis*phosphate carboxylase, characteristic of all autotrophs.

2. *Reductive extraction of oxygen,* catalyzed by the substrate-level oxphos unit of the glycolytic chain (Chapter 7) acting in reverse, but with NAD replaced by NADP (in plant cells, not in bacteria). Oxygen is extracted by high-energy electrons of NADPH, with additional help from ATP hydrolysis. The resulting water is contained in the products of ATP splitting. (It will appear explicitly after regeneration of the ATP by photophosphorylation.

3. Rearrangement leading to the *formation of glucose,* the main product of photosynthesis (which is stored in the form of its polymer, starch). A complex enzyme system, known as the pentose-phosphate-pathway system, catalyzes this rearrangement, which allows twelve molecules of the three-carbon compound phosphoglyceraldehyde (36 C) to be rearranged into six molecules of the five-carbon ribulose-5-phosphate (30 C) and one molecule of glucose (6 C). This system is also present, with only minor modifications, in most heterotrophic organisms, where it supports important metabolic functions.

4. *Phosphorylation of ribulose-5-phosphate,* making it ready for participation in step 1.

Note that, for each turn of the cycle, six molecules of CO_2 are condensed reductively into one molecule of glucose, with the help of twelve electron pairs furnished by NADPH and of additional energy provided by the hydrolysis of eighteen molecules of ATP.

to NADP$^+$, in which case the efficiency of the process would be much higher than estimated above.

Even the lowest estimates represent excellent quantum yields, compared with other photochemical systems. They should be pondered by all those who are trying to harness solar energy. To do better than the multilayered shroud of countless microphototransducers that cyanobacteria and their descendants have woven around the earth will not be an easy task. Perhaps we should try to find better ways to make them work for us, rather than attempting to rival them.

The Dark Reaction

The only light-powered reaction in photosynthesis is the two-step photoelectric transduction that extracts electrons from water and raises their potential by about 1,200 mV. Its products are NADPH and ATP. These suffice to support all the autotrophic biosynthetic mechanisms; they can do so perfectly well in the absence of light, as they do in the nonphotosynthetic chemolithotrophic organisms.

Before we leave the chloroplasts, we should take at least a brief look at the most famous of these "dark reactions," namely CO_2 fixation. Tracing this pathway is one of the triumphs of radioisotope technology used in conjunction with chromatographic separation (Chapter 1).

By illuminating leaves in the presence of radioactive $^{14}CO_2$ for shorter and shorter durations (as little as a few seconds) and then extracting and separating the labeled compounds, workers were able to identify the earliest product of CO_2 fixation as phosphoglyceric acid. With this clue, an appropriate reaction was sought and duly found. It is a remarkable process, in which a *bis*-phosphorylated five-carbon sugar, ribulose-1,5-*bis*phosphate, reacts with CO_2 and water to give two molecules of phosphoglyceric acid. The enzyme catalyzing this reaction is bound to the thylakoid membrane.

The substance made by this enzyme, phosphoglyceric acid, is not unknown to us. We met it in Chapter 7 as the product of the central phosphorylating oxidative step of glycolysis. Indeed, photosynthesis has borrowed this primeval oxphos unit for its key reductive step—first, as such and, later, with NAD replaced by NADP as coenzyme. Taking advantage of the high energy potential that it maintains in the NADH or NADPH reservoir, it causes the electrons to flow in the reverse direction through this oxphos unit, consuming ATP in the process. And, in this manner, phosphoglycer*aldehyde* is made from phosphoglyceric *acid* with the help of one pair of electrons supplied by NADH or NADPH and of the free energy of hydrolysis of ATP.

These are the key reactions of photosynthesis. They are incorporated within a complex cyclic process known as the Calvin cycle, named for the American chemist who first unraveled it. We will not look at the details, but only at the end result. A molecule of glucose is made from 6 molecules of CO_2 and 6 molecules of water. The 12 electron pairs required to extract the 12 excessive oxygen atoms are supplied by NADPH, and the additional energy needs are covered by the hydrolysis of a total of 18 molecules of ATP.

What does this make in terms of energy yield? On the debit side, we have the oxidation of 12 NADPH molecules (660 kcal) plus the breakdown of 18 ATPs (252 kcal), or a total of 912 kcal per gram-molecule of glucose made. On the credit side, we have 686 kcal per gram-molecule, the free energy of oxidation of glucose, for an overall efficiency of 75 per cent. To achieve this, a minimum of 12×164 kcal of light energy is required. Thus the maximum efficiency of photosynthesis is 35 per cent. It could be somewhat lower, depending on the number and location of the oxphos units in the electron-transport chain.

Animals lack the key enzymes of the Calvin cycle. But they do share with plants the use of NADPH as main electron donor in reductive syntheses—for instance, in the conversion of carbohydrate into fat. Like plants, they maintain a relatively high energy level in their NADPH reservoir, of the order of 55 kcal per pair of electron-equivalents, or 1,200 mV. They do this with the help of certain selected substrates acting as high-level electron donors: glucose-6-phosphate is the main one. If need be, they can send electrons from NADH to $NADP^+$ by an energy-dependent process in the mitochondria. Conversely, there is a strictly regulated overflow mechanism, whereby excess electrons in the NADPH reservoir are allowed to fall down to NAD^+.

Growth and Multiplication of Chloroplasts

What has been said about the biogenesis of mitochondria applies similarly to the chloroplasts. These particles also possess a complete genetic machinery. It is actually richer than that of the mitochondria, but nevertheless controls the synthesis of only a small portion of the total chloroplast proteins. It has the same bacterial characters as the mitochondrial system and presumably likewise represents a vestigial remnant of the ancestral endosymbiont, in this case cyanobacteria. Chloroplasts, like mitochondria, display genetic continuity and can undergo mutations that are transmitted via the cytoplasm.

The autonomy of chloroplasts is, however, limited, as is that of mitochondria. Most of their constituents are manufactured by cytoplasmic ribosomes under the control of nuclear genes. As in mitochondria, the manner in which these constituents are transferred across the particle membranes and inserted into their proper location is not well understood.

11 Peroxisomes and Sundry Other Microbodies

The first microbody was spotted in mouse kidney in the early 1950s by a Swedish anatomist, who found it to be of such nondescript character that he could think of no better name for it. Soon after, similar objects were seen in rat liver and, later, in a variety of cells in both the plant and animal kingdoms. Although widespread, microbodies remain restricted to certain cell types. In mammals, they have been detected mainly in liver and kidney.

Wherever they are found, microbodies have a similar appearance. They are roughly spherical structures, from 0.5 to 1.0 μm in diameter, which makes them slightly smaller than mitochondria. They are surrounded by a membrane and are most often filled with a fairly compact, amorphous matrix. In some cells, this matrix hides a gem—a dense, crystalloid core, or nucleoid, of beautifully delicate texture. This kind of purely morphological information leaves much to the imagination, and it did indeed provide the grounds for some rather fanciful constructions. When biochemical clues eventually put investigators on the right track, the truth turned out to be even stranger than the fictions, revealing that not one, but several, distinct types of microbodies exist, each concerned with what looks like a primitive, if not primeval, set of metabolic reactions.

Peroxisomes and Glyoxysomes

The peroxisomes are the most widespread form of microbody. They derive their name from hydrogen peroxide, H_2O_2, a key intermediate in their oxida-

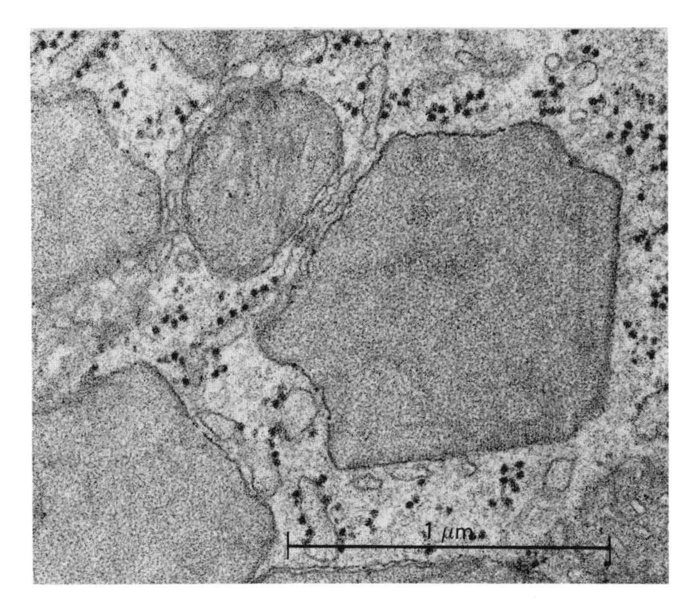

Peroxisomes in a rat kidney proximal tubule cell. Three irregularly shaped microbodies surround a small mitochondrion.

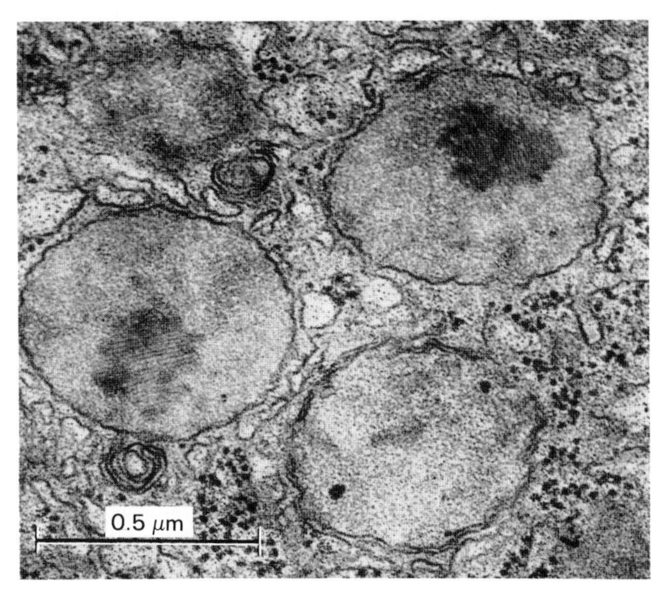

Two peroxisomes in rat liver. The regularly structured core contains an enzyme that oxidizes uric acid with the formation of H_2O_2.

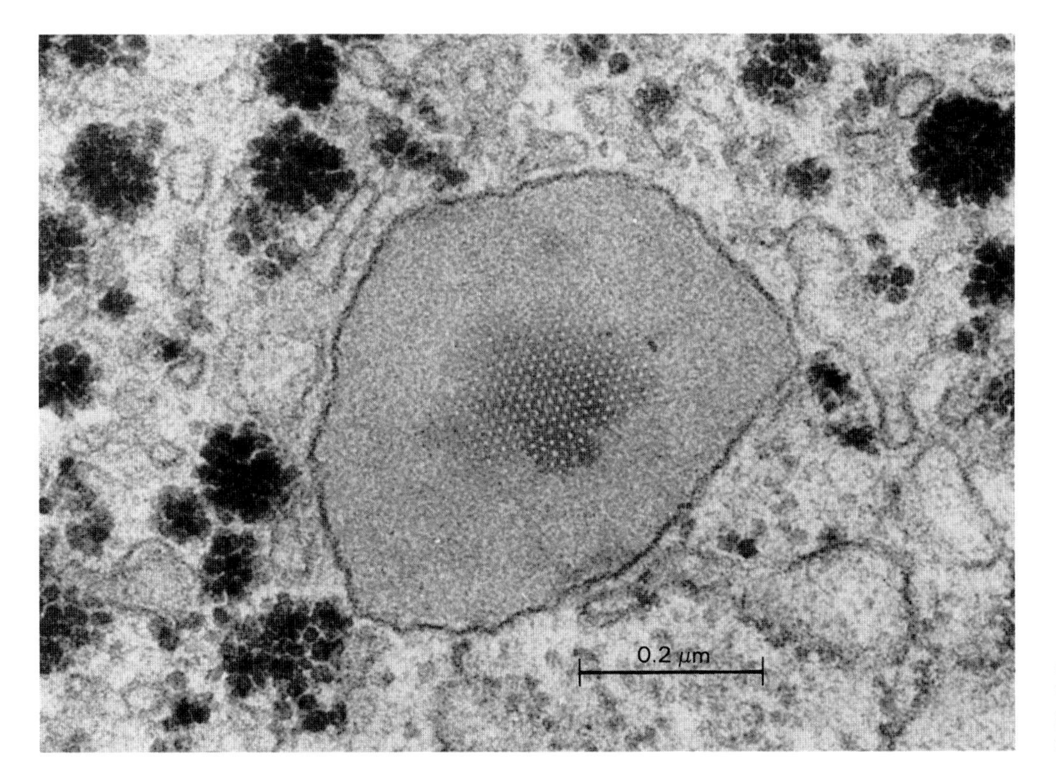

A peroxisome in guinea pig liver. The crystalloid core, which, as in rat liver peroxisomes, contains urate oxidase, shows a regular hexagonal lattice.

tive metabolism. Hydrogen peroxide is produced by a family of enzymes called type II oxidases, mostly flavoproteins, sometimes also copper proteins, that use molecular oxygen as electron acceptor and reduce it to H_2O_2:

$$RH_2 + O_2 \longrightarrow R + H_2O_2$$

The electron donors in these reactions include amino acids, fatty acyl-coenzyme A derivatives, purines, and some products of carbohydrate metabolism, such as lactic acid—in short, representatives of all major classes of foodstuffs.

The hydrogen peroxide produced in the peroxisomes is further metabolized through the action of catalase, a green hemoprotein that reduces H_2O_2 to water with, as electron donor, certain small organic molecules, such as ethanol, methanol, or formic acid, and, in the absence of a suitable donor, hydrogen peroxide itself:

$$R'H_2 + H_2O_2 \longrightarrow R' + 2\ H_2O$$

or

$$H_2O_2 + H_2O_2 \longrightarrow O_2 + 2\ H_2O$$

The latter reaction is called a dismutation. In it, one molecule of hydrogen peroxide is reduced and the other oxidized. The net result is destruction of H_2O_2 with evolution of oxygen. You can readily see catalase in action by dabbing some hydrogen peroxide on a wound: watch for the frothing oxygen. Catalase is one of the fastest-acting enzymes. It was detected as early as 1818 by the French chemist Jacques Thenard, the discoverer of H_2O_2.

Acting together, the peroxisomal oxidases and catalase allow oxidations to take place through the following mechanism:

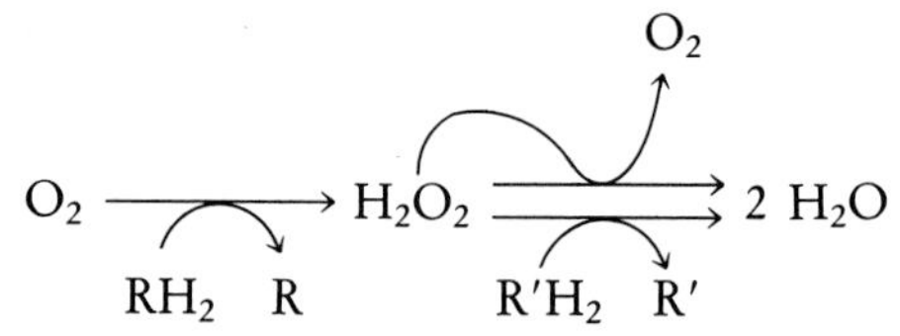

Compare this "respiratory chain" with that of mitochondria and you have all the difference between a reckless profligate and a prudent economizer. Both achieve the same result, bringing about the oxidation of all sorts of foodstuffs, with reduction of oxygen to water. But, whereas in mitochondria much of the free energy of combustion is retrieved in the form of usable ATP, in the peroxisomes it is all dissipated as heat. This drawback is compensated by a remarkable simplicity of design. It seems likely that the peroxisomal type of respiration arose long before the delicate mitochondrial microcircuits were put together; it may represent one of the earliest adaptations of living organisms to oxygen, as was pointed out in Chapter 9.

In addition to oxidizing enzymes, peroxisomes may, depending on cell type, contain a variety of other systems. Prominent among these is the glyoxylate cycle, a Krebs cycle variant that plays an essential role in the conversion of fat into carbohydrate. This is an important biochemical process; among other functions, it makes it possible for fatty seedlings, such as castor beans, to utilize their oily stores upon germination. Hence, the peroxisomes that contain this system have been named glyoxysomes. In animals, the glyoxylate cycle was an early victim of evolution. It is present in lower forms but not in many higher vertebrates, including mammals. We can make fat out of carbohydrate—a privilege many of us would be happy to forsake—but we are unable to reverse the process because our peroxisomes are not glyoxysomes.

Microbodies identifiable as peroxisomes, while not present in every type of cell, are universally distributed in nature. They are found in a large variety of plants and animals, in molds, fungi, and protozoa. This, together with the primitive character of their respiratory machinery, has suggested that all peroxisomes may be evolutionary descendants of a single ancestral particle that was present in the first eukaryotic cell from which all plants and animals are believed to have originated. Possibly this ancestral peroxisome was already part of the primitive phagocyte in premitochondrial times and fulfilled, crudely but efficiently, the essential function of guarding against oxygen.

This hypothesis leaves the historian with two perplexing questions. First, why were peroxisomes not elimi-

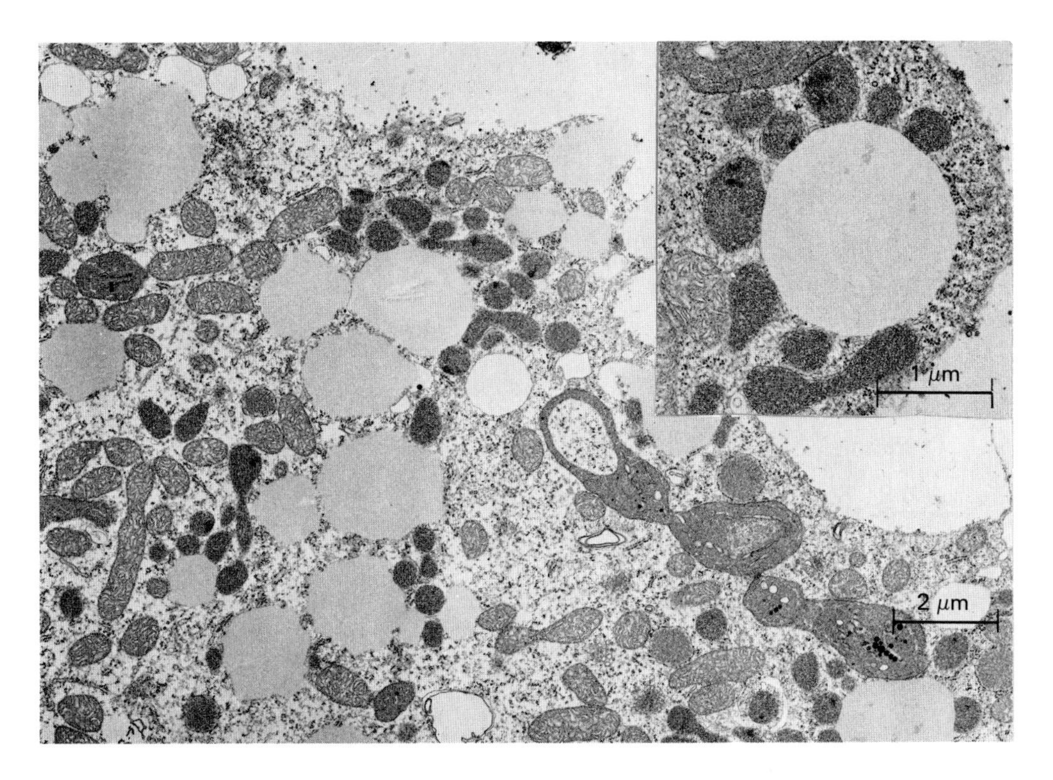

This electron micrograph of castor bean endosperm shows numerous lipid droplets (appearing as holes), mitochondria (bodies of moderate density with internal membranes), and glyoxysomes (darker microbodies with coarsely granular matrix). The inset illustrates the characteristic clustering of glyoxysomes around lipid, which they convert into carbohydrate.

nated by natural selection after they were joined by the better-equipped mitochondria? This may well have happened in many cells, although even this possibility is debatable. Many of the mammalian cells that lack regular microbodies possess tiny, membrane-bounded particles that, on the basis of their content of catalase and sometimes of other typical peroxisomal enzymes, have been identified as microperoxisomes.

In any case, in spite of a considerable attrition of their metabolic potential—the loss of the glyoxylate cycle is an example—mammalian peroxisomes are by no means "fossil organelles" of no more than vestigial interest. They undoubtedly fulfill important functions. They are involved in several ways in the metabolism of lipids, possibly also of cholesterol; in the breakdown of amino acids, including the D-amino acids, which occur only in bacteria; and in the catabolism of purines. So far as is known, they do not provide the cells with utilizable energy, but the heat they produce may sometimes be physiologically significant. It is said that the special brown-fat tissue which helps the Norwegian rat survive the rigors of winter owes part of its thermogenic capacity to peroxisomes. There are also indications that peroxisomes play an essential role in the synthesis of certain phospholipids called plasmalogens. Perhaps the best proof of the importance of these organelles in mammals is provided by pathology. There is a rare human genetic deficiency, known as Zellweger's disease, in which no morphologically detectable microbodies are seen in the liver or kidneys, where they normally are present in large numbers. Infants afflicted with this condition do not survive more than a few months.

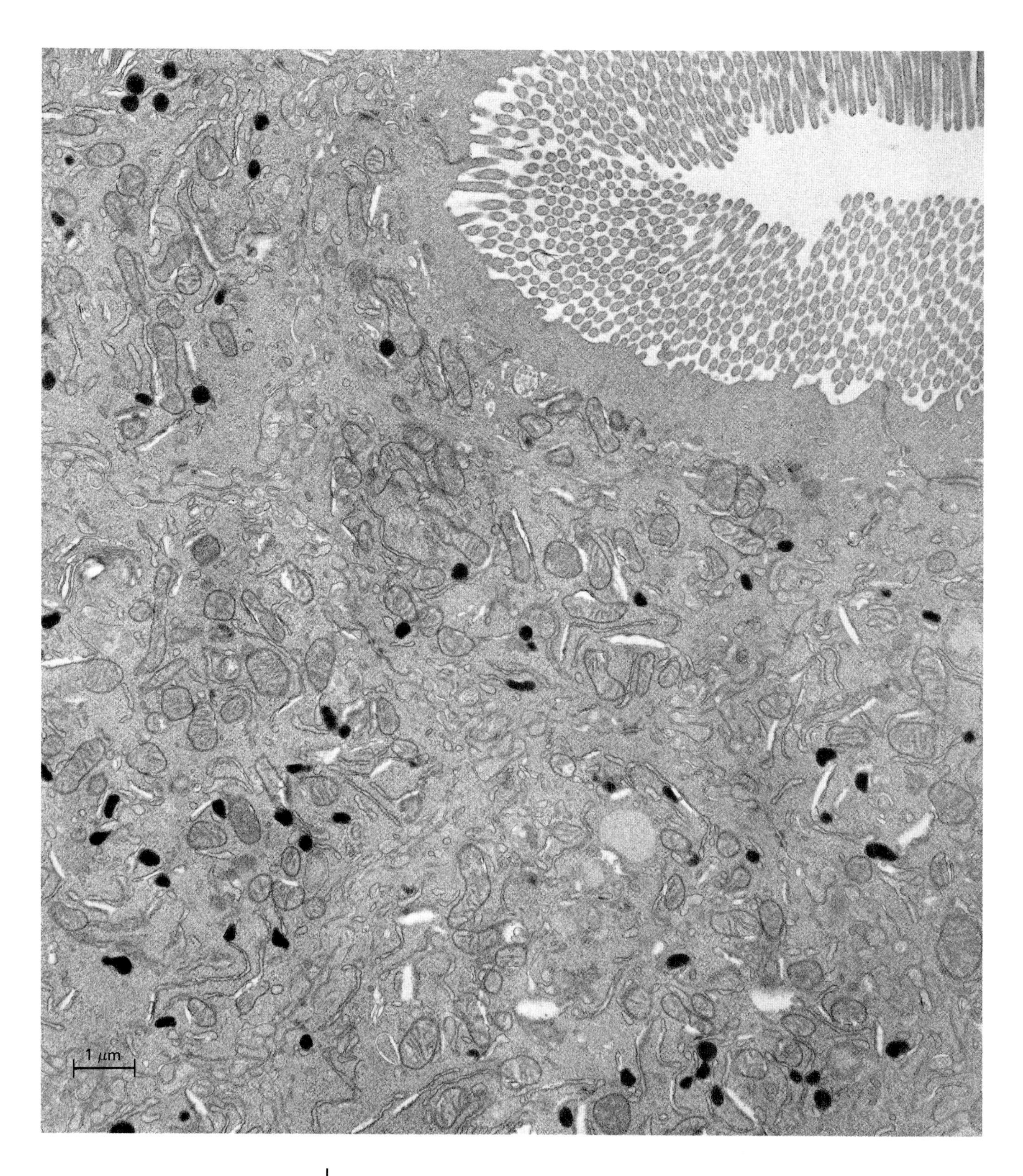
1 μm

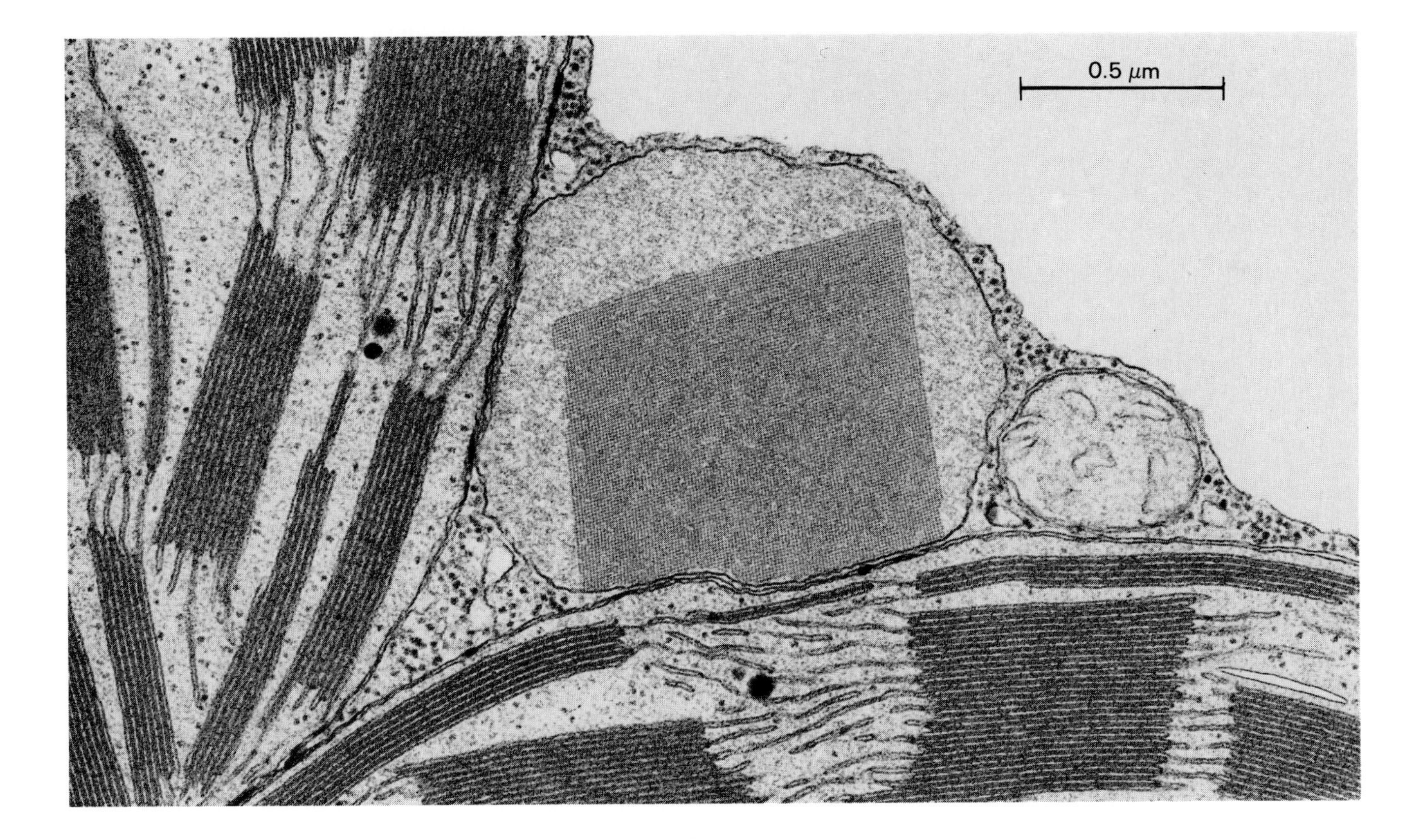

A peroxisome containing a beautiful crystalline structure believed to consist of catalase is seen wedged between two chloroplasts, and adjacent to a mitochondrion, in this electron micrograph of a tobacco-leaf cell. Localization reflects the close metabolic collaboration that exists between the three organelles.

On the facing page: Microperoxisomes in guinea pig ileum have been stained by means of a cytochemical reaction based on the oxidation of diaminobenzidine by catalase in the presence of H_2O_2. Note the numerous mitochondria in the cytoplasm and the sections through microvilli in the upper right-hand corner.

The functions of peroxisomes are even more important and varied in lower animals and protozoa, and especially in the plant world. We have already seen the key metabolic role of those peroxisomes that qualify as glyoxysomes in the conversion of fat into carbohydrate. Innumerable plants that pack their seeds with lipid reserves would be unable to reproduce without this function. Green leaves, which contain peroxisomes of uncommon beauty, depend on an intricate trilateral collaboration between these particles, chloroplasts, and mitochondria to salvage photosynthetic products that would otherwise be lost through photorespiration—a form of light-induced oxidation that results from a peculiar subversion of the central enzyme of the Calvin cycle by oxygen. These interactions are of great economic importance, as they affect the net photosynthetic yield of many crop plants. A particularly remarkable adaptation of peroxisomes is observed in certain strains of yeasts. When grown on metha-

These pictures show the remarkable structure and packing of peroxisomes in cells of yeast *Hansenula polymorpha* adapted to growing in a medium containing methanol as the sole source of carbon.

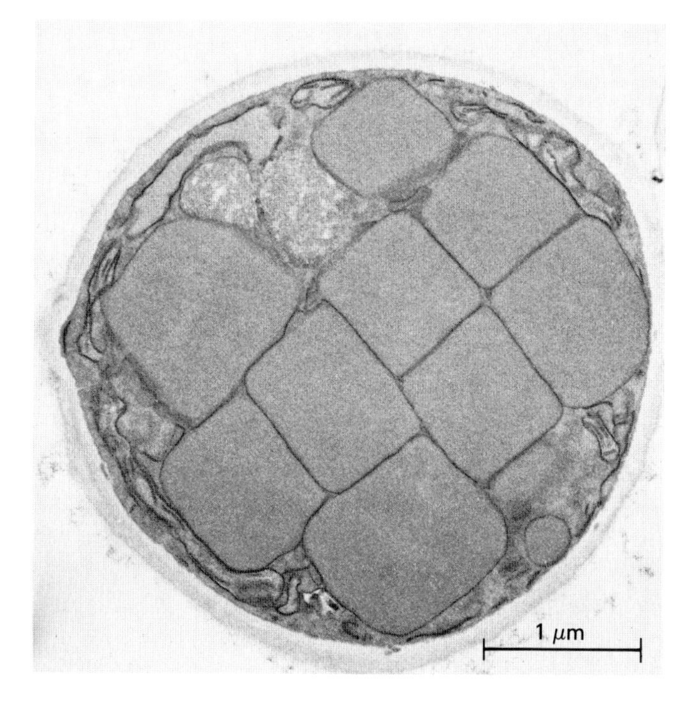

A. A thin section shows cytoplasm occupied mostly by box-shaped peroxisomes.

B. A similar view, but with relief emphasized, is offered by a carbon replica of a preparation that has been fractured in the frozen state and then shadowed with platinum.

nol, on alkanes (which are saturated hydrocarbons such as are found in petroleum), or on some other outlandish nutrient as sole source of carbon, these cells respond by an enormous increase in the level of some of the peroxisomal enzymes needed for the breakdown of those substrates. To accommodate these enzymes, the peroxisomes develop greatly in size and in number. Sometimes the cells become so packed with peroxisomes that the particles assume the shape of square boxes.

When one comes to think of it, it is not really so surprising that cells should retain peroxisomes, even after acquiring mitochondria. Right from the beginning, peroxisomes may have possessed useful attributes that were lacking in the mitochondria, and therefore their retention was favored. Or some of their characters may have become essential at a later stage—for instance, after deletion of a mitochondrial property by a mutation. Nevertheless, peroxisomes are not what they used to be—or so it seems.

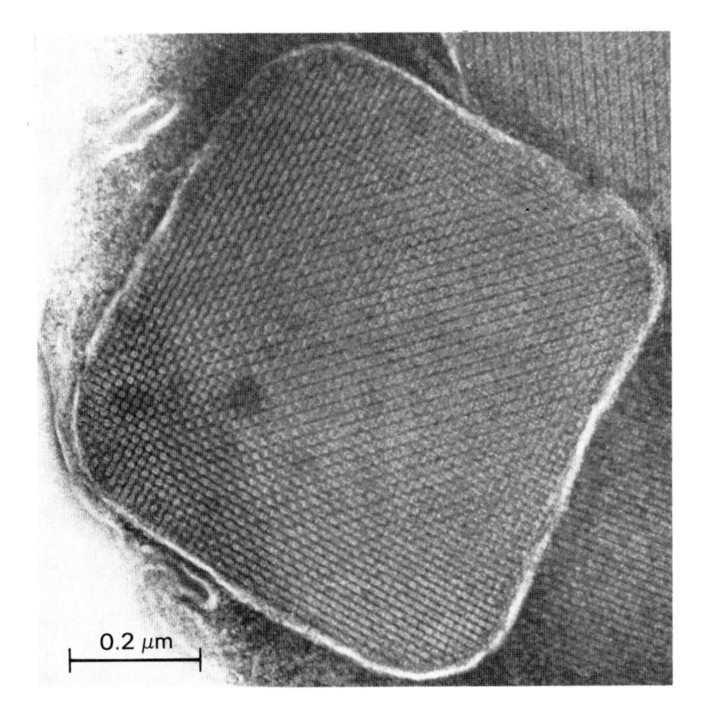

C. A higher magnification shows that a peroxisome is largely filled by a crystalline structure. This structure is made up of molecules of an enzyme that oxidizes methanol, with the formation of H_2O_2.

Putting together all the individual functions that are found today in the different types of peroxisomes, and assuming that they are all inherited from the same ancestor, one arrives at the picture of a mighty metabolizer of considerable versatility, of which today's descendants are but pale replicas.

Which brings us to the second question the historian is likely to ask: How did the primitive eukaryote—or, perhaps, its prokaryotic phagocytic precursor—manage to acquire such a powerful organelle in the first place? A few years ago, this question would hardly have been asked, since it was widely believed that peroxisomes grow out of the ER in the form of bulging buds and may even, in some cells, remain attached to that structure by membranous stalks. But this has become doubtful. Investigators looking for nascent peroxisomal proteins in ER cisternae have been disappointed. They have found instead that these proteins are made by free polysomes and released into the cytosol, from which they subsequently gain access to the peroxisomes by an unknown mechanism. As to the purported connections between peroxisomes and ER, they may have to be re-examined. We must remember that all membranes look more or less alike in the electron microscope and that investigators examining thin sections do not have our advantage of a three-dimensional view of the cellular structures. Unfortunately, the picture that we see is not as clear as we would have it. It does, however, look as though peroxisomes form clusters of interconnected particles, but perhaps separate from the ER.

If peroxisomes are not offshoots of the ER—which, incidentally, they still could be phylogenetically, if not ontogenetically—the origin of these mysterious microbodies must be sought elsewhere. Some cell detectives, their imaginations fired by the tales that circulate about mitochondria and chloroplasts, have raised the possibility that the peroxisomes may be evolutionary descendants of yet other bacterial endosymbionts, adopted at a more remote time. The primitive character of their respiratory mechanisms is consistent with this possibility. Nevertheless, it must be emphasized that, in contrast with mitochondria and chloroplasts, peroxisomes offer not a shred of evidence in support of such speculations. As far as is known, they are entirely lacking in DNA, ribosomes, or other parts of a genetic machinery. But lack of evidence does not invalidate the hypothesis. After all, if an endosymbiont can become integrated more than 90 per cent, as are mitochondria and chloroplasts, why not 100 per cent? Some day we may get an answer to these questions by interrogating the microbody proteins and extracting from their molecular structure some vestigial mementos of their evolutionary history (see Chapter 18).

Hydrogenosomes

On numerous occasions, the need for a suitable acceptor to collect electrons at their exit from oxphos units has been emphasized. But only passing mention has been made of what would appear to be the simplest solution to this problem—namely, the use of protons as acceptors:

$$2\,e^- + 2\,H^+ \longrightarrow H_2$$

Protons are available everywhere. Why, then, do we not all breathe out hydrogen, instead of having to inhale oxygen? Some organisms actually do just that. But they are very few, probably because protons are about the most unprofitable electron acceptors a cell can use. Hydrogen formation takes place at a very high potential level—of the order of 50 or more kcal above water/oxygen level. This means that, if usable energy is to be gained from the transaction through an oxphos unit intercalated between donor and acceptor, the substrate must deliver its electrons at a level of at least 64 kcal per pair of electron-equivalents. Only a handful of substances (pyruvic acid is one) can muster this kind of energy.

In the bacterial world, there is a small group of obligatory anaerobes, known as clostridia, that produce hydrogen. Among them are the pathogenic organisms that cause gaseous gangrene, an infection that develops in wounds that are not properly "aired." These bacteria seem to have an exceptionally old evolutionary history and to have developed as an independent branch in those very early days when life was still exclusively anaerobic. They never learned to adapt to the presence of oxygen. Or perhaps they lost the art after learning it. Remember, all this is conjecture.

Remarkably, the capacity to produce hydrogen is shared in the animal world by a tiny subgroup of protozoa, the trichomonads, most of which are parasites of the genital tracts of both animals and humans. They are one of the main agents of sexually transmitted diseases. These parasites can live both in the presence and in the absence of air. Anaerobically, they support their energy needs by a special reaction in which pyruvic acid is oxidized to acetic acid and CO_2 and the liberated electrons are channeled to hydrogen by way of a substrate-level oxphos unit:

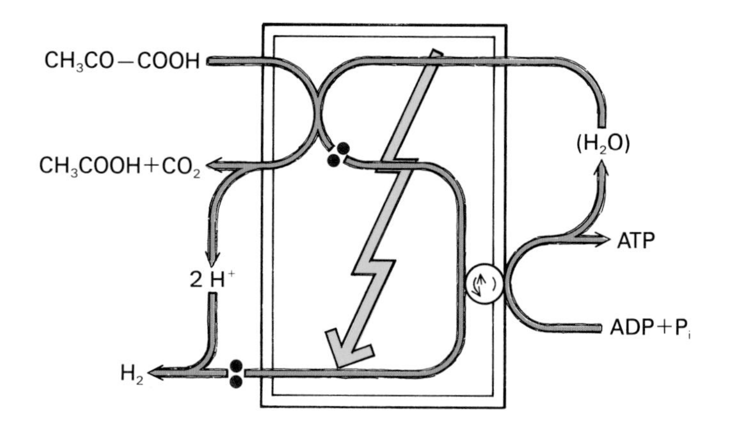

In the oxidation of pyruvic acid to acetic acid and CO_2 by hydrogenosomes, electrons are transferred to protons (with the formation of molecular hydrogen) through a substrate-level oxphos unit. As in the substrate-level phosphorylation of the glycolytic chain (see the illustration on p. 111), the oxygen added to the substrate arises from a cryptic molecule of water generated by the condensation of ADP + P_i.

$$CH_3\text{—}CO\text{—}COOH + (H_2O) \longrightarrow CH_3\text{—}COOH + CO_2 + 2\,e^- + 2\,H^+$$

$$ADP + P_i \longrightarrow ATP + (H_2O)$$
$$2\,e^- + 2\,H^+ \longrightarrow H_2$$

Aerobically the organisms shut off their hydrogen production and switch over to an oxygen-dependent metabolism, though one of a very simple kind since they do not contain mitochondria.

They do have prominent microbodies, however—a fact that, some years ago, raised the exciting possibility that here, perhaps, was a direct descendant of the primitive phagocyte, the offspring of a cell that had never captured a mitochondrial endosymbiont, constrained unto this day to depend on the original ancestral peroxisome for its oxidative metabolism. As often happens in science, when this possibility was put to the acid test of experi-

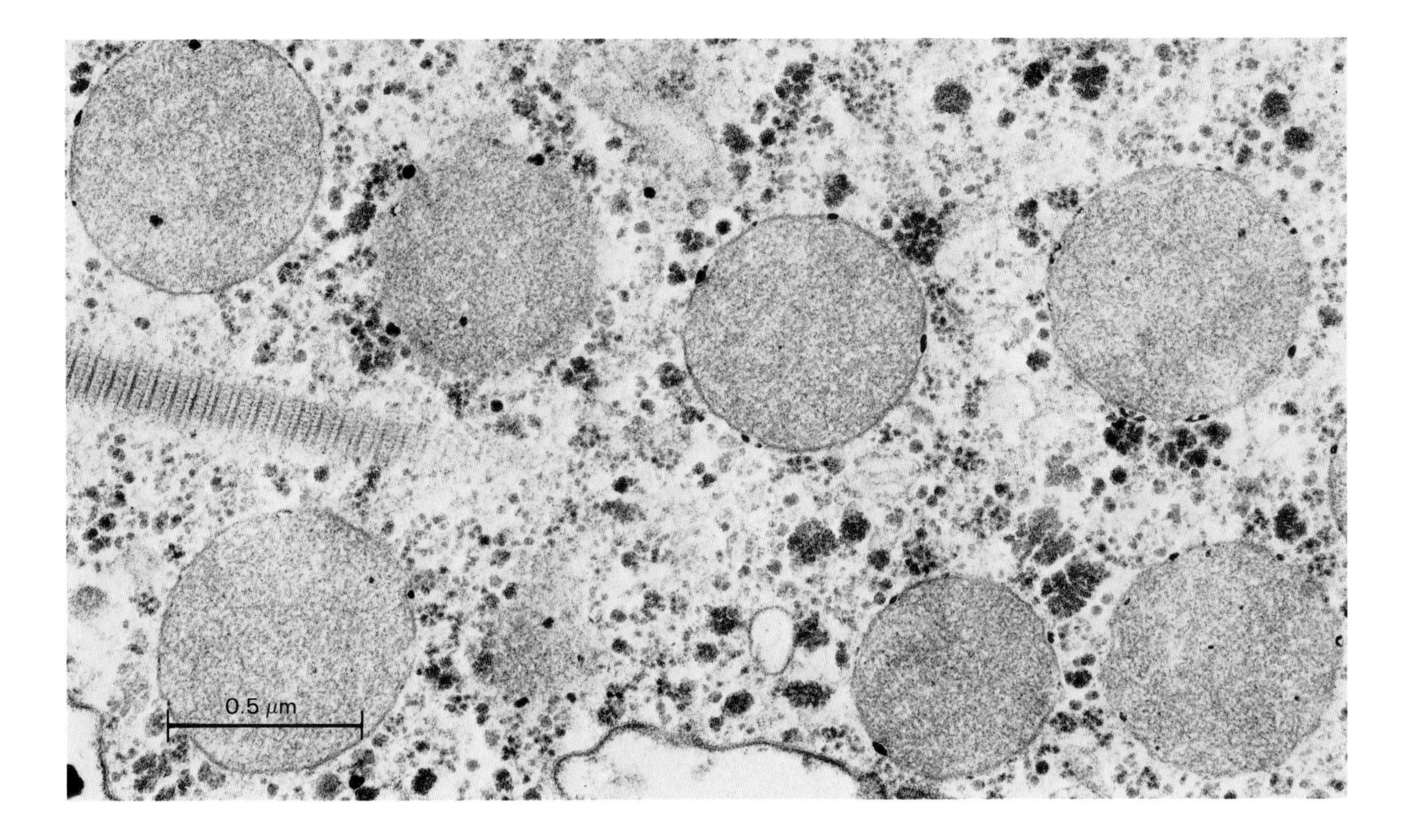

Hydrogenosomes in the human parasite *Trichomonas vaginalis.* The striated object on the left is a riblike structure, the costa.

ment, the answer was a disappointing "No." But, at the same time, it opened a door into the entirely unexpected. The microbodies of trichomonads are hydrogenosomes; they contain the whole system responsible for the oxphos-linked, hydrogen-producing breakdown of pyruvic acid.

Like all discoveries, this finding has raised new questions. The most provocative one is: Where does the hydrogenosome come from? Is it simply another kind of bud growing out of the ER? Or did it originate in some primitive clostridium that was caught and domesticated in bygone days by a member of the voracious phagocyte family? There are as yet no clues to this question. But some of the hydrogenosome proteins are being probed to reveal their ancestry.

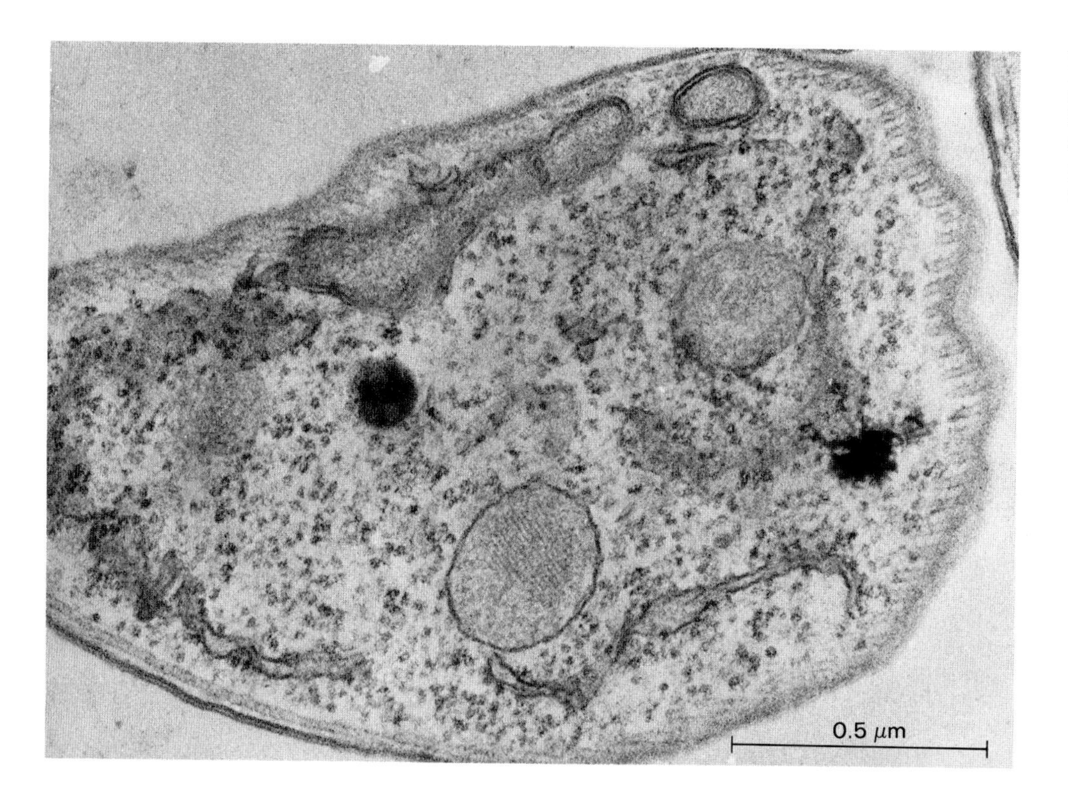

This electron micrograph shows two glycosomes in the cytoplasm of *Trypanosoma brucei*, the agent of sleeping sickness of cattle. One glycosome contains a crystalline core of unknown chemical composition.

Glycosomes

Not long ago, investigators exploring a subgroup of protozoa called trypanosomes discovered yet another species of microbody, which, amazingly, was found to contain a big segment of the organism's glycolytic chain. In every other cell type that has been analyzed in this respect, the glycolytic system has invariably been found in the cytosol. That is where we met it for the first time, earlier in our tour. How, then, did it manage to become segregated within the confines of a membrane-bounded microbody in the trypanosome cytoplasm?

Here, again, imagination is lured to fill the void created by ignorance. Could it be, we cannot help wondering, that the glycosome—as it is called for obvious reasons—is also derived from an endosymbiont, this time a primitive fermenter? Could it perhaps even be that its captor was saved by its prey from the consequences of what would otherwise have been a lethally crippling mutation of its own glycolytic apparatus? If so, the event is still making waves—and rather unpleasant ones—more than one billion years after it happened. For among the trypanosomes are some of the nastiest parasites of animals and humans, causing such severe conditions as African sleeping sickness and the dreaded South American Chagas' disease.

12 Cytobones and Cytomuscles

The shapes of living cells, including those of their infinitely varied surface folds, protrusions, and invaginations so dramatically revealed by scanning electron microscopy, are determined largely by an intricate inner scaffolding of crisscrossed solid fibers and hollow tubes, which make up what is known as the cytoskeleton. These cytoskeletal elements are the girders and cables that we noticed when we first entered the cytosol. But our view of them was obstructed by all the balloon-shaped objects that fill the cytoplasm. Should we be able to remove these objects, we would notice that the components of the cytoskeleton may reach impressive lengths, sometimes extending right across the cell, and that they are linked together in a variety of patterns of astounding beauty and often remarkable regularity. Empty a cluttered museum of its contents and only then will you appreciate the graceful and cleverly engineered architecture of the exhibition halls.

The Cytoskeleton

We cannot empty a cluttered cell in this way, but we can use antibodies directed against cytoskeletal proteins to coat in a very specific manner the structural elements of which these proteins are a part. If these antibodies are made to carry molecules of a fluorescent dye, the structures to which they attach will be as though covered with a coat of fluorescent paint. Illuminate such a decorated cell with ultraviolet light, and the whole cytoskeletal framework coated by the antibody will light up

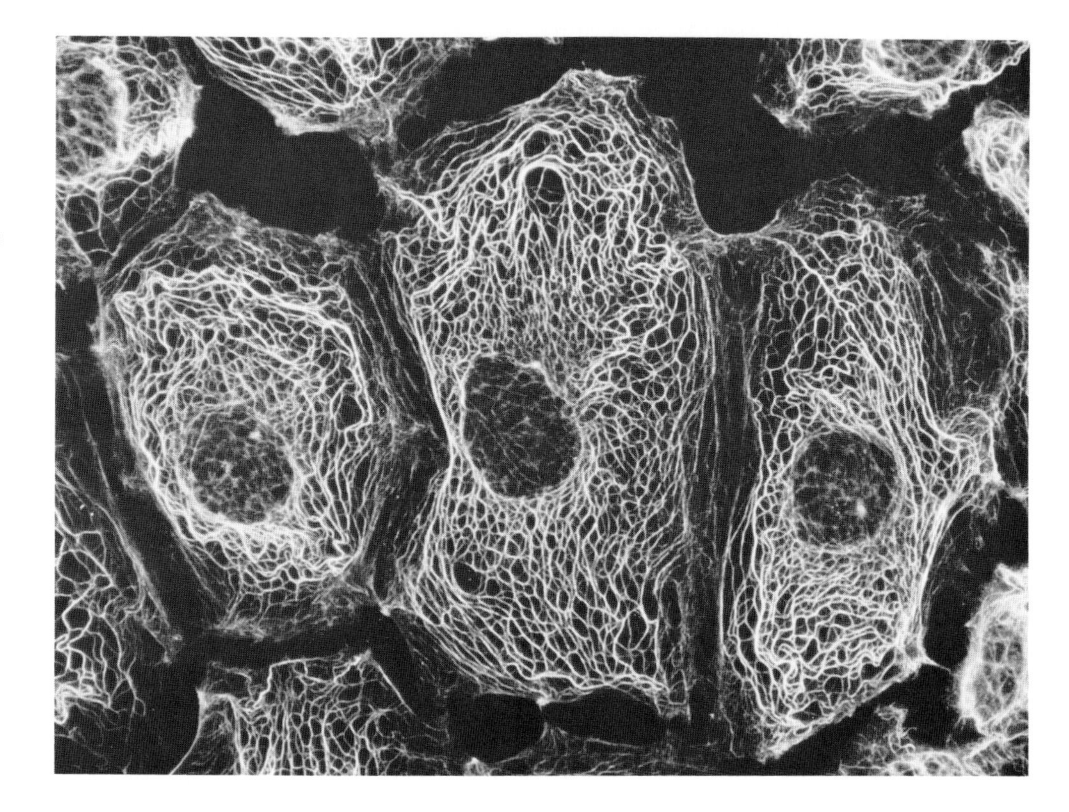

Keratin network of rat kangaroo cells (PtK_2) is revealed by immunofluorescence technique.

brightly on a dark background, as though delineated by neon tubing.

Called immunofluorescence, this elegant technique requires a great deal of prior biochemical research, since the proteins must be purified before they can be used to elicit antibodies. An additional difficulty is that several of the cytoskeletal proteins are poor immunogens—that is, they do not readily induce the production of antibodies when injected into an animal of a different species. The reason is that homologous cytoskeletal proteins, even of widely different species, have closely similar chemical structures and thereby fail to be recognized as foreign by the immune system. This kind of evolutionary conservation of structure is a strong indication that the functional properties of the proteins depend on specific amino-acid sequences that can suffer few modifications without the properties being lost. Most of the mutations that affect such proteins are therefore incompatible with maintenance of function and are eliminated by natural selection.

Beautiful as they are, the pictures revealed by immunofluorescence provide only frozen "stills" of what are constantly changing patterns. Cells alter their shapes all the time; they reshuffle their contents, generate cytoplasmic streams, propel some of their granules in saltatory motion, bend and distort membranes. They move around, creep, crawl, swim, crouch, contract, stretch out, flatten themselves on surfaces, or squeeze through narrow openings. They catch, surround, and engulf bulky objects, push out and retract pseudopods, spew out the contents of stored granules. They wave undulating veils, sweep spiraling flagella, and create currents around themselves by means of beating cilia. What of the cytoskeleton in all this frenzy of movement?

The answer to this question, as might be expected, is that the cytoskeleton is not a rigid framework. In fact, it is not even an articulated framework of the type its name might suggest. It is a much more versatile and complex arrangement of structural elements, only some of which are true fixtures. Others have the remarkable ability of rapidly disassembling into small building blocks and reassembling from them into a different shape, thus explaining the protean metamorphoses of which cells are capable.

Keratinization. This drawing illustrates the progressive differentiation of an epidermal cell as it is pushed toward the skin surface by newly generated cells arising from the germinating layer.

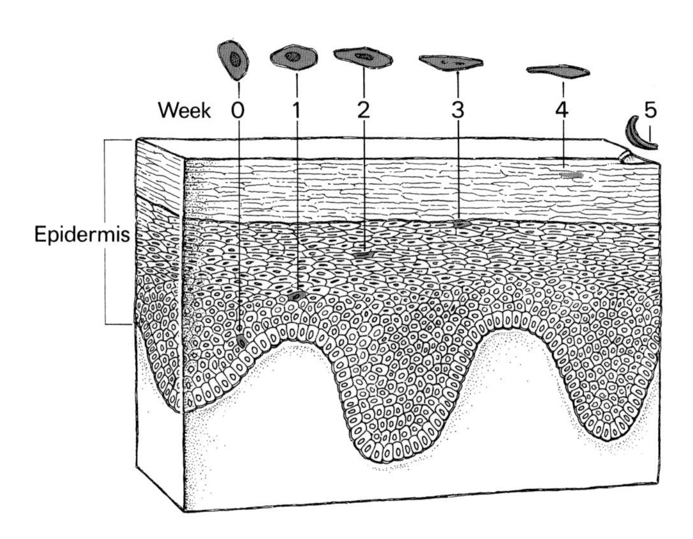

As to the more orderly forms of movement accomplished by cells or by their parts, they seem to depend mostly on the sliding of one structural element on another, elicited by ATP-powered cross-bridges between the two.

The various parts of this machinery are built entirely of protein molecules of various kinds, naturally endowed with the ability to interact either with their congeners or with other cytoskeletal proteins so as to fall together automatically into all those delicate scaffoldings and laceworks that are rendered visible in the ultraviolet light microscope by fluorescent antibodies. Only rarely does the electron microscope reveal these patterns in all their beauty and complexity. In most cases, it allows only glimpses of individual skeletal elements, which appear either as hollow microtubules or as solid filaments. The latter are divided according to their diameters into thin (6–7 nm), intermediate (8–10 nm), and thick (15–20 nm) filaments. Exceptionally, as in the fibrils of muscle cells, or in cilia and other regularly built surface projections, the molecular architecture of the cytoskeleton attains such a remarkable degree of orderliness as to appear clearly in a single, appropriately oriented, ultrathin section. Such pictures have helped considerably in the interpretation of the more random-looking patterns seen most commonly. They have also provided morphologists with some of their greatest esthetic delights.

Functionally, certain cytoskeletal elements serve only to provide the cells with an inner, essentially static, framework and to tie them together by various types of junctions. Most intermediate filaments are of this kind. In conformity with their role in structural differentiation, they tend to be different in different cell types. Examples are the keratin of epithelial cells, the vimentin of mesenchymal cells, the desmin of muscle cells, and the neurofilaments of nerve cells. Other cytoskeletal elements are present in all cells, although their arrangements may vary greatly from one cell type to another; they serve a dynamic, as well as a structural, function. Typical of these are the actin-myosin system, the microtubule-dynein system, and clathrin. We will take a quick look at each of these within the inevitable constraints of time and space that restrict the scope and depth of our visit.

Keratin, the Stuff of Toughness

From the Greek *keras*, horn, the term keratin designates a family of sulfur-rich, fibrous proteins that are the main constituents of skin, hide, hair, horns, hooves, nails, claws, scales, feathers, beaks—that is, of all the outer coverings and appendages with which vertebrates arm themselves against the assaults of the outside world. Unlike the protective coverings of plants and lower animals, which are built extracellularly from secretory products, those of the vertebrates are constructed intracellularly as a result of a very remarkable differentiation process.

The process takes place in epithelial cells. These originate from stem cells that form a layer deep beneath the skin and that divide asymmetrically into undifferentiated stem cells, which remain in the germinating layer, and differentiating daughter cells. As superficial skin layers slough off, daughter cells move slowly to replace them, pushed toward the surface by more newly generated daughter cells.

Should we visit such a young epithelial cell at the beginning of its journey toward the periphery, we would find its cytoplasm traversed here and there by sturdy fibers, about 8 nm thick (one-third of an inch at our millionfold magnification). On closer inspection, the fibers are seen to be bundles of filaments, themselves made up of

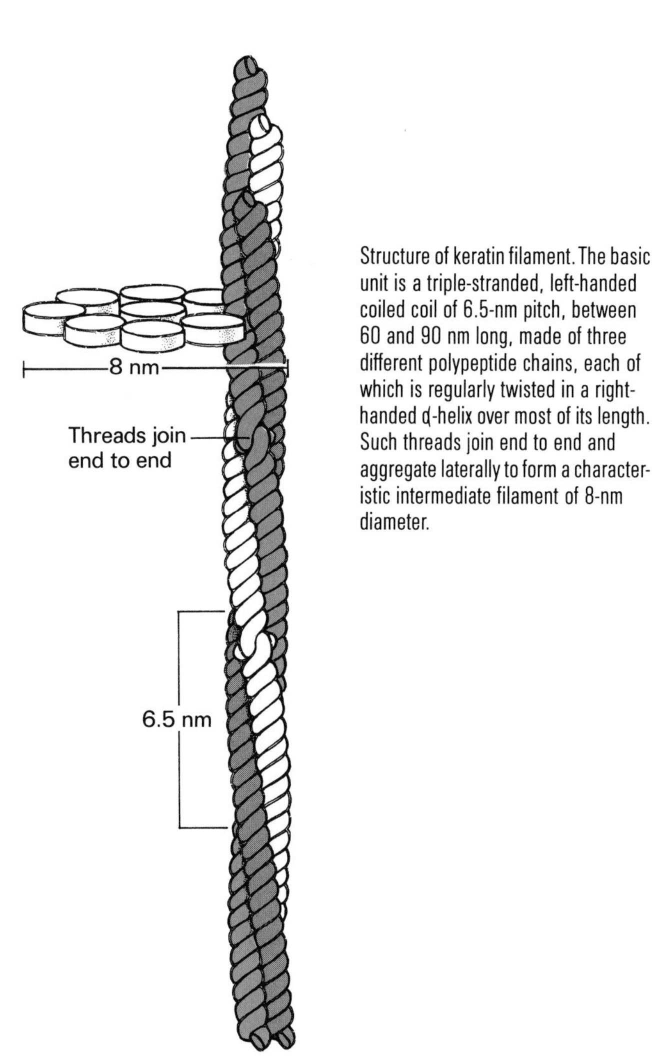

Structure of keratin filament. The basic unit is a triple-stranded, left-handed coiled coil of 6.5-nm pitch, between 60 and 90 nm long, made of three different polypeptide chains, each of which is regularly twisted in a right-handed α-helix over most of its length. Such threads join end to end and aggregate laterally to form a characteristic intermediate filament of 8-nm diameter.

thinner threads that, at first sight, remind us of the structure of tropocollagen. Like the basic unit of connective-tissue fibers, the elementary keratin subunit is a triple-stranded coiled coil made of polypeptide chains that are themselves helically twisted. The pitch of the coiled coil is similar to that of tropocollagen, about 6.5 nm. Closer inspection, however, reveals major differences between the two structures. In keratin, the triple-stranded coil is left-handed instead of right-handed. Its three constitutive polypeptide chains are usually different, and each is twisted into a typical right-handed α-helix, of the kind seen in many proteins (Chapter 2). We will encounter remarkable samples of this structure a little later, in myosin and tropomyosin. The kind of helical structure exhibited by collagen, on the other hand, with its left-handed turn and 1-nm pitch, is unique and explained by this molecule's very unusual amino-acid composition and sequence. Keratin subunits are also shorter than tropocollagen: from 60 to 90 nm, as against 300 nm. They aggregate laterally and longitudinally to form the characteristic intermediate filaments of 8-nm diameter.

Together, the keratin bundles form a loose, three-dimensional network that envelops the nucleus within a basketlike arrangement and is strung between a number of reinforced plates distributed over the plasma membrane. As a rule, these plates are cemented by means of some sort of dense adhesive material to similar plates on the surface of neighboring cells, forming the adhering junctions, or desmosomes, that attracted our attention when we first entered a blood vessel (Chapter 2). The anchoring keratin fibers are the tonofilaments, which are seen to radiate transversely from the desmosomes into the depth of each cell and to straddle the desmosomes so as to establish direct links between the keratin networks of the two cells.

As epithelial cells move slowly from the germinating stem-cell layer to the skin surface, they devote themselves increasingly to the production of keratin, which they support by the massive autophagic destruction of their contents. At the same time, the keratin fibers become increasingly cross-linked with each other and with amorphous matrix components by means of disulfide bonds. By the time the cells reach the skin surface, they are all shriveled and dried out, lifeless and inert, but very, very tough. Firmly riveted to each other by the desmosomes, they form a single protective sheet, the horny layer, which is continuously shed by surface desquamation (flaking off) and replaced by newly differentiated epithelial cells.

If sloughing off is prevented by welding together of superimposed cell layers, the horny layer will thicken to a callus. Change the pattern of growth somewhat, modify the nature or proportion of the various polypeptides that make up the keratin, change their degree of cross-linking, and the resulting structure may be a scale, a nail, a claw, a horn, or a beak. If the differentiating cells grow into a tube instead of a sheet, they end up as all sorts of hairs and spines and, through more complex designs, as the intricate combinations of quills, barbs, and down that cover feathery creatures. There is really no end to the number of architectural variations that evolution has constructed from the central keratinization theme. For millenia, some of these variations have provided mankind with clothing and tools. Even in our contemporary plastic age, no man-made fiber can yet compete with the keratin produced by an Angora goat.

As mentioned above, keratin is a characteristic product of epithelial cells. In other cell types, the supporting intermediate filaments are made from other proteins, which assemble differently. Time does not permit their detailed inventory, and we must return to the "generalized cell" that forms the main object of our tour.

The Actin-Myosin System

Actin and myosin are two proteins that form a very remarkable locomotor combination, first discovered in muscle cells but now known to exist in all cells, in which thin actin filaments make up what may be called the "cytobones," and thick myosin filaments the "cytomuscles."

The actin filaments are usually grouped into slender bundles, which may be seen stretched like telephone wires across the cytoplasm (stress fibers) or meshed together in the form of variously structured cables, belts, felts, or webs, which most often serve as backing for the plasma membrane. They also make up the axial shafts that provide support to such surface protrusions as microvilli. Each actin filament is from 6 to 7 nm thick (one-fourth of an inch at our millionfold magnification) and consists of two intertwined threads.

Unlike the other filamentous structures that we have encountered so far, such as collagen and keratin, these threads are not built from fibrillar proteins but rather from small protein balls of the usual curled-up type, linked by complementary binding sites situated at their two poles. An actin thread thus resembles one of those indefinitely growing trains or chains that can be built out of identical interlocking pieces often put in the hands of

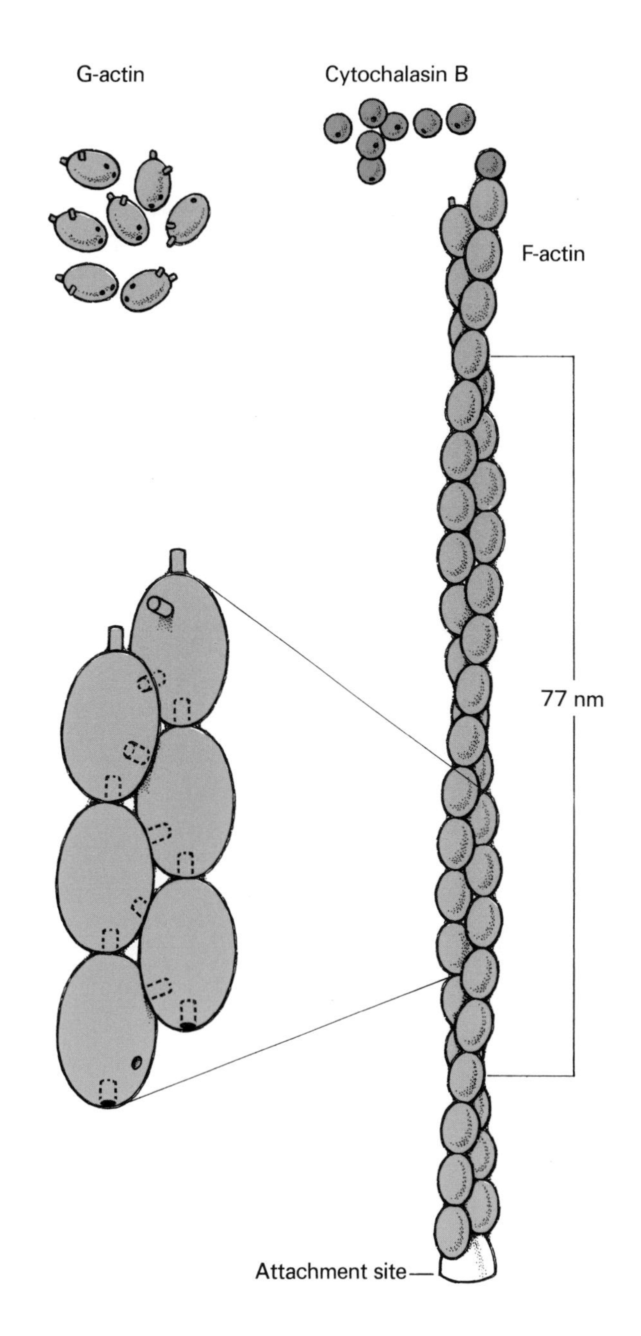

Structure of actin filament. Globular, or G, actin is an egg-shaped protein molecule fitted with two pairs of complementary (lock-key) attachment sites, one polar, the other lateral, which allow the molecules to polymerize into a twisted double-stranded string (fibrous, or F, actin) containing fourteen pairs of monomeric molecules per complete turn (77 nm). Polymerization is directional, away from the filament's attachment point, and reversible. It requires ATP and is blocked by cytochalasin B, a fungal poison.

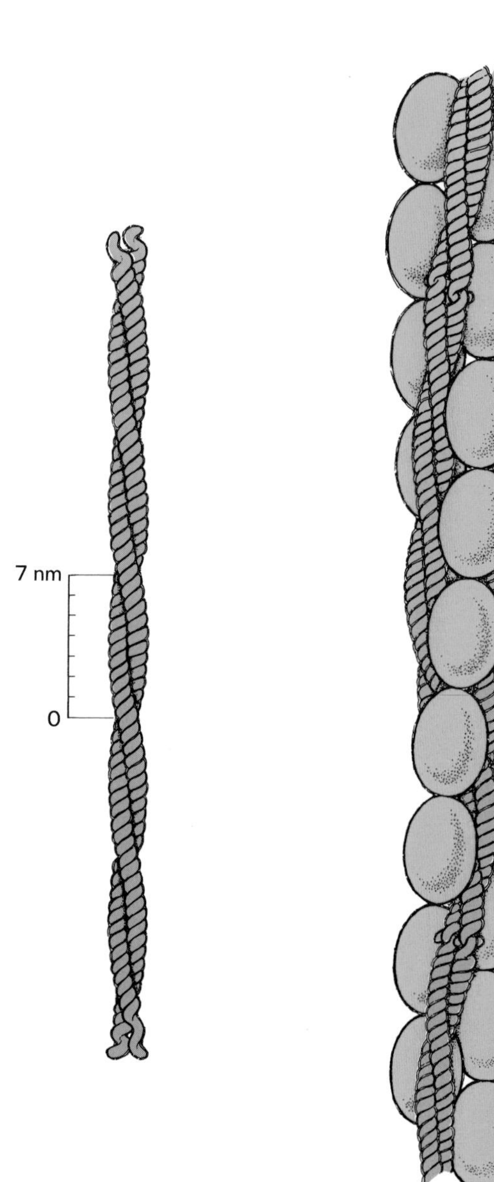

Structure of tropomyosin. This double-stranded coiled coil is 41 nm long and consists of two identical polypeptide chains, α-helical over most of their length, and twisted around each other with a 7-nm pitch.

Arrangement of tropomyosin around actin filament.

young children to test their combinatorial skills. Like such a chain, it has a polarity, defined by the direction of the "lock-key" axis of the pieces. In the case of actin, the chain is not single but double, because the pieces have a second lock-key set situated laterally. The binding of two parallel threads by this set occurs with a slight right-handed twist, generating an elongated double-stranded helix with a 77-nm pitch (about 3 inches at our magnification) containing fourteen pairs of subunits per turn.

The subunits of the actin filaments are called globular, or G, actin. The product of their polymerization is fibrous, or F, actin. The interconvertibility of these two forms is what enables living cells to dismantle certain parts of their actin skeleton and to reassemble them into a different pattern. These operations are controlled by a complex set of directions, so far largely undeciphered, involving the active participation of ATP. There is a special binding site for ATP on the G-actin molecule. Upon polymerization, this ATP is hydrolyzed, and the resulting ADP remains bound to the F-actin. ADP has to be displaced by ATP before the structure can be disassembled.

As a rule, the spiral grooves separating the two strands of an actin filament are occupied by a thin thread made up of another protein molecule, called tropomyosin (Greek tropê, turn; *mys*, muscle). This molecule is a left-handed coiled coil, about 41 nm long, made of two identical polypeptide chains, α-helical over most of their length and twisted around each other with a 7-nm pitch. It provides us with our second example of the α-helix, which we saw a short while ago when we looked at the keratin fibers. Note, however, that tropomyosin is double stranded, whereas keratin is triple stranded. In the actin filament, each tropomyosin molecule extends over seven G-actin subunits, or exactly one-half turn of the helix. So, for each 77-nm turn in F-actin, there are two pairs of dimeric tropomyosin molecules twisting together with fourteen pairs of G-actin subunits. In striated muscle cells but not in others, this repeating structure bears four additional molecules of another protein, called troponin, attached to the actin thread near the junction between successive tropomyosin molecules. Troponin, as we shall see, plays a key role in the regulation of muscle contraction.

Actin filaments are attached by one of their extremities—always the same one with respect to the polarity of the filament—to a sort of flat, disklike structure in which the end of the filament is firmly anchored. A protein called α-actinin has been identified as a component of this structure, but there are several others. These anchoring points remain when filaments are disassembled and provide nucleation sites from which new filaments can grow. Cytochalasin B, a poison of fungal origin, has the property of binding to the growing ends of actin filaments and of inhibiting their further elongation. It thereby interferes with all cellular activities that require remodeling of the actin cytoskeleton. It has become an invaluable tool in the identification and analysis of such activities.

The anchoring disks of the actin filaments are themselves attached to special patches on the inner face of the plasma membrane, or to each other, or to other intracellular organelles. In this way, they form the wide variety of arrangements that are seen in different cells or in the same cell in different functional states. Their attachments are mediated by a number of structural proteins that bear such suggestive names as vinculin (Latin *vincula*, bond), ankyrin (Greek *ankyra*, hook), or spectrin (so named because it was first isolated from erythrocyte "ghosts"). In muscle cells, numerous attachment disks are knit together by a protein called desmin (Greek *desmos*, bond) to form what looks like two brushes glued back to back, with their bristles pointing in opposite directions.

We will see how these bristles function in muscle contraction. But before that, it is helpful to explore the whole length of a single bundle of half a dozen actin filaments, starting from their anchoring point. At first the filaments form a rather loose skein and seem to be held together only by their common anchorage. But eventually they tighten into a cylinder in which the six filaments are grouped around a long, central shaft about 15 nm thick. This shaft extends beyond the ends of the filaments and continues over a distance of several hundred nanometers—sometimes as much as 1 μm—to end up as the core of a second bundle of filaments, symmetrically oriented. What we see at our millionfold magnification is a rigid rod, as much as several feet long and three-fifths of an inch

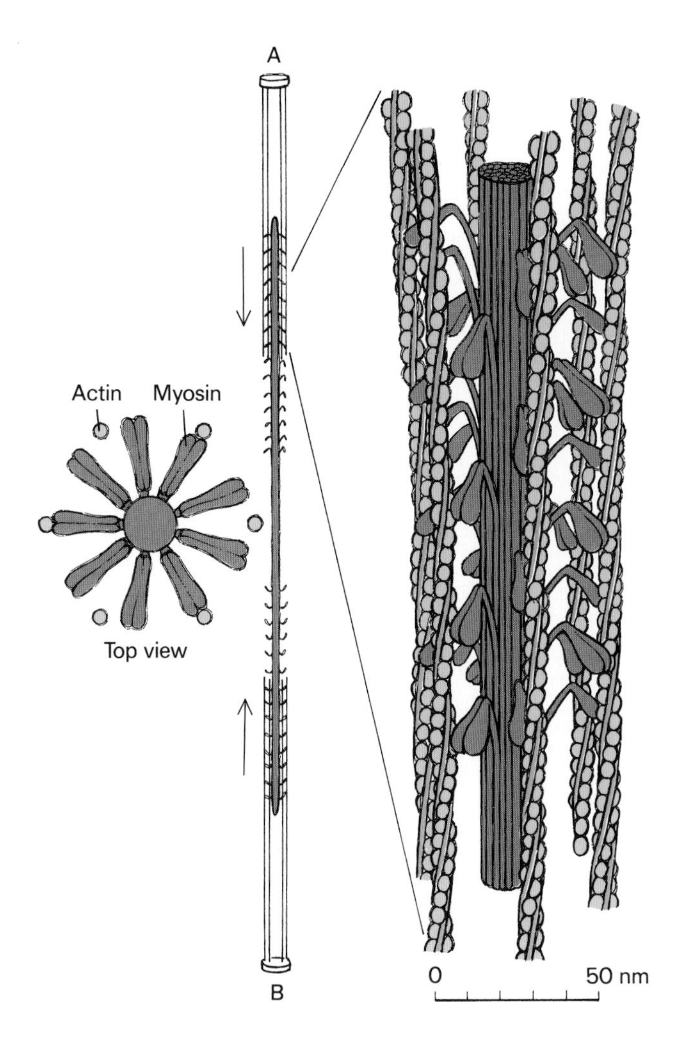

Elementary cytomuscle. A thick myosin filament connects two opposed bundles of six thin actin filaments. The forcible pulling of the actin filaments toward the center of the myosin shaft draws anchoring points A and B together.

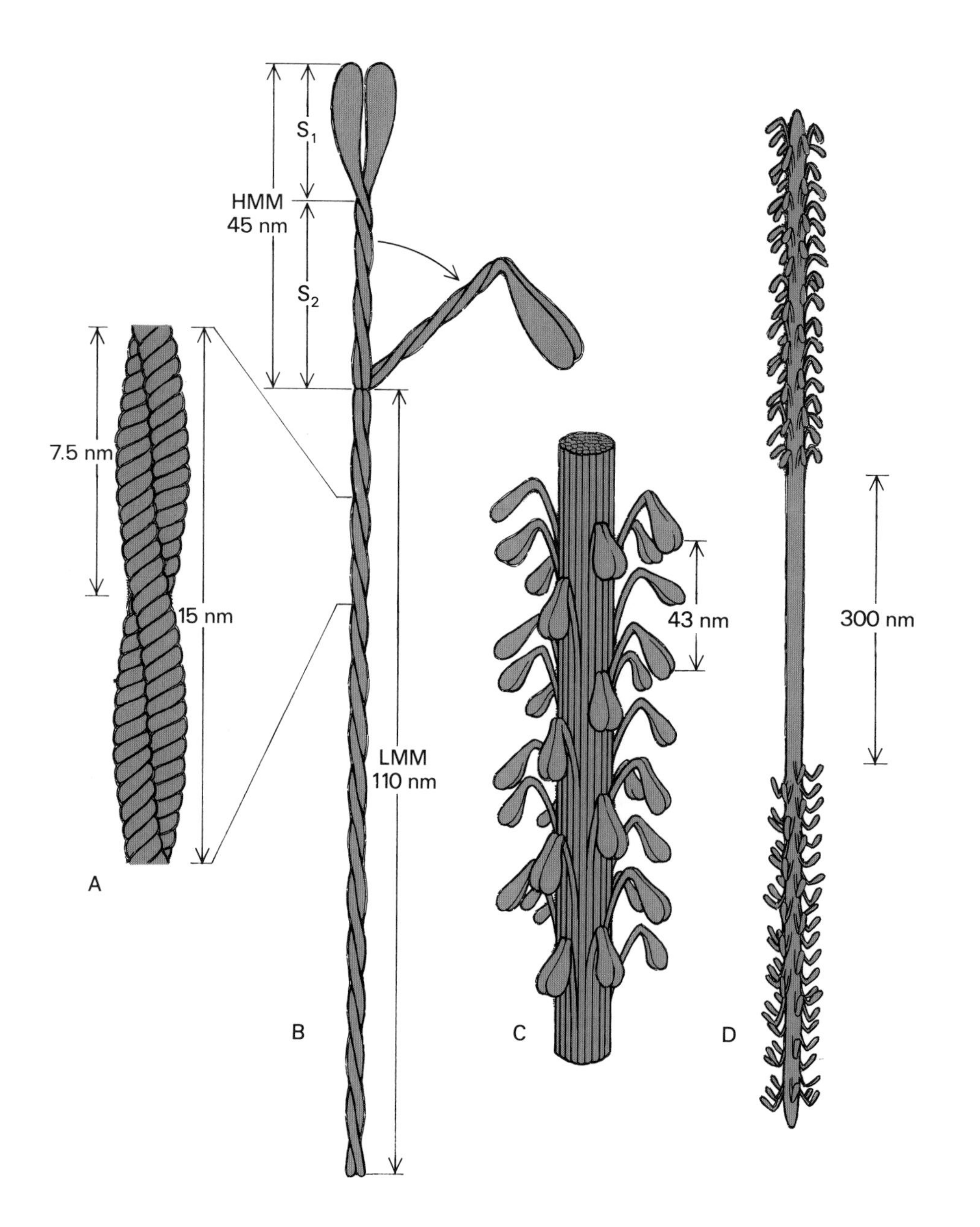

Structure of thick filament. The elementary unit is the myosin molecule (B), a long, duplex, protein molecule, consisting of a 135-nm long tail bearing twin, bulbous ends, or heads. The tail is a coiled coil of two identical, almost perfectly α-helical strands (shown in detail in part A). One irregularity in this structure provides a hinge between a hydrophobic stem, 110 nm long, and a more hydrophilic neck bearing the heads. This hinge is easily cut by proteolytic enzymes, to give heavy (HMM) and light (LMM) meromyosins. The hinge between the head and neck in HMM is also cut easily, with the formation of subfragments S_1 and S_2. Myosin molecules have the remarkable property of assembling into thick rods, from 15 to 20 nm thick and from 1.5 to more than 5 μm long. Molecules join by their hydrophobic LMM tails and have their hydrophobic HMM stalks and heads protruding on the surface of the rod in a regular helical arrangement. This arrangement is symmetrical (as shown in part D). In the middle of the rod is a bare segment, 300 nm long, made of the interdigitated stems that bear the heads of the first row on each side.

Polarity of actin filament determined by "decoration" with isolated myosin heads (S_1 fragments): (below) electron micrograph; (right) model.

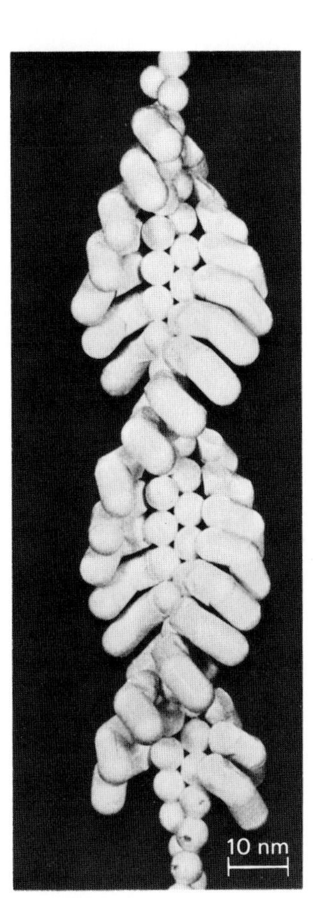

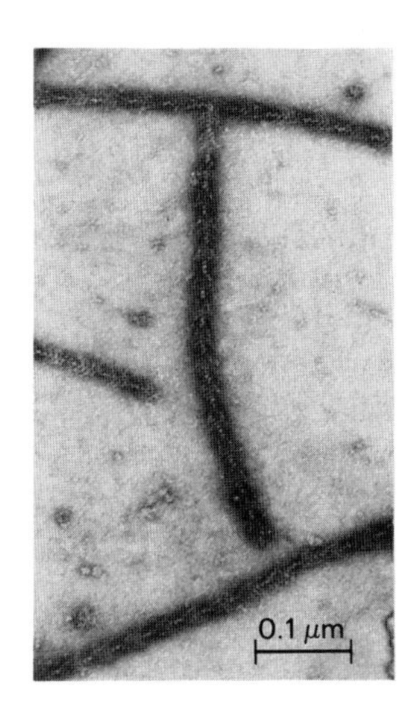

thick, connecting two opposite bundles of six identical double-stranded wires about one-quarter of an inch thick.

This connecting rod is made of myosin, a unique kind of protein that, like tropomyosin, derives its name from the Greek word *mys*. It is a long, duplex molecule, 155 nm in length (a little more than half a foot, at our magnification). Its shape has been likened to that of a golf club (with a split head) or, more romantically, to that of a long-stemmed, twin flower. The stem, about 135 nm long and 2 nm thick, is constructed very much like tropomyosin. It is a left-handed, double-stranded coiled coil, with a pitch of about 7.5 nm, made of two identical, almost perfectly α-helical, polypeptide chains. The myosin stem is the longest such structure known in nature. At its upper end, the two strands separate into distinct flexible stalks that progressively curl up into twin, globular "flowers," or "heads," each of which consists of the terminal portion of one of the stem's strands, or heavy chains, intertwined with two additional light chains. There is one irregularity in the structure of the myosin stem. It provides a hinge between a hydrophobic distal portion, about 110 nm long, and a more hydrophilic proximal segment bearing the heads. This irregularity allows cutting by the proteolytic enzyme trypsin, which sections myosin into two parts, named light meromyosin (LMM) and heavy meromyosin (HMM). A second hinge exists between the heads and the stem in HMM. It is susceptible to splitting by another proteolytic enzyme, papain, which cuts HMM into two subfragments, S_1 and S_2.

Myosin molecules have the remarkable property of associating spontaneously into bundles, which look like miniature exotic trees elegantly festooned with flowers. The "trunk" of the tree is made of the stems of the myosin molecules, joined lengthwise by their hydrophobic LMM portions and staggered helically so as to allow indefinite lengthening over constant thickness (except at the top, which is tapered). The heads, or flowers, protrude from this trunk on the more hydrophilic LMM portions and form a garland that spirals around the trunk in a regular helical fashion. The exact configuration of this arrangement is still a matter of debate among experts and may vary according to cell type and animal species. What seems to be constant is a longitudinal spacing of the heads by 14.3 nm along the trunk axis. An additional, but absolutely essential, feature of this assembly process is that it is symmetrical. Two such trees are always joined by their roots, so that each myosin filament is made of two garlanded shafts of opposite polarity, linked together by a naked central portion about 300 nm long.

The flowers on the myosin tree—in more technical terms, the heads of the myosin molecules—are the hooks to which the actin filaments are attached. Their disposition around the trunk is such that six parallel actin filaments can bind to them, thus surrounding the myosin shaft with a sheath of hexagonal symmetry. One such sheath can form at each extremity of the myosin shaft, which thereby can connect two opposed actin bundles.

Such positioning requires the actin filaments to bear appropriately oriented binding sites for myosin heads. This property can be used to identify actin filaments in tissue sections and to determine their polarity. Myosin

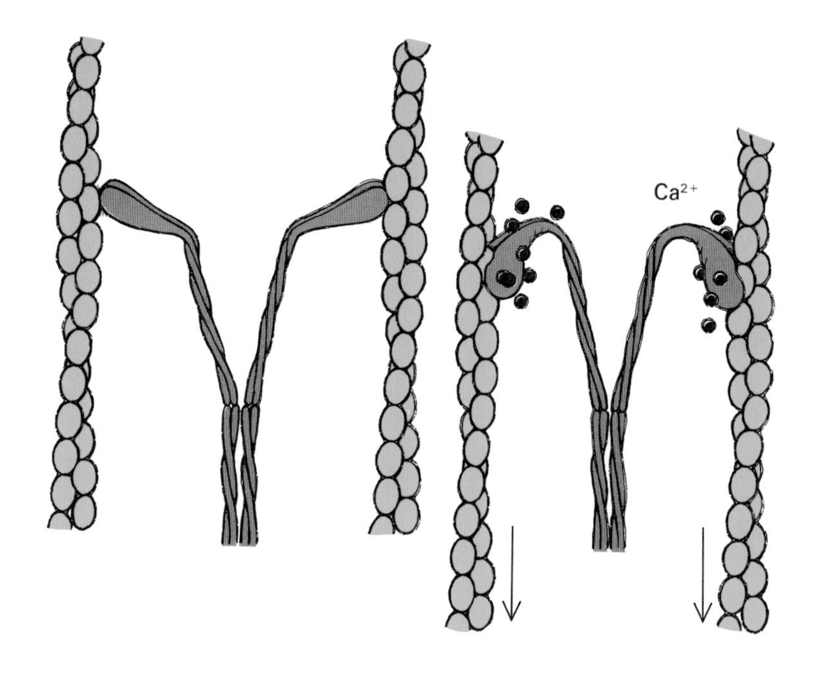

Sliding-filament mechanism of actin-myosin contractile system. In the presence of calcium ions, the ATP-bearing myosin heads bind to actin and become active as ATP-hydrolyzing enzymes. ATP hydrolysis is obligatorily coupled to the bending of the myosin heads and the longitudinal displacement of the actin filaments with respect to the central myosin filament.

heads, cut off enzymatically from their stalk and purified, are used as reagents. They bind specifically to the actin filaments and "decorate" them in a typical arrowhead pattern, with the arrows pointing in the direction of the free end of the actin filament.

So far, we have seen the myosin shaft serving simply as a connecting rod between two bundles of actin filaments. The sight is graceful but static. But let just a few calcium ions reach the system, and you will be offered one of the most dazzling displays of molecular pyrotechnics to entertain a cell tourist. With dramatic suddenness, the myosin heads come alive. Wherever they are attached to a binding site on an actin filament, they bend violently on their stalks, tugging the filament some 10 nm inward, in the direction of the middle of the shaft. This done, they relinquish their hold. By that time, however, other myosin heads have come into register with actin binding sites and exert a further 10-nm pull. This goes on as long as calcium ions (and ATP, see below) are present. With all actin filaments at each end of the myosin shaft being pulled inward in the same way, the net result is to make the two opposing actin bundles that are joined by the myosin shaft slide into each other and drag their respective anchoring points closer together. Thus the distance between these anchoring points contracts, although none of the filaments that link them actually shorten. They slide along each other by means of what may be described as a molecular ratchet.

This process accomplishes mechanical work and therefore requires energy, which, you will hardly be surprised to learn, is supplied by the hydrolysis of ATP into ADP and P_i. Myosin heads are really ATP-hydrolyzing enzymes (ATPases) of a very special kind. They can exert their catalytic activity only if (1) they are bound to actin; (2) they are activated by calcium ions (in a way that we shall consider later); and (3) they are permitted to bend at the same time. The obligatory coupling between the chemical event of ATP hydrolysis and the conformational change that forces the actin-linked myosin head to bend on its stalk and drag along the actin filament is the fundamental property that allows the actin-myosin system to convert the free energy of hydrolysis of ATP into mechanical work.

The kind of work that is accomplished depends on the cellular localization of the two anchoring points that are being pulled toward each other and on their topological relationships. Often, one point remains fixed and the pull is exerted entirely on the other. Take cell creeping, for example. A patch of plasma membrane to which an actin bundle is anchored is made to stick firmly to the substrate (adhesion plaque), after which the actin-myosin complex shortens and pulls whatever is attached to it—which

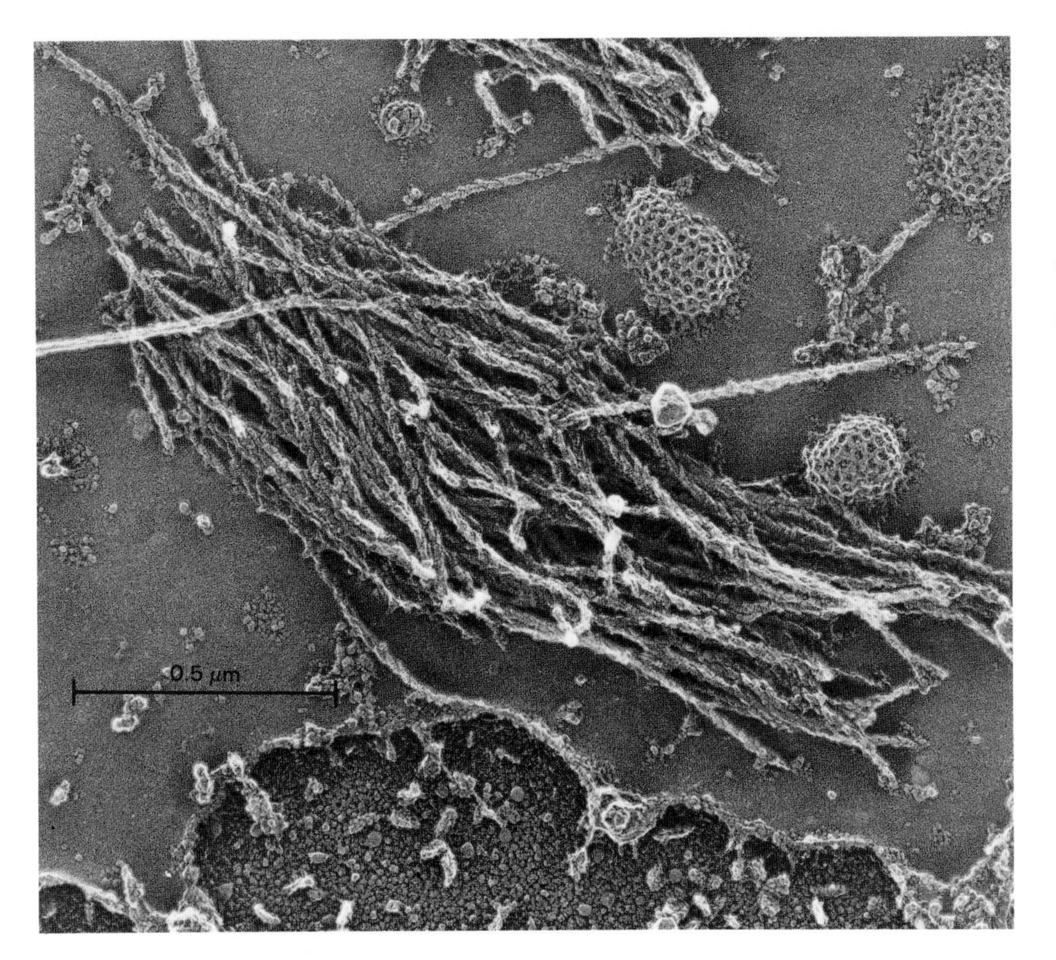

Special quick-freeze, deep-etch technique allows startling visualization of bundles of actin fibers lining the cytoplasmic face of the plasma membrane of a fibroblast. The actin fibers have been "decorated" with isolated myosin heads (S_1 fragments). Structures with a chicken-wire appearance are clathrin baskets over coated pits (see pp. 218–222).

may be the whole cell through its interconnected cytoskeleton—toward the adhesion site. When the movement is over, the adhesion plaque is "unstuck," and new ones form in front through the operation of a ruffling extension, called the lamellipodium (Latin *lamella*, small, thin plate; Greek *pous*, gen. *podos*, foot), which sends out sticky, fingerlike protrusions called filopodia (Latin *filum*, thread). This, according to some observers, could be the basic mechanism of amoeboid movement, whereby cells propel themselves toward (positive chemotaxis) or away from (negative chemotaxis) certain objects that send out chemical signals.

Actin fibers, as we have seen, often occur just below the plasma membrane in the form of belts or webs of various shapes, sometimes connected to axial strings in the inner core of microvilli and other cellular protrusions. One can see how tightening the belt may bring about cellular constriction; how retracting the web may purse, wrinkle, pucker, round up, evaginate, invaginate, or otherwise deform the surface of the cell; how pulling the strings may bend a microvillus. And, if the anchoring points are attached to intracellular organelles, one can see how these can be displaced with respect to each other and with respect to the cell surface.

Indeed, one can visualize, at least as a purely theoretical exercise, how the immense variety of movements displayed by living cells can be brought about by the coordinated operation of hundreds of tiny myosin cytomuscles busily pulling miniature actin cytobones in various directions, just as the complex movements of our own body can be explained by the manner in which our muscles pull our bones or fold our skin. But it will take an enormous amount of ultramicroscopic exploration and dissection of a kind that is still largely beyond our present technical

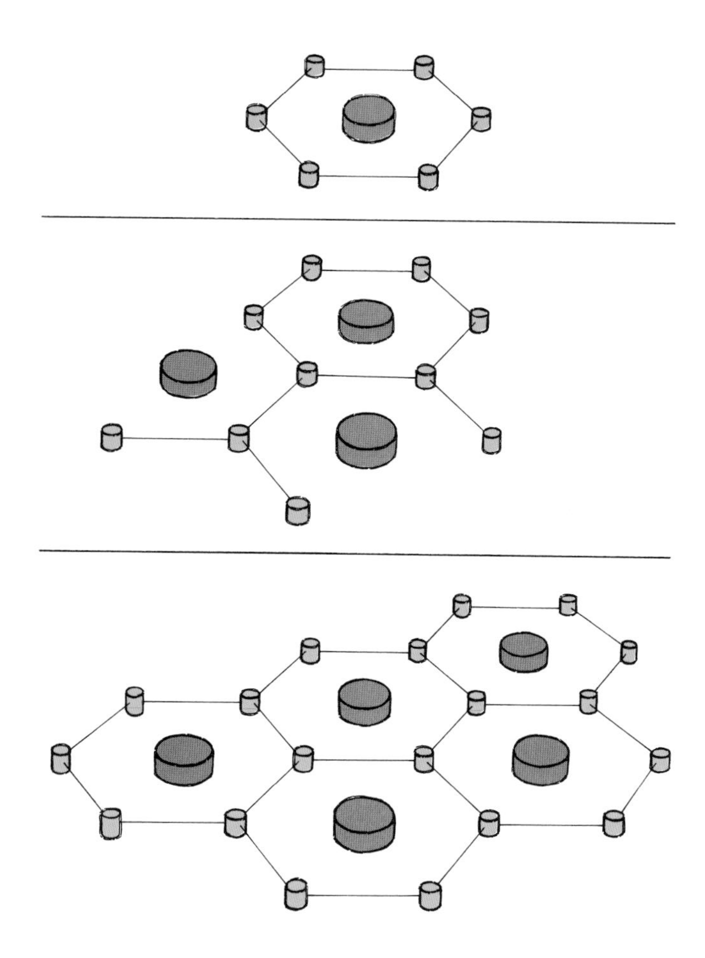

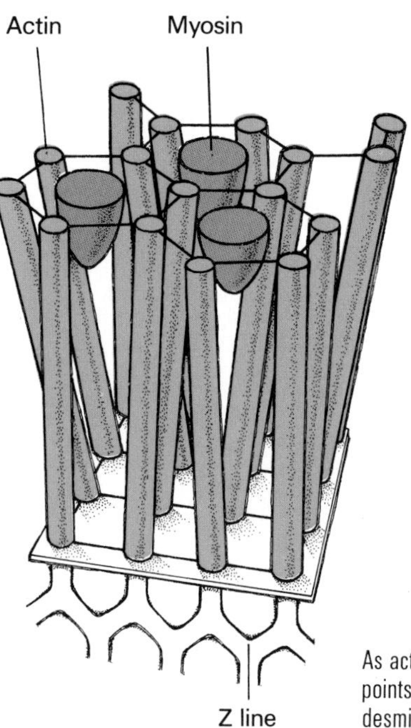

As actin filaments approach their points of attachment to the actinin-desmin meshwork (Z line), their arrangement changes from hexagonal to rectangular.

Diagram showing how the lateral joining of actin and myosin filaments leads to the assembly of a myofibril with regular hexagonal arrangement.

means before the detailed anatomy of the cell's bones and muscles can be mapped out and understood in dynamic terms. These difficulties are further compounded by the fact that the shapes and interconnections of many cytobones change, as do the sites of many cytomuscular insertions. Furthermore, as we shall see, cells contain a second cytoskeletal system made of microtubules and a second type of cytomuscle, dynein, associated with it.

The force exerted by an actin-myosin cytomuscle depends on the number of myosin heads that are pulling together at any time—that is, on the length over which filaments interact—and on the number of parallel filaments that join in this effort. If we examine the single actin-myosin connection that we have explored so far, we see that each of the six actin filaments that surround the myosin shaft has unoccupied myosin-binding sites on its exposed surface. It follows from this property that the six filaments of the first bundle can bind additional myosin shafts, which in turn can surround themselves with more filaments, and so on. Repeating such an arrangement, we end up with a structure of hexagonal symmetry, in which each myosin filament is surrounded by six actin filaments and each actin filament is flanked by three myosin shafts alternating with three actin filaments. Should you immobilize the actin filaments in such a structure by gluing together their α-actinin roots and then slide out the intercalated myosin rods, you would end up with a bristle of thin actin filaments planted from 15 to 25 nm apart at every angle of a regular hexagonal lattice, in which each hexagon surrounds a hole capable of precisely accommodating a thick myosin filament. This is what desmin, the connecting protein referred to earlier, manages to achieve in the muscle cell, except that it actually forms a rectangular network, which changes to a hexagonal arrangement upon insertion of myosin shafts into the actin bristle.

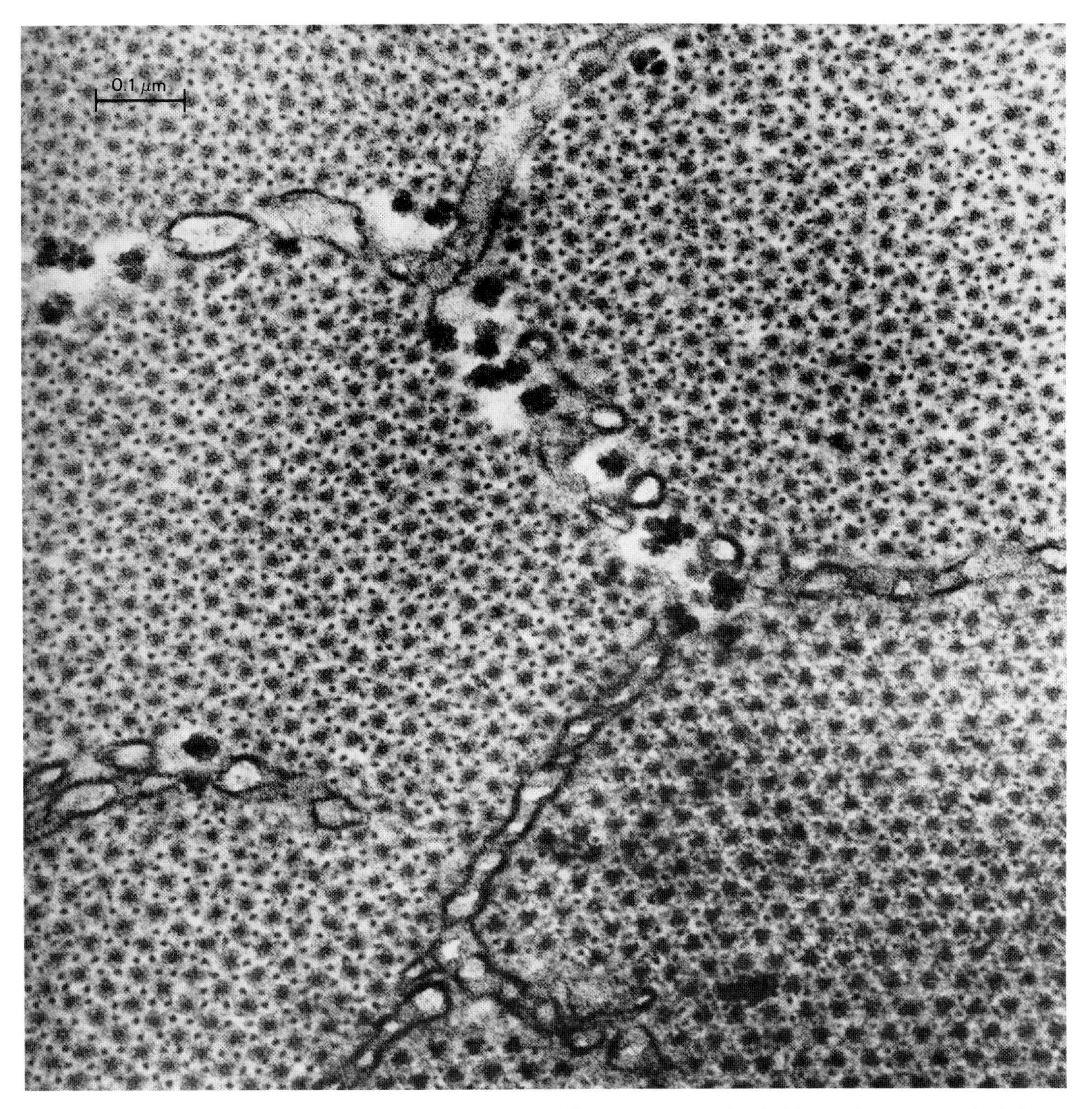

Hexagonal arrangement of thin (actin) and thick (myosin) filaments appears clearly in this electron micrograph of a transverse section through a striated muscle fiber. This picture and that of a longitudinal section shown on the next page are historical documents. They were taken more than thirty years ago and played a major role in the development of the sliding-filament theory of muscular contraction.

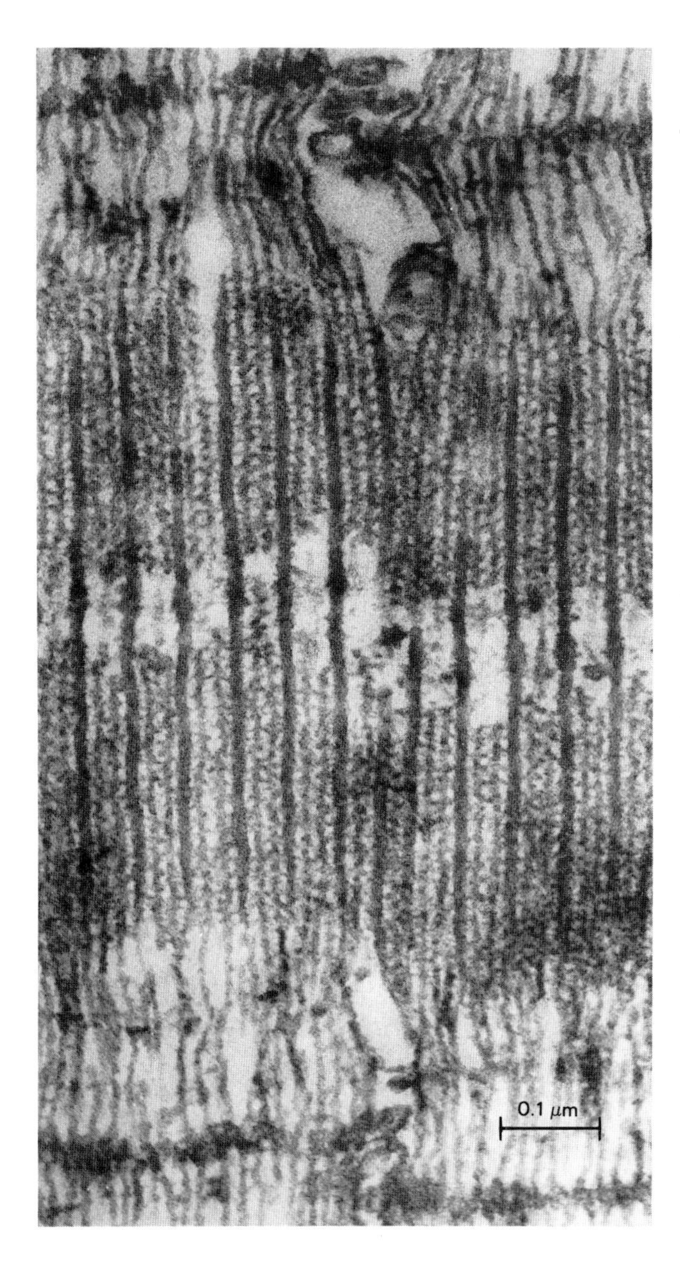

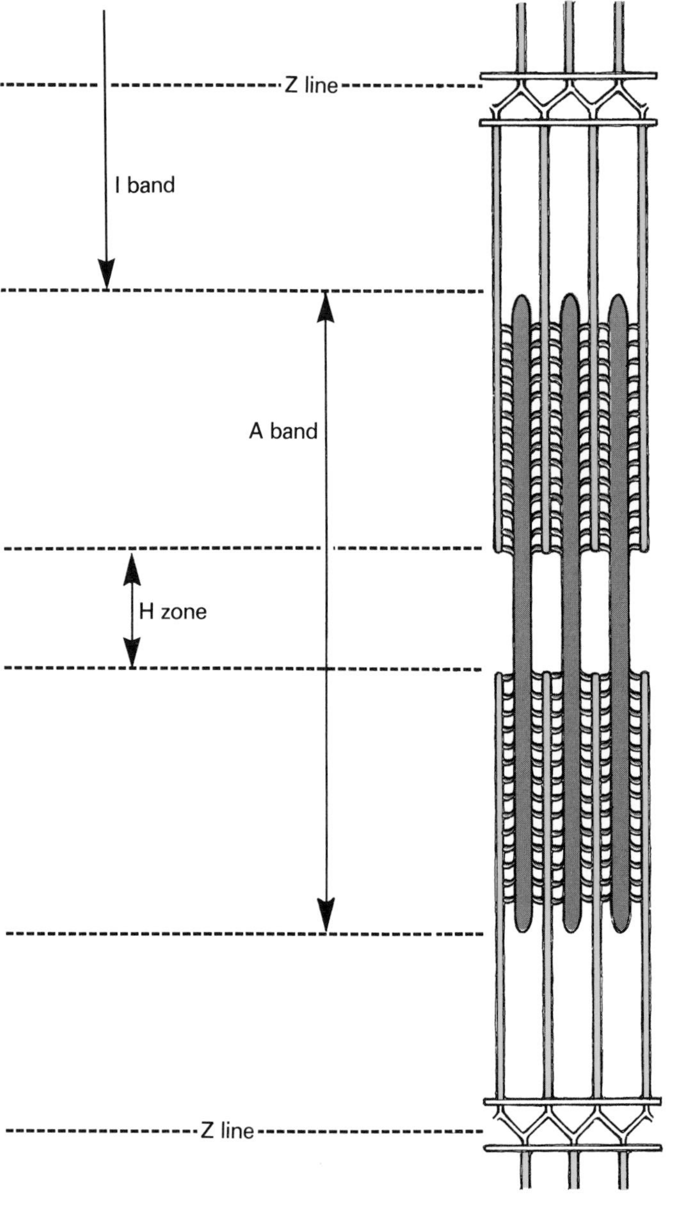

This electron micrograph of a longitudinal section through a striated muscle fiber illustrates the parallel arrangement of thin and thick filaments. The picture shows a complete sarcomere limited at both ends by a Z line (top and bottom) in which the actin filaments are joined to the actin filaments of adjacent sarcomeres by an actinin-desmin meshwork. The shape of the myosin filaments, with beveled extremities, lateral attachments (heads) to actin filaments, and bare midpiece, is clearly recognizable. This myofibril is in a partly contracted state, with thin and thick filaments having slid past each other over a considerable distance. There are two, not one, actin filaments between myosin filaments because of the orientation of the section (see the diagram on p. 202).

Diagrammatic representation at the right above illustrates the main features of cross-striation. The A (anisotropic) band, of constant length, is made of myosin filaments interdigitating at both ends with variable lengths of actin filaments, depending on the degree of contraction. The I (isotropic) band contains only actin; its length depends on the degree of contraction. The I band is bisected by a Z line, where the actinin-desmin meshwork knits the ends of two sets of actin filaments together. Midpieces of the myosin filaments make up the H zone in the middle of the A band.

In vertebrate striated muscle, all the actin filaments have the same length of 1 μm and all the myosin shafts are 1.5 μm long. The complete unit of two opposed actin bristles, connected by the central myosin shafts, is called a sarcomere (Greek *sarx*, flesh; *meros*, part). Its length varies from about 3.5 μm (actins almost entirely pulled away from the myosins) to 1.5 μm (fully interdigitated). In the latter state, the free ends of the actin filaments (250 nm) bundle up in the spaces between the bare middles of the myosin shafts, whereas the tapered ends of the myosin shafts press against the α-actinin–desmin meshwork in which the actin filaments are rooted. It would be wonderful if we could wander freely through a sarcomere, which must look like a tree planter's dream. Just imagine the veined, flower-garlanded myosin trunks alternating with the slender, helically fluted actin shafts, gracefully trimmed with tropomyosin and troponin decorations. Unfortunately, such a tour would require us to shrink at least another hundredfold. Even at our millionfold magnification, the trees of this molecular forest stand less than an inch apart, and this space is further obstructed by their flowery outgrowths. In addition, there is always the danger of a few calcium ions sneaking in and of the forest closing up on us. We must content ourselves with the two-dimensional images revealed by the electron microscope. Even so diminished, their beauty is arresting.

In a muscle fibril, a large number of sarcomeres are linked in series through the binding action of desmin, which not only knits actin anchoring points together to form an appropriately planted bristle, but also, as already mentioned, glues two such bristles back to back. The regular alternation of different zones gives the fibril its characteristic cross-striated appearance. Such fibrils may reach enormous dimensions at the cellular scale. They are built from hundreds of cells that fuse to form a syncytium (Greek *syn*, with). They are surrounded by an elaborate system of membranes that serve primarily in the rapid release and withdrawal of calcium ions whereby muscle contraction is regulated and by rows of mitochondria that provide the necessary ATP fuel.

Such a fibril can adopt three distinct states. In the absence of ATP and calcium, actin and myosin are rigidly

Structure of striated muscle. Each fiber is a giant multinucleated syncytium containing a bundle of myofibrils enveloped by sarcolemma.

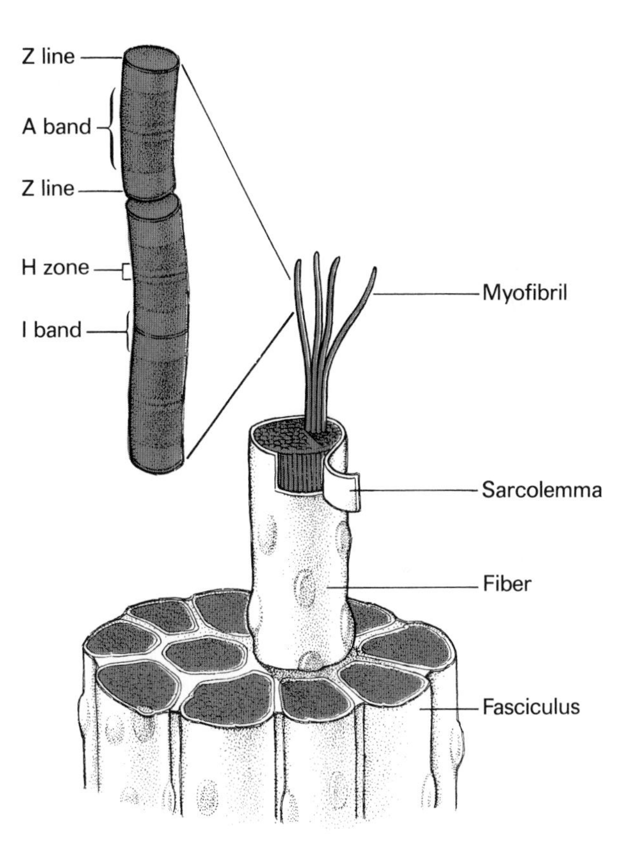

The three states of actin-myosin.

A. *Relaxation:* In the presence of ATP and absence of calcium, the system is plastic. Filaments slide freely along each other.

B. *Contraction:* The addition of calcium ions causes the myosin head to interact with the actin filament (see the next two illustrations). Actin is pulled downward, while ATP is hydrolyzed.

C. *Rigor:* The removal of calcium, with ATP absent, locks the system in a state of rigor. The addition of ATP restores the relaxed state A.

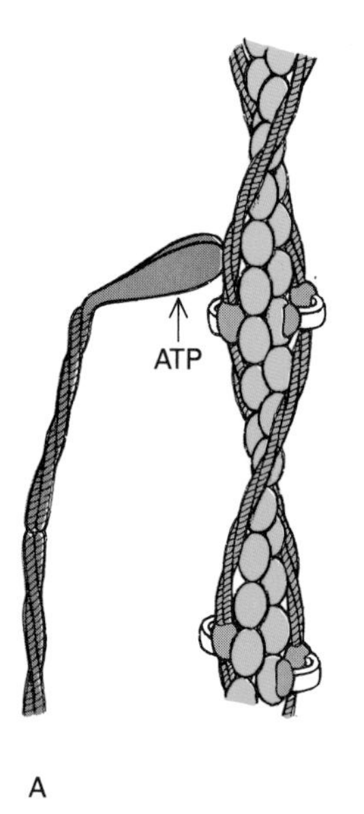

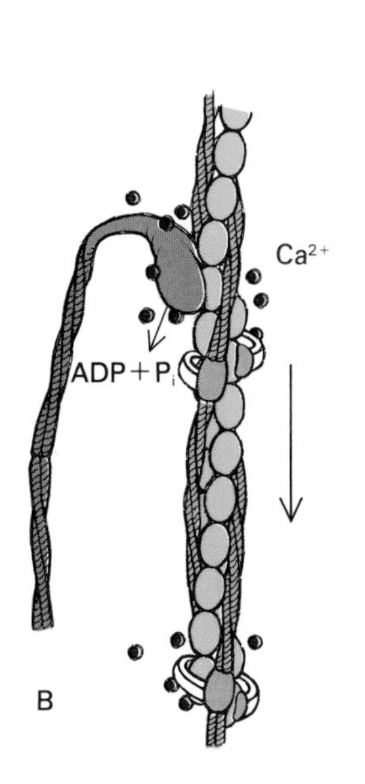

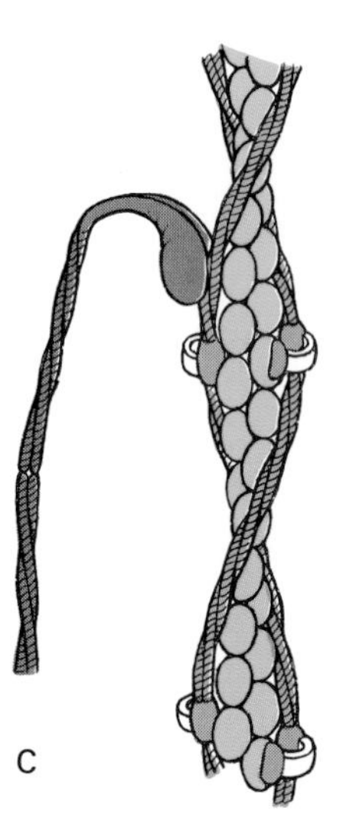

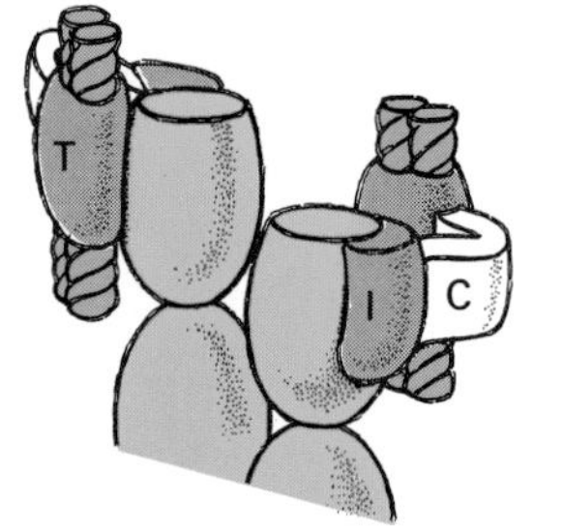

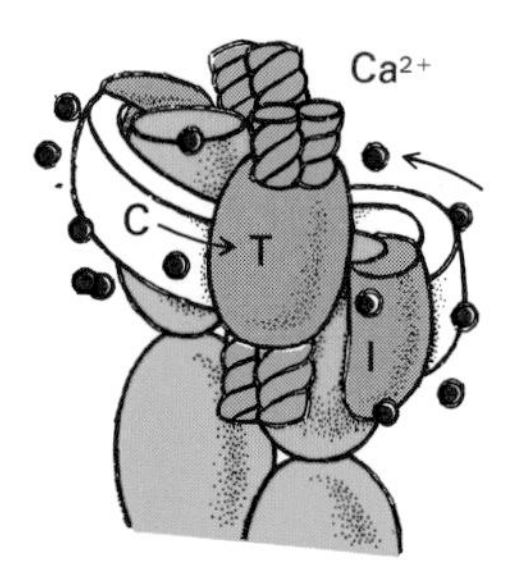

Structure of a troponin molecule. As shown at the left, it is made of a tropomyosin-binding subunit (T) and of an actin-binding subunit (I, because it inhibits the binding of myosin to actin), joined together by a calcium-binding subunit (C). The shape of the C subunit changes depending on whether calcium ions are present or absent. This conformational change alters the orientation of the T and I subunits with respect to each other. This drawing is purely imaginary. The exact shape of the molecule is not known.

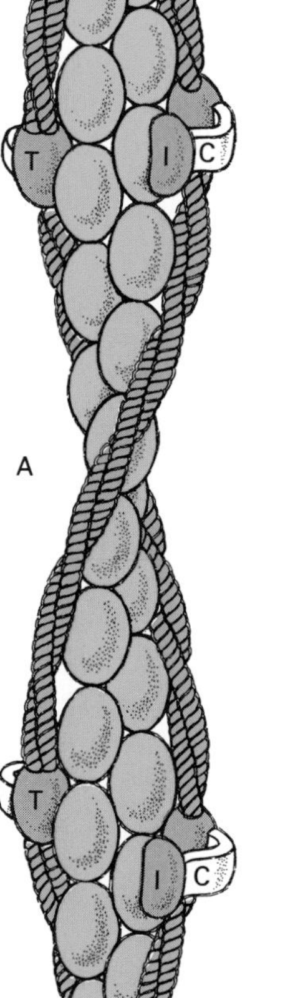

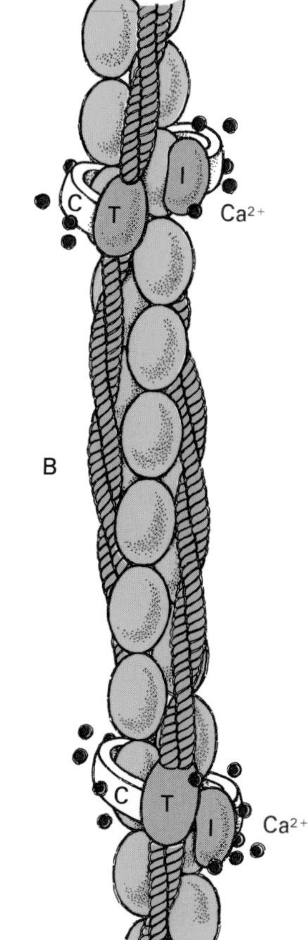

Twin filament (shown at the right), complete with tropomyosin and troponin: (A) in the absence of calcium ions ("off" position); (B) in the presence of calcium ions ("on" position). The displacement of tropomyosin is exaggerated for clarity.

interlocked. It is the state of rigor mortis that forms after death, when ATP production ceases. Add ATP and the structure becomes plastic. There is an ATP-binding site on the myosin head and, when this site is occupied, the head can no longer interact with the actin (in the absence of calcium, see below), and the filaments slide smoothly along each other. Such a fiber offers little resistance to passive stretching until other structural elements intervene to prevent its coming apart completely. This, for instance, is the state of an extensor muscle when the corresponding flexor contracts. The third state is triggered by calcium ions, which act through the troponin molecules that are attached to the actin filaments. Troponin consists of three subunits. One binds to actin and is called I, because its presence inhibits the attachment of myosin. Another, called T, binds to tropomyosin. These two subunits are linked by a third one, called C, which binds calcium ions. The C subunit is the trigger of the muscle machine. In the absence of calcium, its conformation is such that the tropomyosin threads are kept away from the grooves in the actin filament and cover the myosin-binding sites ("off" position). Occupancy of the C troponin subunit by calcium changes its shape in such a way that tropomyosin is shifted toward the grooves of the actin filament ("on" position). This allows myosin to interact with actin, which simultaneously activates ATP hydrolysis and the coupled conformational change that we have witnessed in the single actin-myosin bundle. Note that, in the fibril, each actin filament is pulled by three myosin filaments acting in a concerted fashion. Not all myosin heads are operative at the same time, however; they participate actively in the pulling only when strategically placed with respect to actin binding sites. A very smooth shortening is thereby ensured. If shortening is opposed, tension develops. This goes on for as long as the troponin C subunit is occupied by calcium and ATP is made available to cover the energy cost.

Troponin is found only in striated muscle. In other types of muscles and in nonmuscle cells, contraction is controlled by other means. But the basic principles seem universal. As far as is known, actin filaments of opposite polarity are always made to slide toward each other by a ratchet type of mechanism, powered by the ATP-splitting heads of myosinlike molecules.

Before we move on, we should pause for a last look at the actin molecule, certainly one of the most remarkable assemblages of atoms offered for our contemplation. It is not a giant molecule. It has a molecular mass of 42,000 daltons and is made of 374 amino acids. Yet, the three-dimensional ordering of this chain is such as to produce no less than eight specific, perfectly positioned, binding sites: four for the mutual association of actin molecules in the precise double spiral of F-actin; one for the ATP/ADP involved in the polymerization process; one, perhaps two, for tropomyosin; one for the troponin I subunit; and one, particularly important, for the attachment and simultaneous activation of the ATP-splitting myosin head. In addition, weaker binding sites allow actin fibers to join laterally into regularly structured bundles, or rods. How such extraordinary construction ever came to be is not known. But, once it appeared, evolution could not alter it any more. Between the amoeba and the rabbit, there is virtually no change in the structure of actin.

The Tubulin-Dynein System

More than one billion years ago, a living organism "discovered" the advantage of building scaffoldings with tubular elements, a discovery made only recently by our engineers. Presumably, the evolutionary invention was made by the usual "chance mutation–natural selection" mechanism (see Chapter 18). Perhaps some actinlike globular protein with the property of longitudinal self-assembly suffered a genetic change that altered its faculty of lateral association so that single filaments would tend to form cylindrical sheets rather than helical duplexes.

Indeed, tubulin, the constituent of the microtubules, is, like actin, a small globular protein, about 4 nm in diameter, fitted with a complementary lock-key axis that allows indefinite linear association. However, there are two tubulins, designated α and β, which are undoubtedly

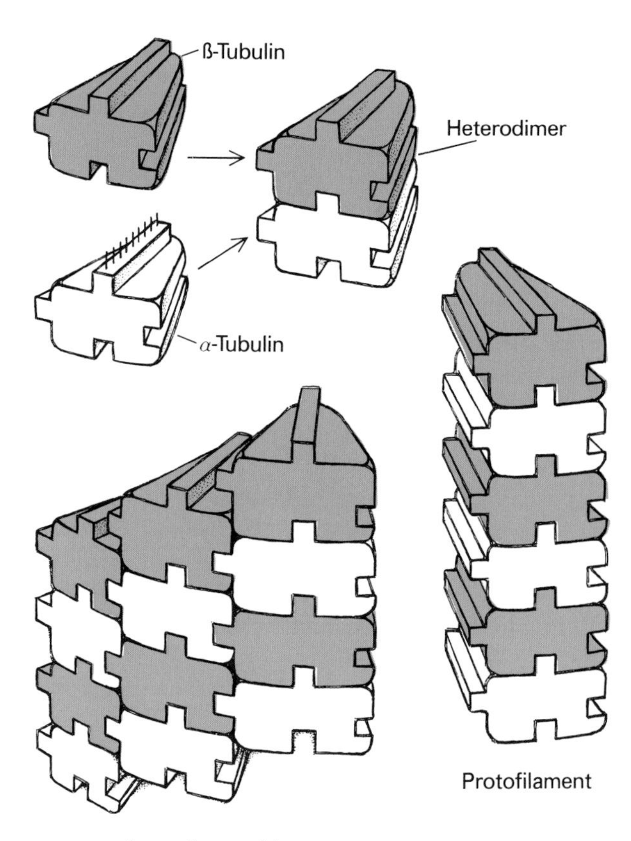

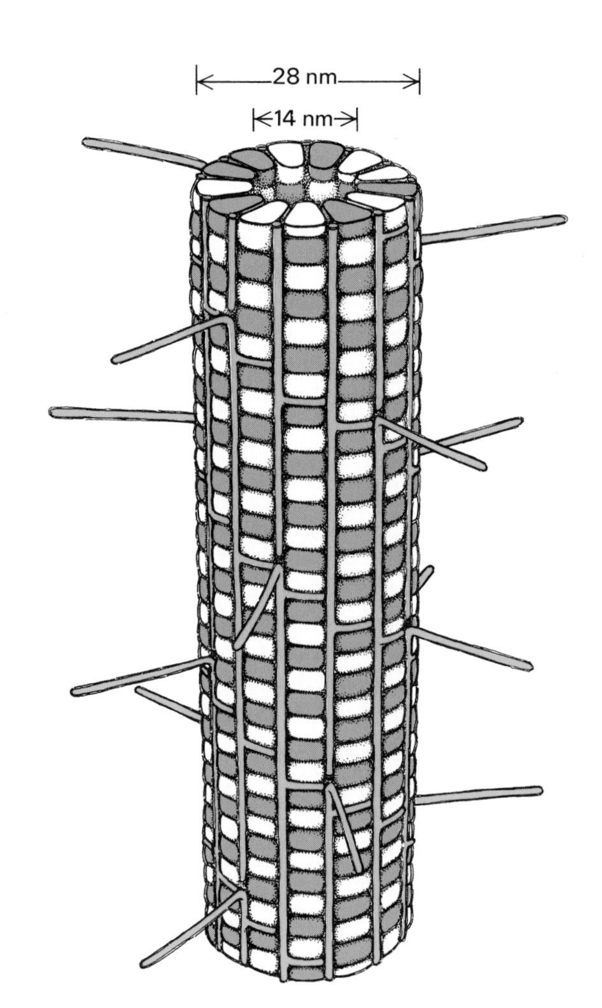

The drawings at the left above show in highly schematic fashion how very similar α- and β-tubulin molecules join preferentially to become stable heterodimers, which themselves assemble longitudinally into protofilaments, by means of a polar lock-and-key arrangement. Another such arrangement, situated laterally, allows the protofilaments to join into a cylindrical structure consisting of thirteen protofilaments, with an outer diameter of 28 nm and a bore of 14 nm. Lateral staggering is such as to create a left-handed helical periodicity of alternating α and β units, with a pitch of 12 nm per turn, and a right-handed helical periodicity of identical units, with a pitch of 40 nm per turn. Microtubule-associated proteins (MAPs) occupy the grooves.

evolutionary siblings, as indicated by extensive similarities in their amino-acid sequences. There is a preferential α–β linkage, which remains stable when microtubules are disassembled, so that the equivalent of G-actin is really an α-β heterodimer, not a monomer. These subunits assemble linearly by reversible β–α linkages to form protofilaments (Greek *prôtos*, first), which associate laterally by means of a second lock-key system, with a staggering such that each α-β subunit is flanked by a β-α couple displaced by about one-fourth of its length. The resulting sheet is not flat but curved and closes into a cylinder when exactly thirteen protofilaments are aligned side by side.

As in the transformation of G-actin into F-actin, the assembly of microtubules is associated with the hydrolysis of a nucleoside triphosphate, in this case GTP (instead of ATP for actin). There is a GTP-binding site on the α-β heterodimer. Upon assembly, GTP is split into GDP,

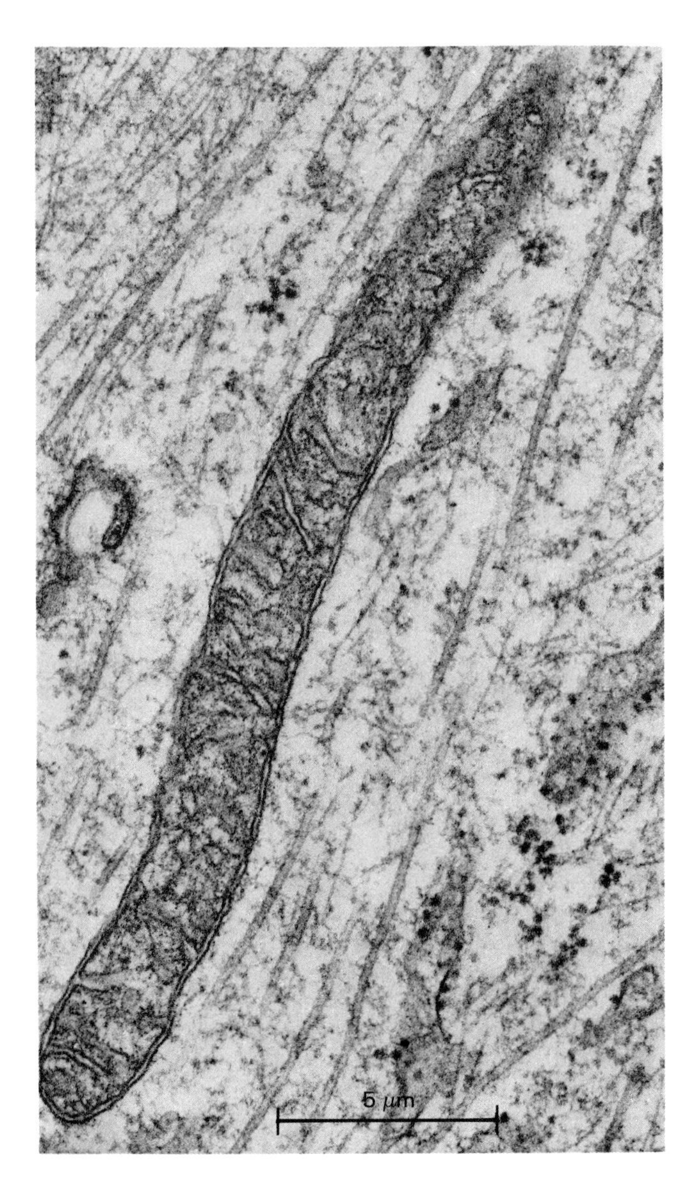

This electron micrograph of a thin section through a cultured human cell shows numerous microtubules running through the cytoplasm, on both sides of the elongated mitochondrion in the middle.

which remains bound, and P_i, which is released. In view of all these similarities, one is tempted to imagine an evolutionary relationship between actin and tubulin, with, as a possible key divergence event, the genetic change that led to duplication of a single gene into two distinct forms (ancestral to the α- and β-tubulin genes). What is known of the amino-acid sequences of actin and the tubulins does not, however, lend any support to this hypothesis.

Whatever their evolutionary origin, the development of microtubules certainly played an important role in the appearance of eukaryotes, in which they are universally present in both the plant and the animal kingdoms. According to one theory, they first appeared in some flagellate microorganism, which subsequently became adopted symbiotically by the common ancestor of all eukaryotes, to provide, among other advantages, the elements of the mitotic spindle (see Chapter 19). Unlike mitochondria and chloroplasts, however, centrioles, which are the organizing centers of the mitotic spindle and which duplicate at each cell division, are not known to contain any DNA.

To the cell tourist equipped with millionfold magnifying glasses, microtubules appear as some sort of thick-walled garden hose with an outer diameter of a little more than one inch (28 nm real size) and a half-inch (actually 14 nm) bore. They have a knobby surface; the knobs, about one-sixth of an inch (4 nm) in diameter, are arranged in thirteen longitudinal rows, in which minor differences in shape reveal the regular alternation of α and β subunits. As a result of the staggered organization of these rows, the knobs form a variety of spiral patterns around the tubing. Among these patterns are a right-handed helix of identical subunits (all α or β), with a pitch of 1.6 inches (40 nm) per turn, and a left-handed helix of alternating α and β subunits, with a pitch of one-half inch (12 nm) per turn. As a rule, this basic skeleton is further decorated with additional proteins (microtubule-associated proteins, or MAPs), partly inserted into the longitudinal grooves and partly projecting freely in the form of a hairy or fuzzy outgrowth, also arranged helically around the microtubule shaft. Altogether, microtubules offer a very pretty sight. Even more arresting are some of the structures that they serve to build.

On the facing page: Electron micrograph of a section through an axopod of the heliozoon *Echinosphaerium nucleofilum.* Note the double-helical arrangement of microtubules with twelvefold symmetry.

Many of these structures are built from labile microtubules. They are transient and changeable, the products of a dynamic equilibrium between two opposing processes that go on more-or-less continually. This equilibrium is readily displaced. For instance, disassembly of microtubules is favored by cooling, by high pressures, by calcium ions, and by drugs such as colchicine or some of the alkaloids (vinblastine, vincristine) extracted from the periwinkle, *Vinca rosea.* These drugs bind to free heterodimers and prevent them from joining with each other. Warming and exposure to heavy water, on the other hand, favor microtubule assembly.

The world of cells abounds in striking illustrations of these phenomena. A favorite of cell tourists is a group of protozoa called Heliozoa (Greek *hêlios*, sun; *zôon*, animal). These unicellular organisms owe their name to the fact that they send out long, thin, rigid spikes, known as axopods, that radiate in all directions up to half a millimeter from the cell body. Just imagine the sight, as seen through your magnifying glasses: a huge sphere, 300 feet or more in diameter, bristling with giant, rod-shaped projections some 3 to 5 feet wide and up to a third of a mile long. Should you wander through one of these projections, you would find it to be supported by a central spine made of hundreds of parallel microtubules, cross-bridged in an admirable double-helical pattern with a twelvefold symmetry. You would see a two-way traffic of particles and molecules moving briskly along this spine, linking the cell body with the very tip of the axopod, where, among other events, endocytic activity is taking place.

Should your heliozoan host be caught in an icy current, a most dramatic change would take place before your eyes. Not only would the axopodal traffic freeze to a stop, but the whole microtubular scaffolding would fall apart. In less than two hours, not a single microtubule would be left, and the axopods would be entirely retracted, leaving the sun a naked ball that has lost its rays. Not all would be lost, however, as might be realized by those who happen to recognize the heterodimeric building blocks of the dismantled microtubules among the swarms of new molecules that now fill the cell. We need only a warm wavelet, and all will be repaired. Reawakened by the gentle heat, the microtubular spine starts growing again from roots, called microtubule organizing centers, that are buried in the depth of the cell body. Soon the cell surface bulges with hundreds of new axopods, and in a matter of hours the sun again flashes its rays in all directions.

Light-micrograph of the heliozoon *Echinosphaerium nucleofilum*. Note the axopods radiating all around the cell body.

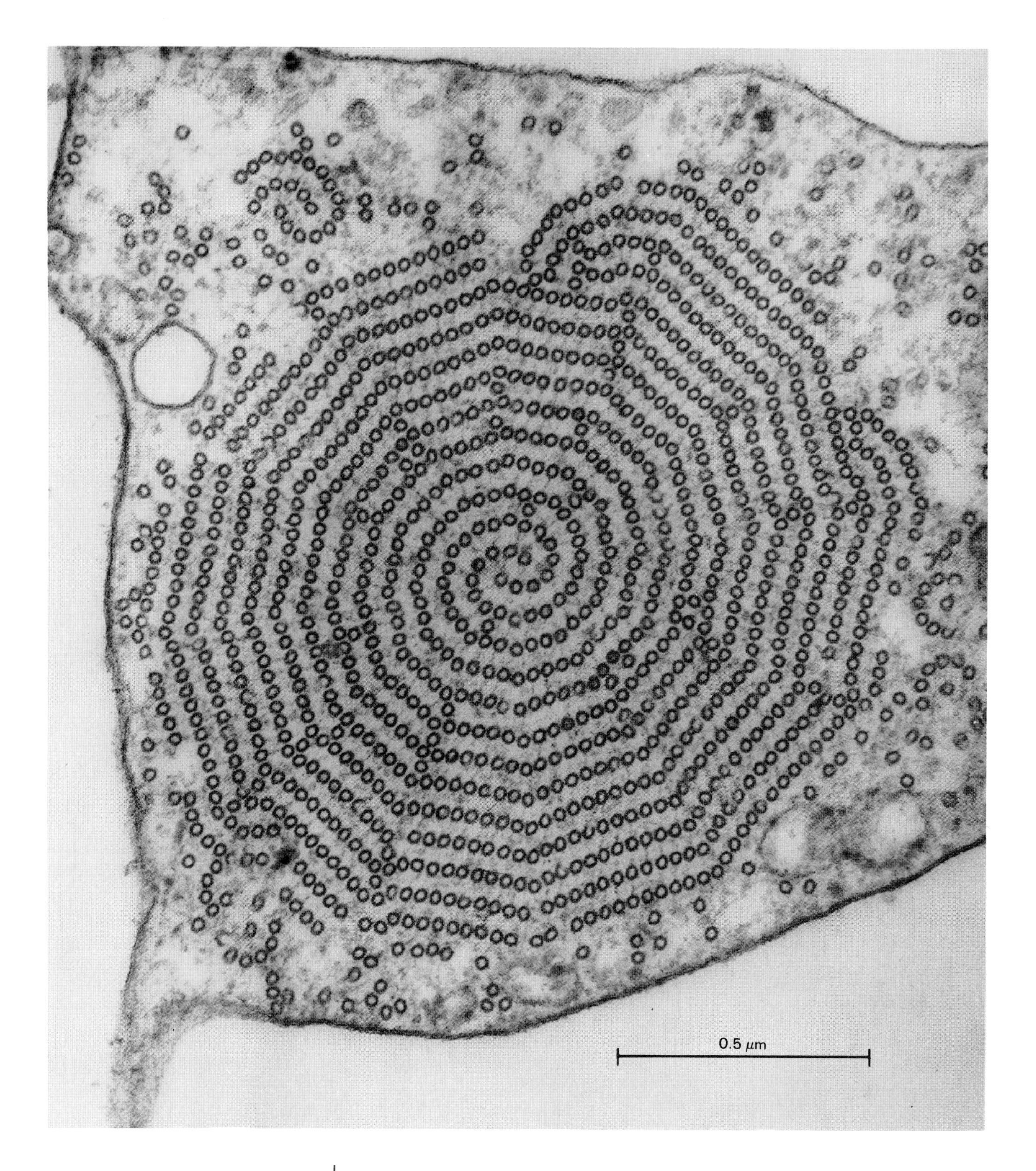
0.5 μm

Light micrograph of a red chromatophore (erythrophore) cell of the fish *Holocentrus rufus*, with pigment granules dispersed.

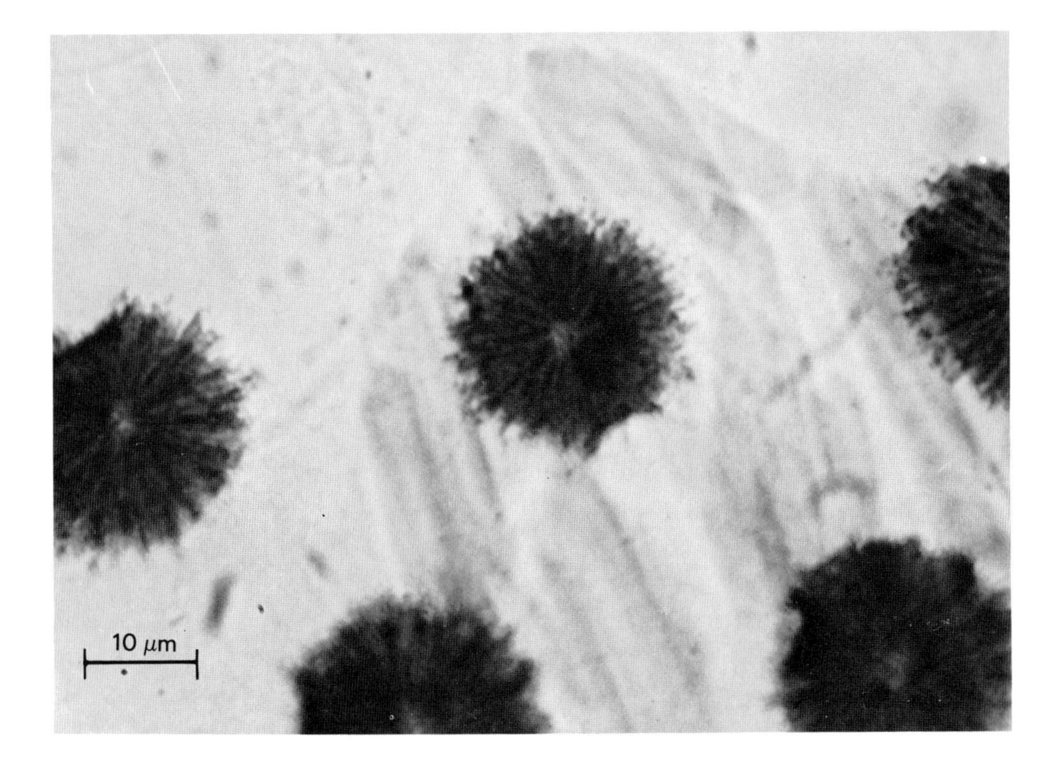

Many crustaceans, fishes, amphibians, and reptiles have in their skin star-shaped pigmented cells, called chromatophores, that send out radiating processes supported by a microtubular skeleton, along which pigment granules move inward or outward with remarkable speed. Under the influence of certain hormones, whose secretion is controlled by light, the granules may, in less than a second, pack centrally and leave the cell largely transparent or invade the processes and cause the cell to darken. In some animals, two or more types of pigment granules of different color participate in these migrations independently. Herein lies the chameleon's secret. Interfere with his microtubules, and there goes his camouflaging skill.

Probably the most extraordinary cellular projections reinforced by microtubules are the axons, the threadlike extensions by which neurons (nerve cells) dispatch their signals and which may, in the nerves of the larger mammals, reach lengths of several meters. Vital communications and exchanges between the main cell body and the terminal branched endings of the axon are maintained over these incredible distances by the so-called axonal flows, of which the fastest component travels at up to 8 μm per second. Magnified a millionfold, such an axon could almost straddle the Atlantic Ocean. It could, in a matter of days, convey products made in the main cell body—say, in New York—to their terminal destination

Electron micrographs of an erythrophore of *Holocentrus rufus* (see the light micrograph on the facing page): (left) with pigment granules dispersed; (right) with pigment granules aggregated.

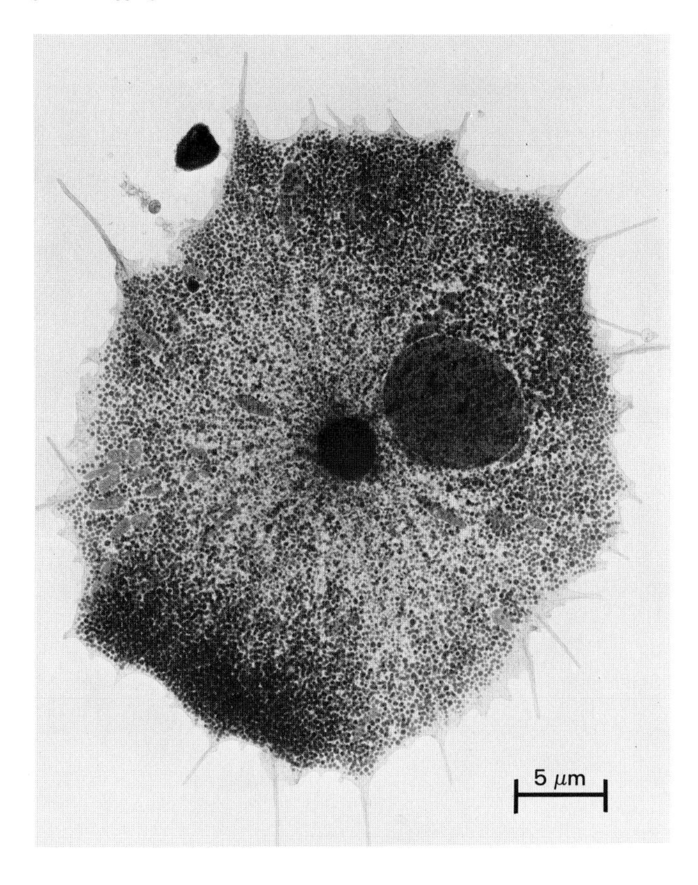

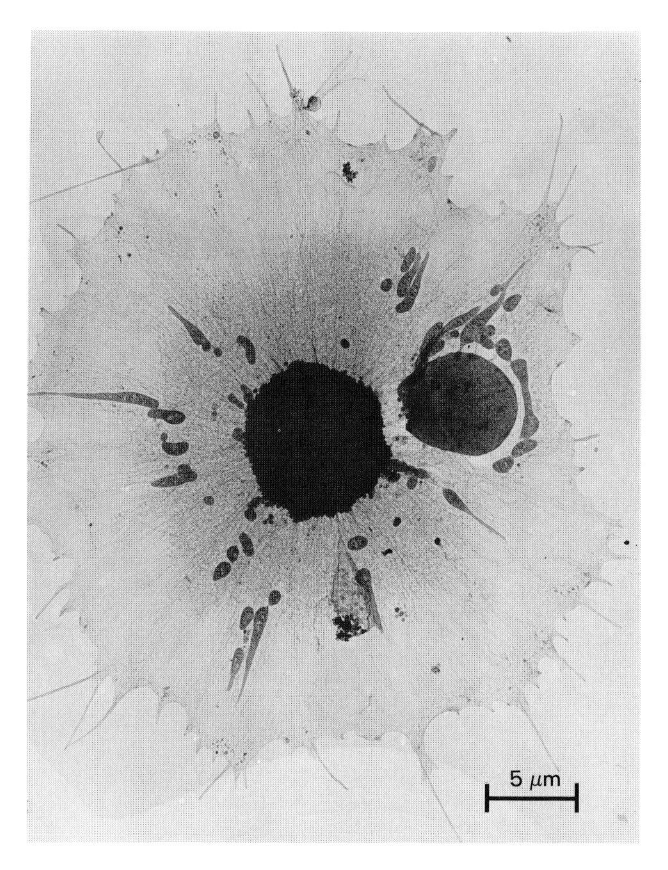

in the outer nerve endings—at some point on the Irish coast.

Microtubules also serve to induce or maintain certain structural asymmetries in cells—for instance, the elongation of growing muscle cells—and to build certain intracellular scaffoldings, generally in support of some directional transport or movement. Near the end of our tour, we will have an opportunity to witness the most grandiose of all such operations, namely the erection and subsequent dismantling of the vast, intricate machinery whereby duplicated chromosomes are disjoined and pulled away from each other to produce two separate sets in the course of mitotic cell division.

In all the examples mentioned, microtubules play an obvious structural role as cytoskeletal elements. In addition, thanks to their labile character, they may also fulfill a morphogenetic function through the alterations in shape that are elicited by their assembly and disassembly. Finally, they often provide leading tracks and bracing props for some form of guided translocation of intracellular objects or materials. The driving force of these transport phenomena could be supplied by actin-myosin cytomuscles. But there is another possibility, suggested by inspection of the more stable microtubular systems that serve to build the two main locomotor organelles of cells—cilia and flagella.

This scanning electron micrograph of the surface of the protozoon *Paramecium tetraurelia* shows rows of cilia arrested in the midst of their synchronous beat.

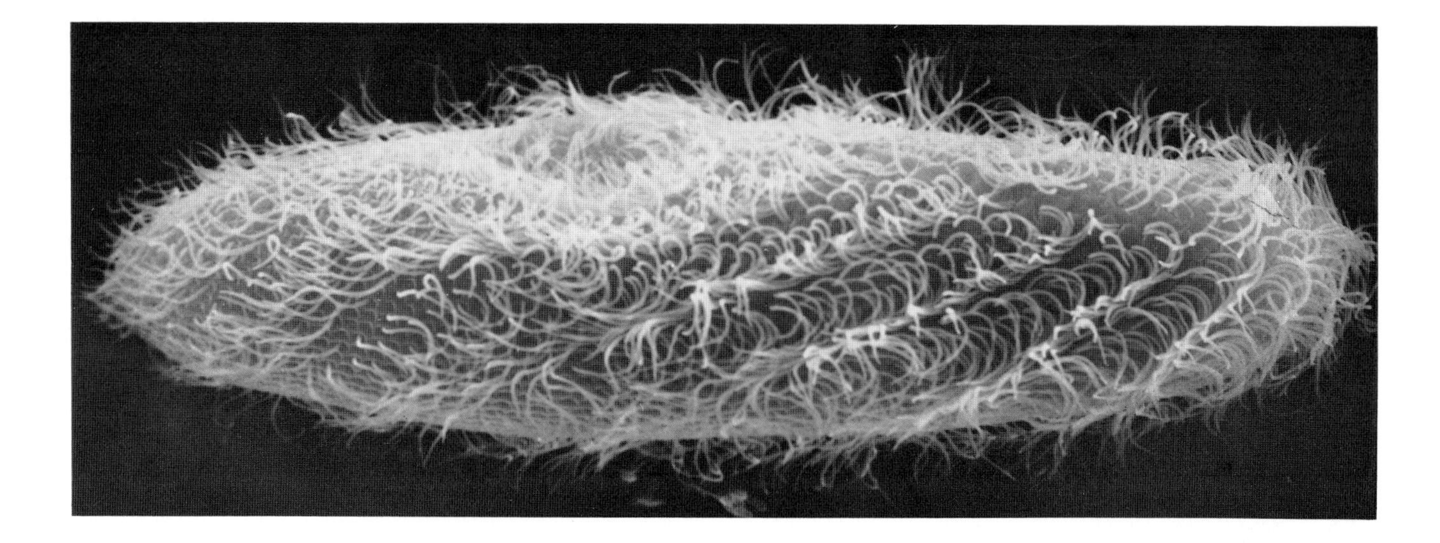

Cilia (eyelashes, in Latin) are beating protrusions. They bend and swing back, remaining essentially in a single plane, with a frequency of oscillation of some ten to forty beats per second. They are usually present in large numbers and are planted in parallel arrays, moved by what looks like waves of excitation signals. For full enjoyment of this display, you should reduce your magnification to some ten-thousandfold, instead of the customary millionfold, and equip yourself with a fast camera and a slow-motion replay system. The cell surface will then appear like a ripe wheatfield bending under a gust of wind. Ciliary beat causes a relative displacement of the cell with respect to the surrounding fluid. If the cell is free, it moves (ciliate protozoa); if fixed, it generates a liquid current. The direction of these movements can be reversed. For instance, when a *Paramecium* (a ciliate) bumps into an obstacle, it immediately swims backward. The agents of this reversal are calcium ions released by the shock.

Flagella (whips, in Latin) are longer than cilia and are usually present singly or in small numbers at the tail end of free-swimming cells, such as flagellate protozoa, gametes of algae, or animal spermatozoa. Like cilia, flagella move in a single plane, but their movement is undulating, not pendular. They serve a propelling function. (Some bacteria also have flagella, but these flagella are constructed differently; tubulin is found only in eukaryotes.)

With only minor variations, all cilia and flagella have the same molecular architecture, based on the so-called 9 + 2 pattern. Axially situated are a pair of microtubules linked laterally by cross-bridges. This central pair is enclosed by a sheath and surrounded further by a cylindrical set of nine parallel microtubular doublets joined to the central pair by radial spokes. Each doublet is made of one complete microtubule (A) of thirteen protofilaments, to which an incomplete microtubule (B) of only ten protofilaments is fused by sharing three of the A microtubule's protofilaments. The A subunit of each doublet sends out pairs of tangentially oriented side arms toward the B subunit of the adjacent doublet in a sort of ring-around-a-rosy pattern, clockwise when viewed from the tip. These side arms, which have been described as dumbbell- or lollipop-shaped, are repeated longitudinally with a spacing of 20 nm. Altogether, more than one hundred distinct proteins participate in the construction of this remarkable

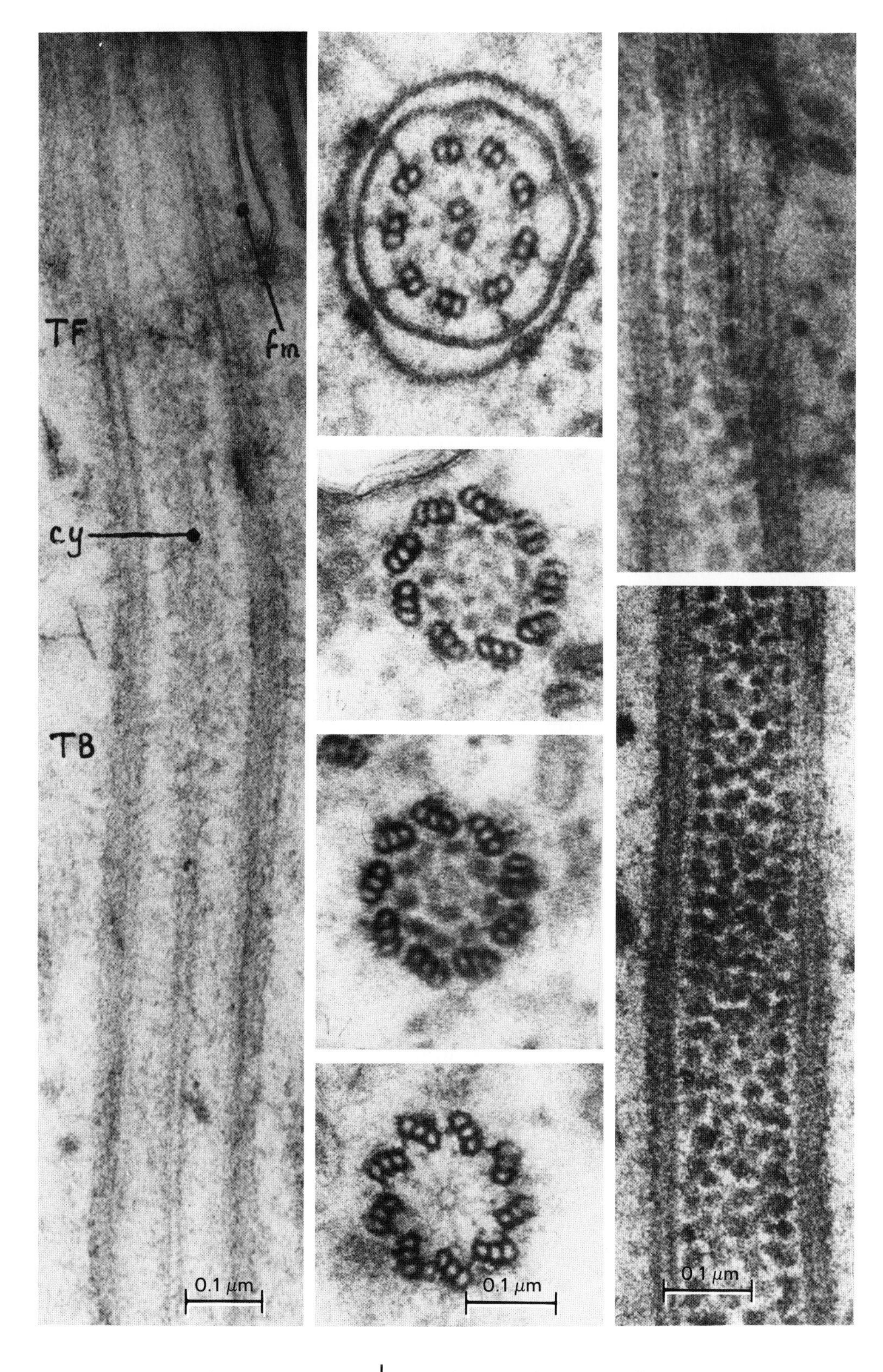

These electron micrographs, taken in the early 1960s, revealed the characteristic structure of flagella in the protozoon *Pseudotrichonympha.* The longitudinal section at the left shows the root of a flagellum on top continuing into the basal body (TF = transition between flagellum and basal body; TB = transition between proximal and distal regions of basal body; cy = central cylinder; fm = flagellar membrane). The transverse sections in the middle illustrate the structure of a flagellum and basal body at different levels (see the diagram on the next page). The longitudinal sections at the right show distal parts of the basal body.

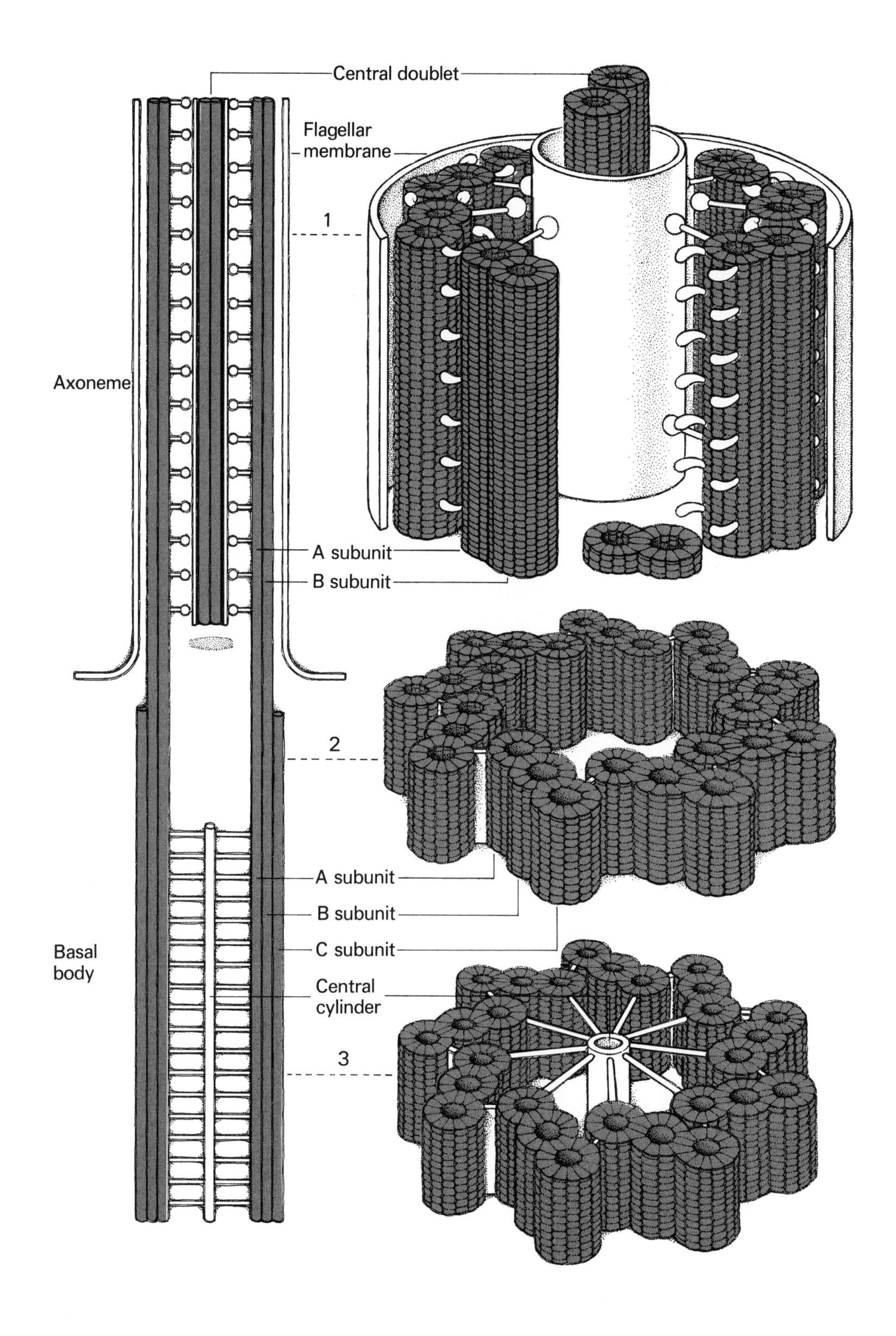

Structure of cilium. The drawing at the left shows a longitudinal section through the lower part of the axoneme and the upper part of the basal body. The drawings at the right illustrate the fine structure at three different levels. Part 1 shows the typical 9 + 2 pattern of the axoneme, with the central microtubule doublet and the ring of nine doublets linked by pairs of dynein side arms. The characteristic cartwheel structure of the basal body, with triplets attached by spokes to a central cylinder, is shown in part 3. The intermediate part connecting the axoneme to the basal body is seen in part 2. (See the electron micrographs on pp. 215 and 217.)

Longitudinal section through a cilium of the protozoon *Tetrahymena thermophilia*, prepared by freeze-etching (see the illustration on the facing page), shows four microtubule doublets connected by rows of dynein arms.

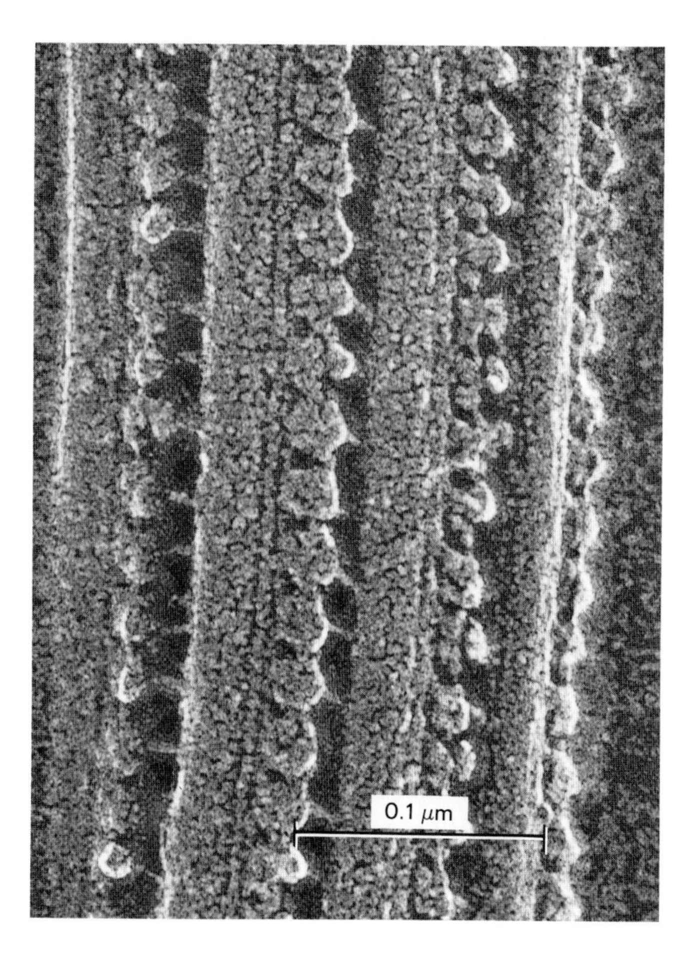

edifice. Guided by their associative properties, they combine to form a delicately sculptured shaft, known as the axoneme (Greek, *nema,* thread). It is about 0.3 μm thick—or 1 foot at our millionfold magnification—and its beauty, like that of the sarcomere, must be enjoyed mostly in cross section.

At the point where the axoneme becomes rooted in the cytoplasm, its structure changes. The central pair of microtubules breaks off and becomes replaced lower down by a single hollow axis. The peripheral microtubular doublets, on the other hand, extend into the cytoplasm, where they are no longer connected by side arms, but instead acquire a third incomplete microtubule (C) fused to the B component of the doublet in the same way as B is fused to A. Radial lamellae connect these triplets to the central axis, replacing the axoneme's spokes. This root is called kinetosome (Greek *kinein,* to move) or basal body. It is endowed with remarkable organizing abilities. For instance, it will sprout a brand-new limb after a cilium or flagellum is amputated. As we shall see in Chapter 19, the centriole, which governs the assembly of the mitotic spindle, is a close relative of the kinetosome.

In the axonemes of cilia and flagella, the microtubules and their radial connections provide a neatly interlocked cytoskeletal framework endowed with just the degree of resilience, flexibility, and elasticity needed to permit, as well as to limit, the bending strains to which these appendages are subjected. The side arms are the power-generating cytomuscles that force the structure to bend. They are made of a protein called dynein (Greek *dynamis,* force), which shares with myosin heads the characteristic property of catalyzing ATP hydrolysis in a manner obligatorily coupled to a change in conformation. The consequence of this mechanochemical event is a relative displacement of the two microtubule doublets connected by the activated dynein side arms, in a manner that recalls the ATP-powered sliding of actin threads along myosin filaments.

This is the principle on which the molecular machinery operates. Details, of course, are considerably more complex. In particular, it is obvious that the nine rows of dynein side arms cannot be activated simultaneously. Those in or near the bending plane (which is perpendicular to the plane defined by the central pair of microtubules) must operate in seesaw fashion, so that those on one side relax their hold while the opposite ones grip, and vice versa. Side arms lying in or near the plane perpendicular to the bending plane, on the other hand, remain largely passive. How this delicate control is exerted has so far eluded the perspicacity of cell explorers.

An intriguing question, as yet unsolved, is whether dynein cytomuscles also function in other forms of intracellular movement for which microtubules provide the cytoskeletal scaffolding or whether this function is fulfilled by myosin filaments pulling appropriately anchored actin threads. In any case, a revealing parallelism is brought out by a comparison of the two main cellular motor systems: both function by a sliding displacement of cytobones under the pull of special ATP-splitting cytomuscles.

Buckminster Fuller's geodesic dome.

Clathrin

In the world of cells, wonders never cease. When the American architect Buckminster Fuller built his celebrated geodesic dome, little did he suspect that hundreds of millions of years ago Nature evolved a protein molecule that could do the same thing entirely on its own. This protein is called clathrin (Greek *klethra,* Latin *clathrum,* lattice bar). Given the opportunity, it will, even in the test tube, assemble spontaneously into basketlike structures formed, like the geodesic dome, by a network of hexagons with a few pentagons and heptagons thrown in for good measure.

We can watch this process and, at the same time, obtain a glimpse of its function by inspecting the inner face of the plasma membrane at a patch where, on the reverse side, receptors occupied by their ligands have just congregated and are about to be engulfed by endocytosis. Alerting us to this phenomenon is the cluster of receptor-molecule tails sticking out through the membrane on our side. Apparently alerted in the same way, clathrin subunits recruited from some cytoplasmic store start polymerizing over the advertised membrane patch, covering it with what looks at first like a ragged piece of chicken wire.

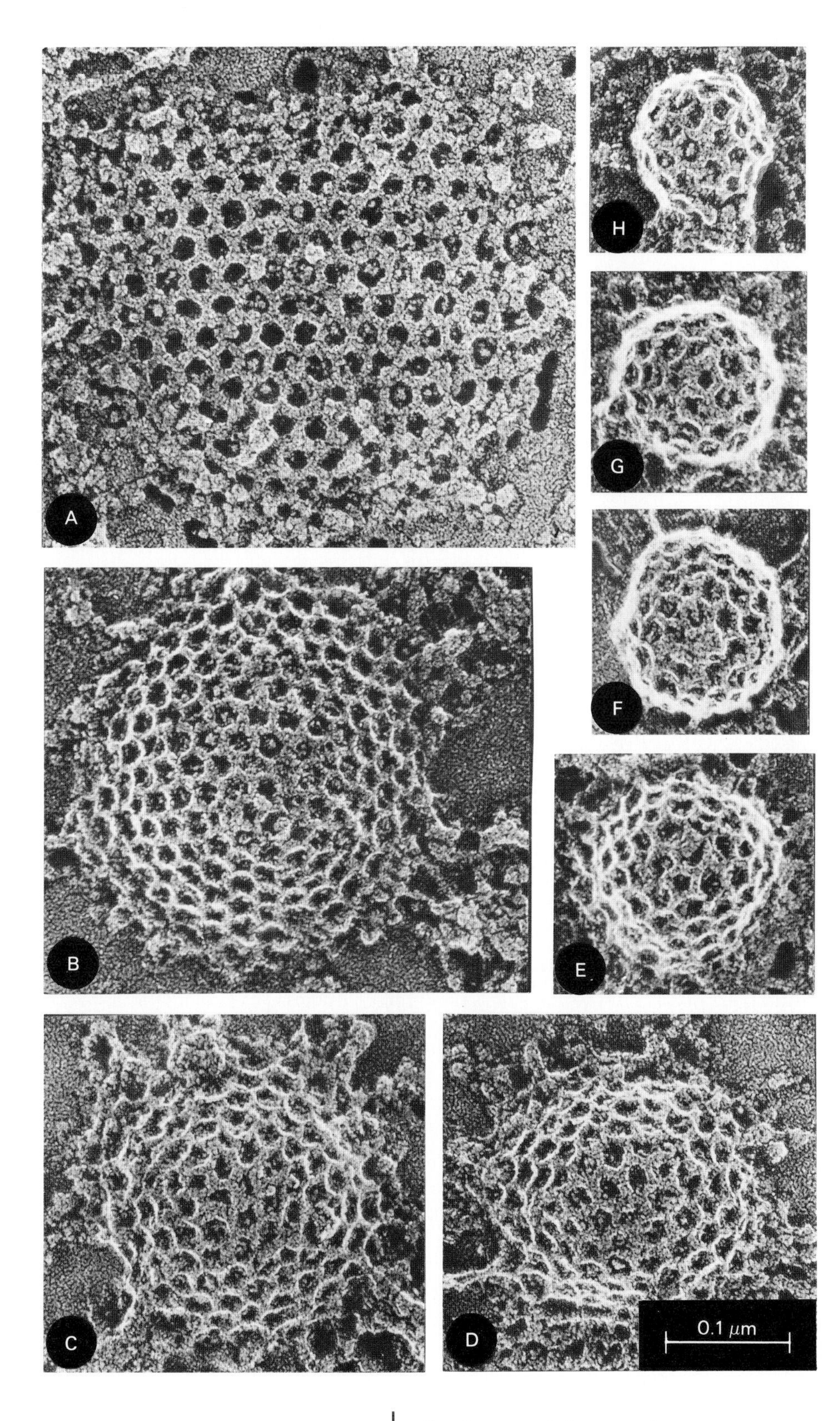

These electron micrographs obtained by freeze-etch technique show clathrin baskets on the cytoplasmic face of the plasma membrane. They cover coated pits and are shown, from A to H, in order of increasing curvature, up to virtually complete closure as a coated vesicle (F to H).

Assembly of clathrin network from triskelion units.

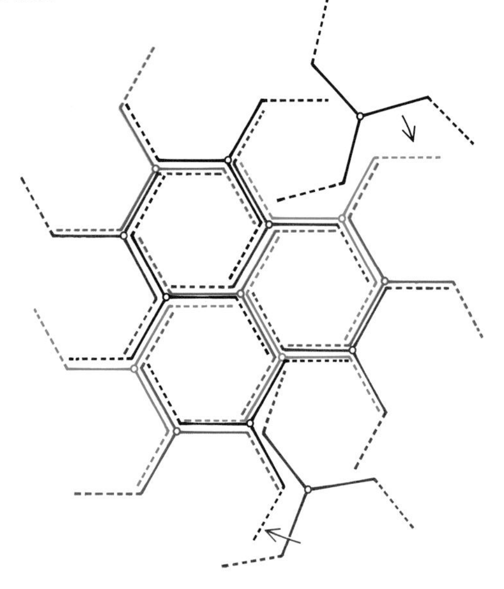

Membrane patch is pulled inward by clathrin cage in endocytosis.

Looking closely, you will notice that each subunit is a trimeric, three-legged structure, or triskelion (Greek *skelos,* leg), with "thighs" and "legs" about 20 nm long (a little less than 1 inch at our magnification). These subunits join by lateral association of their limbs to form a hexagonal mesh in which each vertex is occupied by a triskelion center, and each strut is made of four limbs—two thighs and two legs—aligned side by side. These triskelia, however, are not flat but curved. When "running" clockwise, they fit on a convex surface. Thus, as the structure they form widens, it bulges progressively in typical geodesic-dome shape. To accommodate the increasing curvature, some triskelia fall off, and the regular hexagonal lattice of the network becomes broken in some areas by a few pentagons and heptagons. These are readily assembled by the same mechanism as the hexagons, with only minor conformational stresses of the participating triskelia. As this process continues, the curvature of the dome continues to increase while its outer rim tightens, so that its form changes gradually to that of a pear-shaped basket. Finally, the rim closes, the basket becomes a cage, and the membrane patch turns into a vesicle imprisoned inside the cage. Soon after, the cage disintegrates and disappears, releasing the vesicle.

What we have just witnessed from inside the cell is the characteristic process of receptor-mediated endocytosis.

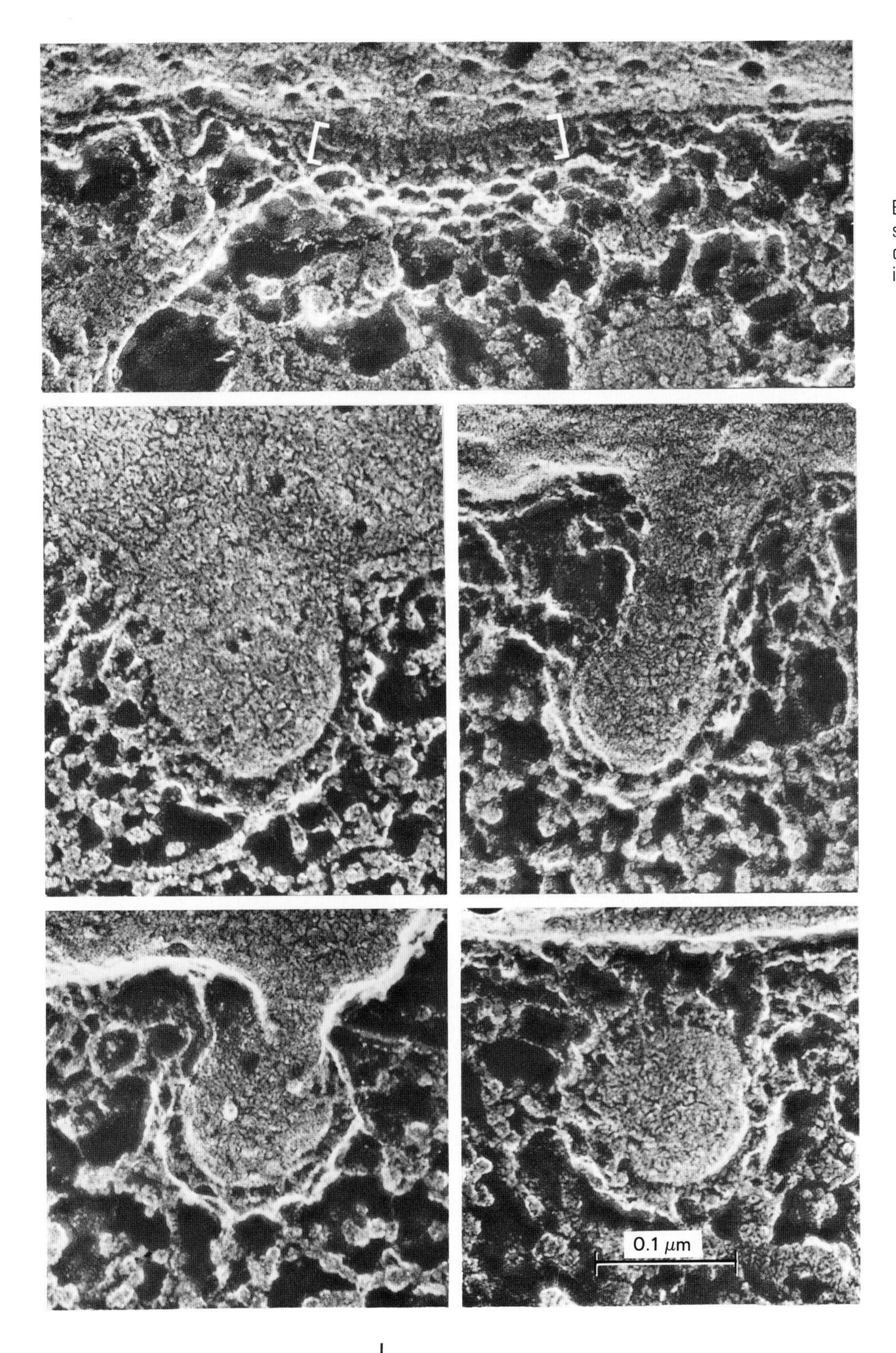

Edge views of freeze-etched coated pits show clathrin baskets at various stages of "strangling" a plasma-membrane invagination into a closed vesicle.

Early electron microscopists were given a two-dimensional hint that an organized structure might participate in this process when they noticed the presence of some sort of fuzzy contour delineating the profiles of certain membrane invaginations and closed vesicles. They invented the terms coated pits and coated vesicles to designate these structures, and put forward the hypothesis that the "coat" somehow plays a role in the process of pinching off vesicles from membranes (p. 55).

Those pioneers were remarkably prescient. Indeed, it appears that clathrin baskets are involved not only in receptor-mediated endocytosis, but also in other processes of vesicular transport, such as the transfer of materials from the ER to the Golgi apparatus. How clathrin operates is, however, far from clear. As we watch the performance of this remarkable molecule, many tantalizing questions come to mind. What, on those little patches of membrane—so diverse in nature and function—where clathrin baskets form, is the nucleating signal that initiates polymerization? Where does the free energy of this process come from, and does it, as we suspect, cover the work of squeezing out a vesicle from a flat membrane? If so, how does endocytosis without the help of clathrin take place? What causes a clathrin basket to fall apart again after it has performed its function?

For the moment, we do not know the answers to these questions, as to countless others. But, if history can be trusted, we can safely predict that some day a cell tourist will find them so intriguing that he will turn into a cell explorer, not resting until he uncovers the answers. Already, there are hints that calcium ions may assist polymerization and that ATP helps in reversing this process.

Microtrabeculae

Explorers who have gone through cells with the most incisive instruments currently available, such as the million-volt electron microscope or the quick-freeze, deep-etch method of preparing samples for three-dimensional electron microscopy, have detected a fine meshwork of filaments pervading the whole of the cytosol and attached by multiple anchoring points to all cytoplasmic structural constituents present. The name microtrabeculae—literally, miniature miniature beams—has been given to the elements of this lattice. Actually, the microtrabeculae are not as miniature as their name implies: they are somewhere between 10 and 15 nm, which makes them definitely thicker than actin filaments and perhaps as thick as myosin filaments. The reason they were not seen before resides in their short length and random orientation in the meshwork. A minimum of three-dimensional vision was required to discern them as part of a lattice.

The pictures of the microtrabecular meshwork that have been brought back so far endow the cytosol with considerably more structural rigidity and organization than was evident to us when we first entered this part of the cell. Admittedly, our progress at that time was rather heavy and clumsy, and we could have missed this fine web. Nevertheless, there is some worry about the reality of the microtrabecular meshwork. It makes the cell look almost "overorganized," and so far has received no chemical identification, which is surprising in view of its very extensive development. The problem remains that, even with their highly sophisticated tools, the explorers who have described the meshwork have to look at cells that are dead, not at live ones. How much of the meshwork could be due to a fixation artifact or to some other postmortem change remains a debated question. On the other hand, the fact that different cell parts are often seen to move in a concerted fashion does support the view of a connecting network between them.

Calcium and Cell Motility

During our brief excursion through striated muscle, we discovered the electrifying effect of calcium ions, which, by combining with troponin C, cause the conformation of this protein subunit to change in such a way that the attached tropomyosin is displaced and no longer prevents myosin-actin interaction and ATP hydrolysis.

As it happens, this is only one of the numerous important triggering functions performed by calcium. These ions also stimulate smooth-muscle fibers, perhaps all types of actin-myosin systems; they inhibit microtubule assembly, stop or reverse ciliary beat; they influence clathrin polymerization and may be involved in the membrane-fusion events that determine endocytosis, exocytosis, and the merger and division of intracellular vesicles; finally, they activate a number of enzymes—in particular, certain protein kinases (Chapter 8). These enzymes play important regulatory roles because the biological activity of the proteins on which they act is crucially dependent on their state of phosphorylation, as will be discussed in Chapters 13 and 18. Thus, those proteins that need phosphorylation to be active are turned on by calcium. Smooth-muscle myosin seems to belong in this category. On the other hand, those that are active only in dephosphorylated form are shut off by calcium ions.

In all these functions, calcium ions act indirectly by way of a calcium-binding protein called calmodulin, which resembles troponin C in that it, too, suffers a conformational change upon binding calcium. It is this altered calmodulin that actually exerts the effects attributed to calcium. Its functional resemblance to troponin C is far from fortuitous. These two proteins are closely related, as indicated by extensive amino-acid-sequence homologies, and appear to be evolutionary descendants of a common ancestor.

Such important functions presuppose the existence of intracellular calcium reservoirs fitted with extremely efficient pumps and unloading valves. The most highly regulated of such systems are found in striated muscles—especially the flight muscles of insects, which alternate between resting and active state as many as several hundred times per second. In striated muscle, calcium is stored in a specialized form of ER, called sarcoplasmic reticulum, which is intimately wrapped around the sarcomeres, so that the distance to be traveled by the calcium ions is extremely short. Release of calcium from this reservoir can be elicited almost instantaneously, thanks to deep indentations of the plasma membrane (T system), closely connected with the sarcoplasmic reticulum in the so-called triad. When a muscle fiber is excited, a wave of depolarization is propagated along the plasma membrane and temporarily induces leakage of calcium ions out of the sarcoplasmic reservoirs as it passes through the triad system. A powerful, ATP-driven pump in the sarcoplasmic membrane sucks back the calcium ions as quickly as they are released.

Specialized ER domains comparable to the sarcoplasmic reticulum may exist in a few other cells. However, the more common calcium reservoirs are, on one hand, the extracellular fluid and, on the other hand, the mitochondria. Each is fitted with a calcium pump. That in the plasma membrane drives calcium out of the cell. That in the mitochondrial inner membrane, which, as we have seen, is operated by the protonmotive force with the help of a natural calcium ionophore, concentrates calcium within the mitochondrial matrix. Thanks to these two pumps, calcium is maintained at a very low level in the cytosol. Any local perturbation of the plasma or inner mitochondrial membranes that either slows down the calcium pump or increases calcium leakage will cause the local cytosolic concentration of calcium to rise and, by way of an enhanced occupancy of the calcium-binding sites of calmodulin, elicit some of the effects that have been mentioned. Some nervous impulses and a number of hormones and other surface ligands produce such changes on the plasma membrane. Intracellular messengers induced by metabolic changes or perhaps, as in the case of cyclic AMP, in response to surface effects act similarly on the inner mitochondrial membrane.

Sources of Illustrations

page 7 (left)
Scanning electron micrograph courtesy of Dr. Brian J. Ford, Cardiff University.

page 7 (right)
Photograph courtesy of Carl Zeiss Inc., Thornwood, New York.

page 10
Reproduction of a portrait by Louis David, *Lavoisier and His Family.* All rights reserved, The Metropolitan Museum of Art.

page 11
Reproduced from *The Journal of Experimental Medicine,* 1945, vol. 81, p. 233, fig. 2, copyright permission of The Rockefeller University Press.

page 13
Electrophoretic patterns courtesy of Dr. Yukio Fujiki.

page 14 (upper right)
Electrophoretic patterns courtesy of Dr. Yukio Fujiki.

page 14 (bottom)
Autoradiograph courtesy of Dr. James D. Jamieson and Dr. George E. Palade.

page 25
Electron micrograph courtesy of Dr. Eldon H. Newcomb.

page 27
Electron micrograph courtesy of Dr. David Chase.

page 29 (top)
Photography by Etienne Bertrand Weill. Arp, Hans. *Serpent* © by ADAGP, Paris 1984.

page 29 (bottom)
Photography by Edward Leigh/courtesy of Sir John Kendrew.

page 33
Adapted from a diagram in the chapter on "*X-ray Analysis and Protein Structure,*" by R.E. Dickerson, in volume II, of *The Proteins* (H. Neurath, editor). Academic Press Inc. Copyright © 1964.

page 34
Scanning electron micrograph courtesy of Dr. Etienne de Harven and Nina Lampen. Sloan Kettering Institute, New York.

page 37 (top)
Illustration from *The Naturalist on the River Amazons,* by H.W. Bates.

page 38
Electron micrograph courtesy of Dr. Jerome Gross.

page 42
From *Tissues and Organs: A Text-Atlas of Scanning Electron Microscopy,* by Richard G. Kessel and Randy H. Kardon. W.H. Freeman and Company. Copyright © 1979.

page 44
Electron micrograph courtesy of Dr. Pierre Baudhuin and Dr. Pierre Guiot.

page 48 (left)
Picture courtesy of *Proceedings of the Royal Society of London,* vol. 66, plate 6.

page 54
Electron micrographs courtesy of Dr. Samuel C. Silverstein.

page 63
Electron micrograph courtesy of Helen Shio.

page 69
Electron micrograph courtesy of Dr. Pierre Baudhuin.

page 70
Electron micrographs courtesy of Dr. Marilyn Gist Farquhar.

page 73
Electron micrographs courtesy of Dr. Pierre Baudhuin.

page 76
Phase-contrast micrograph courtesy of Helen Shio.

page 83
Electron micrograph courtesy of Dr. George E. Palade.

page 88
Electron micrograph courtesy of Dr. George E. Palade.

page 91
Electron micrograph courtesy of Dr. George E. Palade.

page 92
Electron micrograph courtesy of Dr. Marilyn Gist Farquhar.

page 94
de Chirico, Giorgio. *The Anxious Journey.* 1913. Oil on canvas, $29\frac{1}{4} \times 42$ inches. Collection, the Museum of Modern Art, New York. Acquired through the Lillie P. Bliss Bequest.

page 105
Electron micrograph courtesy of Dr. George E. Palade.

page 107
Drawing by Louis Pasteur reproduced by René Dubos in his chapter "Pasteur and Modern Science," pp. 17–32, in *The Pasteur Fermentation Centennial 1857–1957,* copyright 1958 by Chas. Pfizer & Co., Inc.

page 150
Electron micrograph courtesy of Dr. Robert A. Bloodgood, University of Virginia.

page 151 (top)
Photograph of model courtesy of Dr. Hans-Peter Hoffmann and Dr. Charlotte Avers.

page 151 (bottom)
Reclining Figure (1), 1945, courtesy of Henry Moore.

page 152 (left)
Electron micrograph courtesy of Dr. George E. Palade.

page 157
Electron micrograph courtesy of Dr. Donald F. Parsons.

page 174 (bottom)
Electron micrograph courtesy of Dr. Leo P. Vernon.

page 175 (top)
Electron micrograph taken by W.P. Wergin/courtesy of Dr. Eldon H. Newcomb.

page 181 (top left)
Electron micrograph courtesy of Dr. Arvid B. Maunsbach.

page 181 (top right)
Electron micrograph courtesy of Dr. Pierre Baudhuin.

page 181 (bottom)
Electron micrograph courtesy of Dr. George E. Palade.

page 183
Electron micrograph courtesy of Dr. Eugene Vigil, Agricultural Research Center, Beltsville, Maryland.

page 184
Electron micrograph courtesy of Dr. Phyllis Novikoff and Dr. Alex Novikoff.

page 185
Electron micrograph taken by Sue Ellen Frederick/courtesy of Dr. Eldon H. Newcomb. Published in E.H. Newcomb, "Ultrastructure and Cytochemistry of Plant Peroxisomes and Glyoxysomes," *Annals of The New York Academy of Science,* vol. 386, p. 230 (1982).

page 186
Electron micrographs courtesy of Dr. Marten Veenhuis. Published in J.P. Van Dijken, M. Veenhuis, and W. Harder, "Peroxisomes of Methanol-Grown Yeast," *Annals of The New York Academy of Science,* vol. 386, p. 201 (1982).

page 187
Electron micrograph courtesy of Dr. Marten Veenhuis.

page 189
Electron micrograph courtesy of Helen Shio.

page 190
Electron micrograph taken by Isabelle Coffens courtesy of Dr. Pierre Baudhuin.

page 192
Fluorescent micrograph courtesy of Dr. Werner W. Franke.

page 199
Electron micrograph and model courtesy of Dr. Hugh Huxley.

page 201
Electron micrograph courtesy of Dr. John Heuser.

page 203
Electron micrograph courtesy of Dr. Hugh Huxley.

page 204
Electron micrograph courtesy of Dr. Hugh Huxley.

page 209
Electron micrograph courtesy of Dr. Keith R. Porter.

page 210
Light micrograph courtesy of Dr. L.E. Roth.

page 211
Electron micrograph courtesy of Dr. L.E. Roth and Dr. Y. Shigenaka, University of Tennessee, Knoxville, and Hiroshima University, respectively.

page 212
Light micrograph courtesy of Dr. Keith R. Porter.

page 213
Electron micrographs courtesy of Dr. Keith R. Porter.

page 214
Scanning electron micrograph courtesy of Dr. Sidney L. Tamm.

page 215
Electron micrographs courtesy of Dr. Ian R. Gibbons.

page 217
Electron micrograph courtesy of Dr. John Heuser.

page 218
Photograph courtesy of Buckminster Fuller Archives.

page 219
Electron micrographs courtesy of Dr. John Heuser.

page 221
Electron micrographs courtesy of Dr. John Heuser.

Index

Page numbers in boldface type refer to illustrations; those in italic type refer to definitions and etymologies.